Stem Cell Biology and Regenerative Medicine

Series editor

Kursad Turksen, Ph.D.
e-mail: kursadturksen@gmail.com

More information about this series at http://www.springer.com/series/7896

Kursad Turksen
Editor

Tissue-Specific Stem Cell Niche

Humana Press

Editor
Kursad Turksen
Ottawa, ON
Canada

ISSN 2196-8985 ISSN 2196-8993 (electronic)
Stem Cell Biology and Regenerative Medicine
ISBN 978-3-319-21704-8 ISBN 978-3-319-21705-5 (eBook)
DOI 10.1007/978-3-319-21705-5

Library of Congress Control Number: 2015944740

Springer Cham Heidelberg New York Dordrecht London

Printed on acid-free paper

Humana Press is a brand of Springer
Springer International Publishing AG Switzerland is part of Springer Science+Business Media
(www.springer.com)

Preface

Dramatic developments in cell biology generally and methods for isolation, tagging and imaging cells specifically have resulted in major increases in our understanding of stem cell biology. This includes significant advances in identifying and characterizing tissue-specific stem cells and their specific niches. Given perhaps the surprising nature of multiple niches, i.e., their common components, the search remains intense to understand the feature and factors that result in the specificities of the niches for particular tissue-specific stem cells. To provide a summary of what we now know and what remains to be established, I have solicited contributions from several groups that are actively working on different tissue-specific stem cells. I am grateful to all of them for their contributions and thank them for their efforts in helping to make this an outstanding volume.

I thank Aleta Kalkstein for her support and help in getting this project off the ground.

I am also grateful to Emily Janakiram, who had a keen eye for missing details, and was instrumental in completion of the volume with the highest Springer standards.

Kursad Turksen

Contents

Editor and Contributors

About the Editor

Kursad Turksen received his Ph.D. from the University of Toronto in the area of osteoprogenitor biology and cell selection methodologies. He then did postdoctoral training at the Howard Hughes Medical Institute, University of Chicago, studying epidermal biology through use of genetically-altered mouse models. He joined the Ottawa Hospital Research Institute (previously known as the Loeb Research Institute) where he rose to the position of Senior Scientist in the Division of Regenerative Medicine. His research interests focus on stem cell biology, with a particular interest in the Claudin family of tight junction proteins and their role in epidermal lineage commitment and progression during development in health and disease.

Dr. Turksen is currently Editor in Chief of the Springer journal *Stem Cell Reviews and Reports* in addition to serving as series editor for the *Stem Cell Biology and Regenerative Medicine.*

Contributors

Juan Aréchaga Faculty of Medicine and Dentistry, Laboratory of Stem Cells, Development and Cancer, Department of Cell Biology and Histology, University of the Basque Country, Leioa, Spain; Faculty of Medicine and Dentistry, Analytical and High Resolution Biomedical Microscopy Core Facility, University of the Basque Country, Leioa, Spain

Caitlin M. Braitsch Department of Molecular Biology, University of Texas Southwestern Medical Center, Dallas, TX, USA

Celina Yuemin Chee Program in Cancer and Stem Cell Biology, Duke-NUS Graduate Medical School, Singapore, Singapore

Ondine Cleaver Department of Molecular Biology, University of Texas Southwestern Medical Center, Dallas, TX, USA

Süleyman Coşkun Yale Cardiovascular Research Center, Department of Internal Medicine, Vascular Biology and Therapeutics Program, and Yale Stem Cell Center, Yale University School of Medicine, New Haven, CT, USA; Department of Neurosurgery, Yale Program in Brain Tumor Research, Yale School of Medicine, The Anlyan Center, New Haven, CT, USA

Raquel Costa-Almeida 3B's Research Group, Biomaterials, Biodegradables and Biomimetics, University of Minho, Guimarães, Portugal; ICVS/3B's Government Associate Laboratory, University of Minho, Braga, Portugal

Ali R. Djalilian Department of Ophthalmology and Visual Sciences, University of Illinois at Chicago, Chicago, IL, USA

Nicholas D. Evans Faculty of Medicine, Bone and Joint Research Group, Centre for Human Development, Stem Cells and Regeneration, Southampton General Hospital, Southampton, UK

Franklin Garcia-Godoy Department of Bioscience Research, College of Dentistry, University of Tennessee Health Science Center, Memphis, TN, USA

P. Gershovich 3B's Research Group, Biomaterials, Biodegradables and Biomimetics, University of Minho, Guimarães, Portugal; ICVS/3B's Government Associate Laboratory, University of Minho, Braga, Portugal

M.E. Gomes 3B's Research Group, Biomaterials, Biodegradables and Biomimetics, University of Minho, Guimarães, Portugal; ICVS/3B's Government Associate Laboratory, University of Minho, Braga, Portugal

A.I. Gonçalves 3B's Research Group, Biomaterials, Biodegradables and Biomimetics, University of Minho, Guimarães, Portugal; ICVS/3B's Government Associate Laboratory, University of Minho, Braga, Portugal

Karen K. Hirschi Yale Cardiovascular Research Center, Department of Internal Medicine, Vascular Biology and Therapeutics Program, and Yale Stem Cell Center, Yale University School of Medicine, New Haven, CT, USA

Zhengqing Hu Department of Otolaryngology-Head and Neck Surgery, Wayne State University School of Medicine, Detroit, MI, USA

George T.-J. Huang Department of Bioscience Research, College of Dentistry, University of Tennessee Health Science Center, Memphis, TN, USA

Tohru Itoh Laboratory of Cell Growth and Differentiation, Institute of Molecular and Cellular Biosciences, The University of Tokyo, Bunkyo-ku, Tokyo, Japan

Prachi Jain Department of Molecular Biology and Genetics, Cornell University, Ithaca, NY, USA

Mohamed Jamal Department of Endodontics, Henry M. Goldman School of Dental Medicine, Boston University, Boston, MA, USA

Agnieszka A. Janeczek Faculty of Medicine, Bone and Joint Research Group, Centre for Human Development, Stem Cells and Regeneration, Southampton General Hospital, Southampton, UK

Ingrid Lohmann Centre for Organismal Studies (COS) Heidelberg, Cell Networks – Cluster of Excellence, University of Heidelberg, Heidelberg, Germany

Babita Madan Program in Cancer and Stem Cell Biology, Duke-NUS Graduate Medical School, Singapore, Singapore

Mercede Majdi Department of Ophthalmology and Visual Sciences, University of Illinois at Chicago, Chicago, IL, USA

Pablo Moreno Faculty of Medicine and Dentistry, Laboratory of Stem Cells, Development and Cancer, Department of Cell Biology and Histology, University of the Basque Country, Leioa, Spain

Tracey A. Newman Faculty of Medicine, Bone and Joint Research Group, Centre for Human Development, Stem Cells and Regeneration, Southampton General Hospital, Southampton, UK

Catherin Niemann Center for Biochemistry, Medical Faculty and Center for Molecular Medicine Cologne, University of Cologne, Cologne, Germany; Institute for Biochemistry, Center for Molecular Medicine Cologne, University of Cologne, Cologne, Germany

Richard O.C. Oreffo Faculty of Medicine, Bone and Joint Research Group, Centre for Human Development, Stem Cells and Regeneration, Southampton General Hospital, Southampton, UK

Fani Papagiannouli Centre for Organismal Studies (COS) Heidelberg, Cell Networks – Cluster of Excellence, University of Heidelberg, Heidelberg, Germany

R.L. Reis 3B's Research Group, Biomaterials, Biodegradables and Biomimetics, University of Minho, Guimarães, Portugal; ICVS/3B's Government Associate Laboratory, University of Minho, Braga, Portugal

Karen Reuter Center for Biochemistry, Medical Faculty and Center for Molecular Medicine Cologne, University of Cologne, Cologne, Germany

M.T. Rodrigues 3B's Research Group, Biomaterials, Biodegradables and Biomimetics, University of Minho, Guimarães, Portugal; ICVS/3B's Government Associate Laboratory, University of Minho, Braga, Portugal

Edoardo Scarpa Faculty of Medicine, Bone and Joint Research Group, Centre for Human Development, Stem Cells and Regeneration, Southampton General Hospital, Southampton, UK

Unai Silván Faculty of Medicine and Dentistry, Laboratory of Stem Cells, Development and Cancer, Department of Cell Biology and Histology, University of the Basque Country, Leioa, Spain; Uniklinik Balgrist, Institute for Biomechanics, ETH Zürich, Zürich, Switzerland

Rahul S. Tare Faculty of Medicine, Bone and Joint Research Group, Centre for Human Development, Stem Cells and Regeneration, Southampton General Hospital, Southampton, UK

Tudorita Tumbar Department of Molecular Biology and Genetics, Cornell University, Ithaca, NY, USA

David M. Virshup Program in Cancer and Stem Cell Biology, Duke-NUS Graduate Medical School, Singapore, Singapore

Alex B. Wang Department of Molecular Biology and Genetics, Cornell University, Ithaca, NY, USA

Lisa Wasielewski Rosalind Franklin University, North Chicago, IL, USA

Jinhua Yu Key Laboratory of Oral Diseases of Jiangsu Province, Institute of Stomatology, Nanjing Medical University, Nanjing, Jiangsu, China

The Hair Follicle Stem Cell Niche: The Bulge and Its Environment

Alex B. Wang, Prachi Jain and Tudorita Tumbar

Abbreviations

ASK1	Apoptosis signal-regulating kinase 1
BCC	Basal cell carcinoma
Bmp	Bone morphogenic protein
BrdU	Bromodeoxyuridine
CDKI	Cyclin-dependent kinase inhibitor
DMBA	9,10-dimethyl-1,2-benzanthracene
DP	Dermal papilla
FACS	Fluorescence-activated cell sorting
Fgf	Fibroblast growth factor
GFP	Green fluorescent protein
H2B	Histone 2B
HF	Hair follicle
HH	Hedgehog
^{3}HTdR	Tritiated thymidine
iDTR	Inducible diphtheria toxin receptor
IRS	Inner root sheath
K	Keratin
ORS	Outer root sheath
Lgr5	Leucine rich repeat containing G protein coupled receptor 5
LRC	Label retaining cell
MAP3K	Mitogen-activated protein kinase kinase kinase

A.B. Wang · P. Jain
Department of Molecular Biology and Genetics, Cornell University, 260 Biotechnology Bldg, Ithaca, NY 14853, USA
e-mail: abw85@cornell.edu

P. Jain
e-mail: pj99@cornell.edu

T. Tumbar (✉)
Department of Molecular Biology and Genetics, Cornell University, 258 Biotechnology Bldg, Ithaca, NY 14853, USA
e-mail: tt252@cornell.edu

K. Turksen (ed.), *Tissue-Specific Stem Cell Niche*,
Stem Cell Biology and Regenerative Medicine,
DOI 10.1007/978-3-319-21705-5_1

MHC	Major histocompatibility complex
MSC	Melanocyte stem cell
NFATc1	Nuclear factor of activated T cells, cytoplasmic, calcineurin dependent 1
NSC	Neural stem cell
ROS	Reactive oxygen species
Runx1	Runt-related transcription factor 1
SC	Stem cell
SCC	Squamous cell carcinoma
SG	Sebaceous gland
Shh	Sonic hedgehog
TGF-β	Transforming growth factor, beta
TPA	12-O- tetradecanoylphorbol-13-acetate
VEGF	Vascular endothelial growth factor
Wnt	Wingless-type MMTV integration site family

1 Introduction

The mouse hair follicle (HF) has emerged as a powerful system for studying the behavior of stem cells (SCs) within their native tissue context. A major focus of the field has been to understand the regulatory mechanisms underlying self-renewal, differentiation, and quiescence. In the HF, this effort has greatly benefitted from the fact that its SCs and progeny with different degrees of differentiation are spatially and temporally restricted.

The HF is composed of distinct compartments of SCs, progenitor cells, and terminally differentiated cells. The hair follicle stem cells (HFSCs) are located in a region of the follicle called the bulge. This anatomical feature is an outcropping of the outer root sheath (ORS), the outermost layer surrounding the HF, and is located below the sebum producing structure, known as the sebaceous gland (SG). The bulge behaves as a “storage” niche that contains a large population of clustered SCs. SCs within the bulge may remain in the niche and self-renew for extensive periods of time, or they may be expelled from the bulge to become early progenitor cells in the hair germ at the base of the hair. The hair germ expands by divisions to form the matrix cells, which constitute bona fide, mature progenitor cells that subsequently terminally differentiate to form concentric layers of post-mitotic cells known as the inner root sheath (IRS) and the hair shaft in the center.

The HF cycles through stages of growth (anagen), regression (catagen), and rest (telogen), which occur synchronously across the back of the mouse [1] (Fig. 1). In anagen, the bulge SCs divide on average three times [2] and self-renew while remaining confined to the bulge [3]. In parallel, matrix progenitor cells proliferate and differentiate to fuel the growing hair shaft. Then, during catagen, the follicle below the bulge undergoes extensive apoptosis of the progenitor and differentiated cells, leaving a quiescent bulge to rest during telogen [4–6]. The cycling of the HF

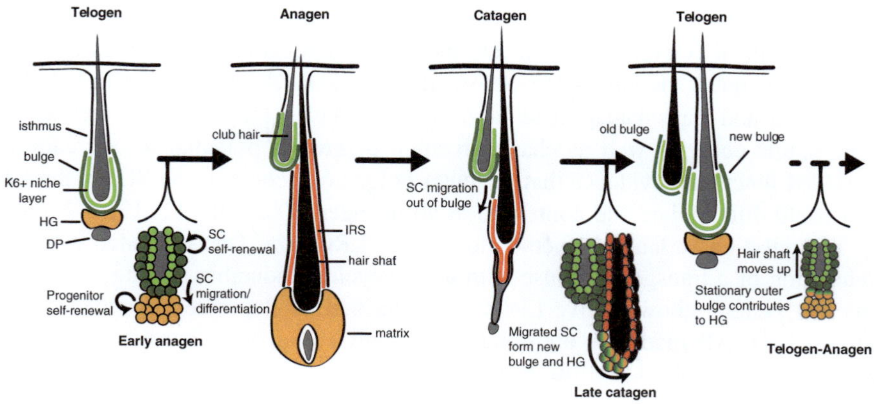

Fig. 1 During the hair cycle, HFs undergo synchronous stages of rest (telogen), growth (anagen), and regression (catagen). Stem cell activation occurs during the transitions between telogen-anagen (self-renewal) and catagen-telogen (differentiation) (highlighted in the offshoots between stages). During the transition from telogen to anagen, progenitor cells of the hair germ proliferate to form the matrix and bulge SCs either migrate into the hair germ to differentiate or remain in the bulge to self-renew. During the transition from catagen to telogen, bulge SCs first migrate out of the bulge into the ORS then as catagen ends, the migrated SCs form a new bulge around the hair shaft and a new hair germ beneath the new bulge. During telogen and into early anagen, the hair shaft continues to be pushed upwards, leaving some outer bulge cells to collapse around its former position and contribute more bulge cells to the new hair germ

depends on signals originating from the dermal papilla (DP), a cluster of mesenchymal cells at the base of the HF, and from the surrounding environment [7–9]. This includes dermal fibroblasts, blood vessels, nerves, fat, immune cells, and the arrector pili muscle, which are non-epithelial cells that surround the HF and influence HF activation and growth. In this review, we will highlight the mechanisms regulating the behavior of HFSCs as well as their interaction with various niche components during the hair cycle, both within the HF as well as non-epithelial components near the follicle. Additionally, at the end of the review we will discuss mechanisms utilized by HFSCs for quiescence, long-term maintenance, and cancer prevention.

2 The Bulge as the Epithelial Hair Follicle Stem Cell Niche

The bulge was originally posited to be the HFSC niche based on the slow-cycling properties of cells located in the outermost layer, as determined by label retention in nucleotide analog pulse-chase experiments [10]. In these experiments, the mouse is pulsed with repeated doses of labeled nucleotides such as tritiated thymidine (^{3}HTdR) or bromodeoxyuridine (BrdU) to label all of the proliferative cells. During the chase period (4–10 weeks), as expected, the short-lived but rapidly dividing progenitor cells of the matrix and the lower ORS dilute the label with each cell division until it is eventually lost. On the other hand, as expected of tissue SCs, bulge cells rarely divide and therefore remain as label retaining cells (LRCs). Dissection and transplantation of lacZ or GFP labeled HF bulges or derivatives from single cells, as well as characterization of growth potential in cell culture, provided functional evidence that the outer bulge layer contains the SCs of the HF and could differentiate and contribute to all lineages of the HF [11, 12].

The pulse-chase labeling method to identify LRCs was later upgraded with the generation of a transgenic mouse with a doxycycline inducible histone H2B-GFP labeling system, allowing live LRCs to be purified by fluorescence-activated cell sorting [13]. Alternatively, the outer bulge cells can also be purified using cell surface markers CD34 and α6-integrin [14] or transgenic labeling with the bulge specific keratin 15 (K15) promoter driving Cre recombinase expression, although K15 expression has also been reported in the hair germ during telogen [15, 16]. Sorting and transcriptional profiling of the LRCs have revealed a variety of transcription factors and signaling pathways that have greatly contributed to our understanding of how HFSC behavior is regulated (discussed in subsequent sections) [9, 17].

In addition to the bulge outer layer that contains the SCs, the bulge also contains an inner keratin 6 (K6)+ niche layer that serves to anchor the old hair (club hair) and promote bulge SC quiescence [18]. K6 is also expressed in the terminally differentiated companion layer between the ORS and IRS during anagen, but unlike this layer, the K6+ inner bulge layer shows expression of HFSC markers such as Sox9, Lhx2, Tcf3, and NFATc1. Unlike the other cells in the bulge, the K6+ inner layer cells make extensive adhesive contacts with the club hair and are post-mitotic [18]. The inner bulge layer constitutes the epithelial component of the HFSC niche and is known to act in synergy with the perifollicular non-epithelial components (discussed further in Sect. 6) to establish a niche environment that regulates HFSC quiescence and activation.

3 Quiescence Within the Bulge

As with some, although not all tissue SC populations, bulge SCs remain quiescent for the majority of their lifetime [19, 20]. Quiescence is maintained both by internal programming of the SCs as well as by external signaling. The major molecular driver of bulge quiescence is BMP signaling. During telogen, the DP, hair germ, and surrounding dermis express Bmp4 [21], and the bulge inner layer expresses Bmp6 [18]. Conditional deletion of Bmpr1a in the bulge causes the quiescent SCs to precociously activate and continuously proliferate without differentiation to form tumor-like cysts [22]. Conversely, introduction of ectopic Bmp by injection of Bmp4-soaked beads into the skin is sufficient to prevent anagen initiation and to maintain quiescence [21]. BMP signaling has been shown to maintain quiescence in the HFSCs in part through the upregulation of the transcription factor NFATc1.

NFATc1 is expressed in the quiescent bulge cells throughout the hair cycle and maintains quiescence at least in part by repressing the cyclin-dependent kinase CDK4 [23].

As mentioned previously, the bulge contains an inner layer of K6+ cells that promotes quiescence in the surrounding outer SC layer. Loss of the K6+ layer by hair plucking or ablation with iDTR in Sox9-CreER mice induces the surrounding CD34+ bulge cells to proliferate and initiate anagen [18, 24]. In both cases, not only the inner K6 layer but also some of the outer, SC containing bulge layer is lost, raising questions about data interpretation attributing loss of SC quiescence solely to damage of the inner layer. Supporting the model in which the K6+ inner layer is required for SC quiescence, loss of cells from the outer layer alone via K15-Cre, led to less robust induction of SC proliferation. The specific mechanisms required in the communication between the outer SC layer of the bulge and the inner K6+ layer remain to be established, although it has been suggested that BMP secretion from the inner layer might play a central role [18].

Quiescence is an essential strategy for long-term SC survival. Over-proliferation of the bulge poses the risk of premature hair loss due to exhaustion of the SC pool [25–27]. To promote quiescence, HFSCs express high levels of cyclin-dependent kinase inhibitors (CDKIs) such as p21, p27, p57, and p15 [28, 29]. While the expression of multiple CDKIs likely provides some redundancy, p21 has been implicated in ensuring that activated HFSCs promptly exit the cell cycle at early catagen after a defined number of divisions. Specifically, the p21 knockout HFSCs show on average one to two additional cell divisions prior to entering quiescence potentially due to increased apoptosis and available space in the SC niche [29]. Because p21 was upregulated in the Runx1 KO, which showed prolonged telogen [30, 31], it was possible that p21 might be responsible for blocking HF proliferation in this genetic background. Surprisingly, in a p21-null genetic background, Runx1 knockout induced quiescence was not overcome, but instead it was even further and excessively prolonged [29]. This raised the puzzling question as to how inhibiting a cell cycle inhibitor (p21) resulted in enhanced quiescence, the opposite effect from the expected. The explanation might reside in the specific upregulation of other CDKIs, in particular p15, which occurs only in the double p21 and Runx1 knockout. P21 has been previously shown to have a dual role as both a CDKI and transcriptional repressor in skin epithelial cells [32], and more recently p27 was shown to work in a similar manner in a different system [33]. Intriguingly, it appears that p21 and Runx1 synergistically repress the p15 promoter at the transcriptional level [29]. This suggests a feedback mechanism in HFSCs that reinforces quiescence by regulating the levels of CDKIs within SCs via the dual function of some CDKIs as cyclin dependent kinase inhibitors and transcriptional repressors [29]. This CDKI mechanism as well as quiescence promoting signaling from the DP and inner bulge layer highlights the variety of methods used to keep HFSCs quiescent. Beyond ensuring that enough HFSCs are maintained to sustain hair growth, quiescence also plays an important role in tumor suppression, which we will discuss later in the review (see Sect. 7).

4 Activation of HFSCs

Bulge activation occurs during two temporally distinct stages: activation for self-renewal occurs during early anagen, whereas activation for differentiation primarily occurs during catagen/telogen [3, 34]. At the beginning of anagen, activation of proliferation occurs in two steps in two HF compartments, first in the hair germ, which is fated to differentiate, and second in the bulge, which self-renews.

The hair germ is a structure located at the base of the telogen HF between the bulge and the DP. The hair germ originates from SCs that migrated out of the bulge into the ORS during early catagen [18, 34, 35]. These cells escape apoptosis during catagen and remain as a cluster of cells beneath the bulge in telogen. These cells are thought of as differentiation “primed” SCs or early progenitor cells, also referred to as activated SCs. When removed from their native tissue environment and plated in cell culture, hair germ cells show a more rapid initial colony formation and short-term expansion than bulge cells, followed by exhaustion and inability to survive multiple passages [34]. Rapid growth and short-term survival are well-established characteristics that define progenitor cells. As of yet, the mechanisms governing this transition from HFSCs to hair germ progenitor cells are not well understood. However, our lab has recently found that the quiescent HFSCs that migrate out of the bulge express high levels of the transcription factor Runx1 as they lose contact with the inner bulge layer [36]. In this study, transgenic overexpression of Runx1 was sufficient to induce catagen and to drive the transition of HFSCs to the progenitor fate, as shown by gene expression profiling and functional studies.

While an initial hair germ forms from bulge cell migration at catagen [18, 34, 35], bulge cells continue to contribute additional cells to the pre-existing hair germ during telogen and very early anagen [3, 37]. This was demonstrated by single bulge cell lineage tracing from our laboratory, in which tracking the distribution of LacZ+ cells within the bulge suggested that during telogen the entire bulge outer layer is shifted downwards with respect to the hair shaft adding new cells to the pre-existing hair germ [3]. The most likely explanation is that the club hair continues to move upwards, dragging the inner bulge layer with it and leaving cells from the outer bulge layer to collapse underneath, thus enlarging the hair germ. Bulge cells that lose contact with the inner layer gain Runx1 expression and as a consequence lose CD34, thus becoming new hair germ cells during telogen/early anagen [3, 36].

Under normal homeostasis, the hair germ is fated to divide in anagen and expand into the matrix, which in turn will terminally differentiate [38]. However, upon injury to the bulge, the hair germ cells are capable of returning to the bulge to replace the lost SCs [35, 38]. This suggests that the hair germ is not fully committed to a progenitor fate, and is consistent with the reversible effects of Runx1 expression on turning the bulge cells to a hair germ phenotype [36]. In accordance with its position directly above the DP, the hair germ is first to receive the activating Wnt and Fgf signals at the start of anagen and begin proliferation [34]. There must also be a special intrinsic state of these cells that allows them to readily respond to activation signals, as suggested by their rapid growth in cell culture. This "activated" state is likely induced by Runx1 through its many target genes implicated in cell size growth and active metabolism, and is confirmed by the rapid growth of keratinocytes with elevated Runx1 levels and by delayed activation of hair germ cell proliferation in the Runx1 epithelial knockout mouse [30, 31, 36, 39].

After the hair germ became activated to divide and for the matrix, a second wave of activation triggers proliferation of the HFSCs in their bulge location. Based on the physical separation of the bulge from the activating signals of the DP and the delayed proliferation following hair germ expansion at anagen initiation, it has been proposed that the proliferating matrix cells actually provide the activating signal for HFSCs rather than the DP. Unlike the hair germ, the bulge does not show activation of WNT and TGF-β pathways, whose signals originate from the DP, but instead the bulge depends on sonic hedgehog (Shh) signaling originating from the matrix [40]. In order to overcome the distance between the bulge and the matrix, the bulge increases its sensitivity to Shh by upregulating hedgehog coreceptors such as Gas1 [40]. Self-renewal of HFSCs occurs in parallel with the differentiation of the matrix cells; the latter is pushed downwards as the HF bulb grows into the dermis, thus distancing the Shh-activating matrix signals from the bulge SCs. After a few divisions the bulge outer layer cells become extremely crowded, likely triggering contact inhibition signals via the Hippo/Yap pathway and p21 [29, 41] to synergize with fading activation signals to reinstate bulge-cell quiescence during mid-anagen.

Bulge HFSCs can also be precociously activated in response to wounding. Upon transplantation or skin wounding, bulge cells display mutipotency and will contribute to neighboring epithelial populations such as the epidermis and the SG [11, 13, 42]. Precociously activated HFSCs migrate out of the bulge and contribute to epidermal wound healing by acting as short-lived progenitors rather than serving as long-term epidermal SCs [43]. Similarly, the SG is maintained independently from the bulge unless disturbed by wounding [44]. Instead, SCs in the isthmus above the bulge may normally contribute to both the epidermis and sebaceous gland during homeostasis [45, 46].

As can be seen, activation of bulge SCs is very tightly controlled, only transiently activating during narrow windows at the beginning of anagen, catagen, and in response to wounding. Interestingly, the bulge niche exclusively houses quiescent and self-renewing SCs, while cells are first exported out of the bulge before differentiation occurs. Presumably signaling within the bulge niche inhibits differentiation, and further investigation into the mechanisms of how the bulge maintains this delineation will be of great interest for the future.

5 Heterogeneity Within the Bulge

Among HFSCs in the bulge there is heterogeneity. Although all HFSCs have the potential to self-renew or differentiate, the two fates do not actually occur at equal frequencies. Single-bulge cell lineage tracing reveals that during early anagen, most

HFSCs remain in the bulge to self-renew through symmetric divisions, as indicated by the orientation of doublets parallel to the basement membrane during early chases, or remain quiescent. On the other hand, only a minority (14 %) of the telogen bulge is exported into the pre-existing hair germ to differentiate [2, 3]. Live-cell imaging has confirmed the orientation of cell divisions relative to the basement membrane and further revealed that fate selection of the anagen bulge depends on location of the HFSC within the vertical axis of the bulge [38]. With this method, Rompolas et al. genetically labeled single bulge cells with *K19-CreER;Rosa-stop-tdTomato* mice, then lineage traced the fate of the labeled cell in vivo through the hair cycle. In general, HFSCs in the upper half of the telogen bulge either remain quiescent or undergo self-renewal and remain in the bulge during anagen and the subsequent telogen. On the other hand, HFSCs in the lower half of the telogen bulge are more likely to exit the bulge into the ORS in late anagen, and become hair germ cells in the subsequent catagen. Labeled hair germ cells exclusively contribute to the differentiated lineages in the subsequent anagen. While the molecular differences between the upper and lower bulge have yet to be characterized, quiescence-related NFATc1 expression is more prominent in the upper bulge while the gene associated with more actively cycling cells, LGR5 (see below), is preferentially detected in the lower bulge. The relationship between location and HFSC fate choice is likely a product of gradients of signaling molecules that pattern the tissue prior to SC activation.

In addition to differences in upper and lower bulge SCs, there is also additional heterogeneity of the bulge cells due to exact relationship to the basement membrane. There are two populations of basal bulge cells that both express CD34 and have label retention ability, but one of them, the so-called "suprabasal" SCs, is more inner relative to the basement membrane and display lower levels of α6-integrin [28]. FACS isolated samples of the two populations display similar colony formation capacity in cell culture and share similar gene expression profiles. However, in vivo, the suprabasal cells are less proliferative based on BrdU labeling and show a corresponding upregulation of the growth inhibitory factors FGF-18 and BMP-6. The significance of this heterogeneity remains unknown.

The basal layer of the lower bulge also contains Lgr5 + SCs [47]. Lgr5+ bulge cells are biased towards the lateral side opposite the old bulge [47], which shows greater proliferation in anagen [2, 37]. Furthermore, separating bulge cells based on their number of divisions as indicated by H2B-GFP levels showed high Lgr5 mRNA expression in the more actively cycling cells and low expression in the cells that divided 0–1 times [3]. Thus, Lgr5+ cells represent the more actively cycling population, and consistent with this conclusion lineage tracing places them mostly in the proliferative hair germ and the anagen ORS [47]. Lineage tracing also shows that the Lgr5+/CD34+ cells in the upper ORS, which are SCs that escaped the old bulge during late anagen/early catagen, are capable of surviving long-term through the hair cycle and contribute to the new bulge SCs, K6+ inner bulge layer, and hair germ [18, 47]. The Lgr5+ cells largely overlap with the Runx1+ cell population, which have been tracked by lineage tracing in a similar manner with Lgr5 [36].

The functional significance of Lgr5 negative cells, which are the more quiescent LRCs that reside in the old bulge or remain on one bulge side after the old bulge is destroyed, remains unclear. Although it has been proposed that they act as a reserve population of SCs [18], selective lineage tracing or killing of the most quiescent LRCs in the bulge has yet to be performed. Some of these LRCs from the old bulge seem to migrate laterally along the basement membrane during catagen, also contributing cells to the new bulge, without having to travel downwards through the upper ORS [37].

In addition to housing the epithelial HFSCs, the bulge is also the niche for the neural crest derived melanocyte SC (MSC) population [48]. MSC activation and differentiation follow a similar spatial-temporal pattern as the epithelial HFSCs [49]. Epithelial targeted mutations that block hair cycle progression also block MSC activation. At the beginning of anagen, MSCs contribute cells to the matrix. These cells proliferate in synchrony with the matrix epithelial cells and generate mature melanocytes in the inner cortex of the HF. These cells produce a pigment called melanin and transfer this pigment to the epithelial matrix cells differentiating into the hair shaft. During catagen, the progenitor and mature melanocytes undergo apoptosis along with the epithelial cells of the matrix. In the bulge, MSCs can be distinguished from epithelial HFSCs by their expression of dopachrome tautomerase (DCT) and tyrosine kinase receptor (KIT), which are specific to the melanocyte lineage [50].

The coordination between MSC activity and hair growth is achieved through activating Wnt and inhibitory TGF-β signaling, and crosstalk between HFSCs and MSCs [51]. In the bulge, TGF-β signaling from HFSCs is critical for maintaining MSC quiescence [50]. Further highlighting the importance of HFSCs on MSCs, conditional ablation of the transcription factor NFIB in HFSCs promotes aberrant MSC differentiation, implicating NFIB's downstream target endothelins as important messengers between the SC populations [51]. Additionally, in anagen, Wnt signaling originating from the hair germ initiates MSC proliferation [52]. While HFSCs are critical for the maintenance of its neighboring MSCs, the converse influence of MSCs on HFSCs seems less critical, as MSCs are lost during aging with minimal effects on hair growth which continues to be independently driven from the epithelial compartment (see also Sect. 7 on aging).

The heterogeneity present in the epithelial bulge population is likely a reflection of localized differences in signaling environments determined by relative proximity to major signaling sources within the hair such as the bulge inner layer as well as from the DP and other non-epithelial neighbors of the bulge. The advent of advanced techniques for studying individual cells, such as single cell sequencing and live imaging is likely to reveal further heterogeneity in the behavior of bulge cells. Furthermore, examination of the heterogeneity in bulge behavior will be essential to our understanding of the mechanisms that dictate HFSC fate selection.

6 Interactions with the Non-epithelial Environment

An emerging area of investigation is how the HF stem cells interact with its surrounding environment. A variety of non-epithelial components are found neighboring the HF and contribute to the niche (Fig. 2). As previously mentioned, the DP is ensconced at the base of the HF and serves as a major signaling hub for the follicle. In addition, the entire HF is surrounded by a dermal sheath with a permanent dermal SC population residing at the base of the hair, a ring of blood vessels and nerves that wrap around the isthmus directly above the bulge, the arrector pili muscle attaches directly to the bulge, and adipocytes and immune cells are found in the surrounding dermis. These neighbors display a range of influences on the bulge cell behavior and on hair growth. The DP, adipocytes, and blood vessels are required for proper HF activation because they provide the follicle with key activation cues or nutrients. On the other hand, neighboring neurons and macrophages appear to allow the HF to adjust HFSC activation in response to the environmental conditions such as wounding.

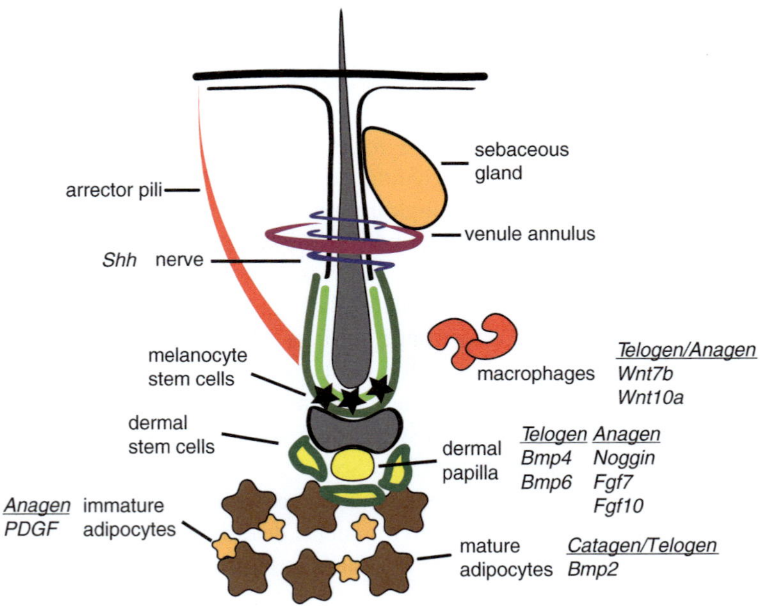

Fig. 2 Schematic of the non-epithelial environment surrounding the hair follicle. A number of these populations influence hair follicle homeostasis through both stimulatory and inhibitory signaling pathways

6.1 Dermal Papilla

As previously noted, the DP is a pocket of cells of dermal origin, which lies at the base of the HF, below the hair germ during telogen and encapsulated by the matrix during anagen. The DP has long been recognized as a critical signaling center regulating HF growth and quiescence. Loss of the DP by laser ablation arrests follicles at telogen, demonstrating the necessity of the DP for anagen progression [53]. Conversely, grafting of DP cells is sufficient for inducing hair growth in hairless skin [8]. Additionally, the number of DP cells effects the size of follicles, with smaller DPs producing shorter thinner hairs associated with aging [54].

FACS purification and gene expression profiling of the DP and its surrounding epithelial populations revealed potential crosstalk between the populations through Bmp, Wnt, Fgf, and Shh signaling, which is being borne out through subsequent studies [55]. The DP typically expresses high levels of HFSC quiescence promoting Bmp4 and Bmp6 [55]. However, to initiate hair germ proliferation at the start of anagen, the DP must also upregulate Bmp inhibitors such as Noggin and Sostdc1 [34, 56]. Following anagen initiation, the DP continues to inhibit Bmp signaling to control the rate of hair growth [57]. The transcription factor Sox2 in the DP has been demonstrated to maintain expression of Bmp inhibitor Sostdc1, and DP specific ablation of Sox2 results in reduced hair growth due to inhibited migration of differentiating matrix progenitors into the hair shaft. Another Bmp inhibitor Noggin forms a regulatory loop with epithelial expressed Shh to maintain both DP function and hair growth [58]. Additionally, Notch-Wnt5a signaling from the DP promotes terminal differentiation of matrix progenitor cells [59].

In addition to Bmp signaling's role in hair cycling, Bmp is also essential within the DP for proper function [60]. When Bmp receptor 1a (Bmpr1a) is knocked out in the DP, the DP loses its ability to initiate hair growth when engrafted onto skin [60]. Similarly for Wnt signaling, HFs with β-catenin ablated in the DP are unable to enter anagen [61]. However, while these studies show that Bmp and Wnt are needed by the DP, the importance of these signals specifically from the epithelium remains unclear.

Although in general the signals seem to flow unidirectionally from the DP towards the bulge, the opposite direction is most likely also true. Early experiments showed that transplanting a bulge into hairless skin is sufficient to initiate formation of a new DP from neighboring dermal cells [11]. Moreover, when activated β-catenin is constitutively expressed in the epithelial bulge, it signals the surrounding dermal cells to form a new DP within each round of hair cycling [62]. These data suggest that in special conditions the bulge cells can regain some embryonic-like capabilities, where mesenchymal-epithelial crosstalk is thought to be essential for de novo HF formation [7, 63].

6.2 Dermal Fibroblasts and the Dermal Sheath

Whereas much is known about the dermal papilla's relationship with the HF, relatively little is known about the HF's relationship with the surrounding dermal fibroblasts and the dermal sheath that outlines the hair. This is surprising because fibroblast have been long recognized as support cells for keratinocytes (skin epithelial cells) in cell culture [64]. These dermal cells are known to provide structural support for the epidermis and HF by producing collagen and other fibers for the extracellular matrix [65]. As mentioned previously, embryonic HF morphogenesis involves the crosstalk between epithelial cells and dermal fibroblasts [7, 63]. In embryonic skin, aggregates of dermal fibroblasts, precursors to the DP, form and induce HF morphogenesis from the developing epidermis [7]. Wnt signaling is believed to be the first signal from the dermal aggregates responsible for hair induction. The transcription factor Runx1 is highly but transiently expressed in the embryonic dermis and is essential for the maturation of emerging HFSCs and their long-term survival in the HF [39]. In this study lineage tracing using the Runx1-CreER knockin mouse showed that this population of Runx1-expressing cells are short-lived and do not survive past birth, perhaps explaining why new hairs are not induced in adulthood.

Nevertheless, dermal fibroblasts continue to influence HFSCs into adulthood. At the end of anagen and into telogen, fibroblasts express high levels of Bmp4, which helps to maintain quiescence in the bulge [21]. Signaling from the bulge to fibroblasts remains unexplored, although epithelial mutants have been shown to have an effect on the dermis. Epithelial activation of β-catenin not only induces dermal cells to form a new DP as stated previously [62], it also specifically reverts dermal fibroblasts around the isthmus to an embryonic state. These fibroblasts display increased proliferation and replaces mature collagen in the extracellular matrix with collagen subtypes found only in developing embryonic skin [66]. Why this effect is specifically seen in fibroblasts near the isthmus is unclear, but perhaps the isthmus maintains these cells in a more plastic state allowing them to be more easily reprogrammed.

A specialized layer of fibroblasts also forms the dermal sheath that lines the outside of the HF. The dermal sheath not only produces supportive collagen layers around the HF, but the lower dermal sheath at the base of the HF serves as a niche for bipotent dermal SCs that can renew the dermal sheath and DP [67, 68]. The dermal sheath cycles through growth and destruction in parallel with the HF [68]. The dermal sheath surrounds the entire HF during anagen, but then undergoes apoptosis during catagen, leaving only the dermal SCs around the DP in catagen. Then during anagen, the dermal SCs proliferate to form a new dermal sheath as well as contributing some cells to the DP. How this synchrony between the dermal sheath and epithelial cells is achieved remains unknown, but it is suggestive of mesenchymal-epithelial crosstalk between the populations. Depletion of the dermal SCs causes delayed entry into anagen and reduced hair length, even though the average number of DP cells remained unchanged, presumably due to compensating proliferation within the DP [68]. This suggests that either the dermal SC derived DP cells serve an additional stimulatory function independent of the resident DP cells, or perhaps the dermal sheath promotes proliferation. Additionally, the bulge has been shown to upregulate connective tissue growth factor, CTGF [13], which may stimulate recruitment of dermal sheath cells to around the HF. Overall, the relationship between dermal fibroblasts, particularly the dermal sheath, and HFSCs is poorly understood, and additional experiments are necessary to determine if these fibroblasts have additional functions in supporting the HFSC niche beyond formation of the extracellular matrix. In addition, it will be interesting to see if the extracellular matrix around the HF is heterogeneous in nature, perhaps dictating some of the spatial heterogeneity observed within the stem cell niche.

6.3 Adipocytes

Below the HF lies an intradermal adipocyte layer: the skin's own fat cells. Immature adipocyte progenitors, derived from the dermal resident adipocyte SCs are necessary and sufficient to drive the proliferation of HFSCs [69]. Using mouse mutants or pharmacological treatments to target various stages of adipogenesis, Festa et al. were able to establish the necessity of the adipocyte progenitor cells for bulge HFSC activation and anagen progression. The adipocyte progenitor cells seem to exert their influence on HFSCs through platelet derived growth factor (PDGF) signaling, which is received by the DP, although a specific knockout of PDGF in adipocytes has yet to be performed. During anagen, this adipocyte layer doubles in thickness from adipogenesis when the progenitor cells generate mature terminally differentiated adipocytes, paralleling the proliferation and differentiation of HFSCs [69]. The mature adipocytes generated in the intradermal adipocyte layer show periodic expression of the inhibitory signal Bmp2 [21]. Bmp2 expression is high in late catagen and early telogen but is absent in late telogen. Bmp2 from adipocytes and Bmp4 from the DP and dermal fibroblasts might be responsible for establishing a refractory period at the beginning of telogen. During this refractory period, the threshold for bulge SC activation is high, and even injury by hair plucking cannot induce anagen. Quiescence is maintained until Bmp levels are again lowered in the surrounding adipocytes and dermis, allowing anagen to be initiated [21].

Signaling from epithelial cells to adypocites remains unknown, but epithelial mutations that block hair cycle progression also block adipogenesis [27, 29, 30]. This strongly suggests that adipocytes wait for a response signal from the epithelial SCs before initiating adipogenesis. It is unclear if the first step in activation is when adipocyte SCs generate the progenitors that in turn must signal to epithelial cells, or whether the epithelial signals are required prior to this step in adipogenesis. In other words, it is currently unclear what the primary signal is that sets in motion the crosstalk between fat and epithelial cells in the skin. In addition, it is tempting to speculate that as the HFSCs begin to proliferate they signal to fat cells for the imminent necessity of vast sources of energy, which are then stored as lipids in mature adipocytes in preparation for rapid hair production/elongation. These lipids would then be subsequently delivered to the matrix cells, thus fueling the generation of the new hair shaft.

6.4 Blood Vessels

Nutrients are delivered to HF and epidermis through a microvascular network that surrounds most of the follicle and undergoes extensive remodeling and angiogenesis throughout the hair cycle [70]. Angiogenesis significantly increases during early anagen, perhaps to support the increased oxygen demands of the rapidly proliferating matrix. Consistent with this possibility, pharmaceutically inhibiting angiogenesis with the drug TNP-470 is capable of delaying anagen onset [70], whereas transgenic overexpression in epithelial cells of vascular endothelial growth factor (VEGF) increases angiogenesis and causes accelerated hair growth [71]. Transplantation assays demonstrate that the bulge and anagen matrix (but not the DP) are capable of stimulating angiogenesis [72, 73], however the signaling mechanism remains unknown. This suggests that rather than associating randomly with the HF, the microvascular network may be directed from the epithelial cells to form around these compartments.

Indeed, three-dimensional confocal imaging has revealed that a ring of venules permanently wraps around the HF isthmus, a compartment between the bulge and sebaceous gland that contains a distinct, rapidly cycling, SC population that renews the upper portion of the HF [74]. HFs can be formed de novo in mouse skin reconstituted from primary keratinocytes and dermal fibroblasts, and when transplanted the association between the isthmus and venules is reestablished. Consistent with recruitment to de novo HFs, the isthmus expresses Egfl6, a member of the epidermal growth factor (EGF) superfamily known to promote endothelial migration and angiogenesis [74]. As yet, whether or not the vascular ring has a role in HFSC function remains unknown. It is interesting to consider the possibility that blood vessels can contribute to oxygen availability and consumption, which in turn would support oxidative phosphorylation, and the generation of reactive oxygen species (ROS). ROS is a hallmark of cells undergoing rapid proliferation and has been shown to act as a differentiation signal in other SC systems [75]. In the HF, ROS signaling is necessary during anagen for matrix cell proliferation and survival and loss of mitochondrial ROS causes follicles to prematurely enter catagen [76].

It has been hypothesized that tissue stem cells in general maintain their quiescence and are protected against oxidative stress by residing in a low oxygen environment [77]. As with the neural, hematopoietic, and mesenchymal niches [77], expression of hypoxia markers suggests that the bulge may also be a low oxygen environment [78]. The blood vessels associated with the upper bulge are venous in nature, meaning the blood is low in oxygen and nutrients, and may therefore establish the hypoxic microenvironment [74]. This suggests that the presence of blood vessels in niche is not only important to provide nutrients to stem cells but also to create gradients of oxygen tension that could play an important role in regulating cell fate. It was demonstrated in the neural stem cell (NSC) niche that hypoxic stimuli lead to expression of hypoxia inducible factor 1 (HIF-1) [79]. HIF-1 in NSCs promotes stemness by upregulating expression of Notch target genes and cyclin D1 to promote cell cycle progression and self-renewal [80]. In the future, it will be interesting to examine the relationship between the blood vessel microenvironment, hypoxia, and HFSC fate. It will be particularly interesting to determine if signals from the bulge are capable of recruiting additional blood vessels in response to the demands of the hair cycle, and whether blood vessels are indeed essential for HF oxygen homeostasis or have additional unknown functions for HFSC survival and cell fate.

6.5 Neurons

Mechanosensory nerves wrap around the HF isthmus and make contacts with the upper bulge allowing the hair to sense when it is touched [81]. Additionally, these nerves form a perineural microenvironment that displays increased Shh signaling [82]. The perineural microenvironment establishes a distinct domain at the bottom of the isthmus directly above the bulge that is K15 negative and expresses the HH mediating transcription factor Gli1. Although hypothesized to contribute to bulge HFSC activity, removing the Shh microenvironment by denervation had no affect on the bulge or normal HF homeostasis. Denervation did however inhibit the ability of isthmus SCs to contribute long-term to epidermal wounds, suggesting a possible role for Shh signaling in this process [82]. While removing the nerves does not affect hair cycling, the converse experiment where nerves are augmented has not been performed yet. It also remains unclear how nerve terminations are guided to the HF, but the bulge cells produce high levels of brain derived neurotrophic factor (BDNF) as well as guiding and repelling signals such as ephrins [13]. These signals might be important in patterning the nerve terminations around the HF.

6.6 Immune Cells

Hair follicles are one of the few sites in the body, along with the brain and the eye, of relative immune privilege. In particular, the anagen bulge and matrix have decreased major histocompatibility complex (MHC) class I expression and increased expression of local immunosuppressants [83, 84]. Although the purpose of establishing immune privilege during anagen remains unclear, its maintenance is critical as loss of immune privilege and autoimmune attacks on the matrix and the bulge are associated with alopecia [85].

Interestingly, macrophages are present in the perifollicular environment and have been found to contribute to the activation of HFSCs during wounding and homeostasis [86, 87]. Following wounding, apoptosis signal-regulating kinase 1 (ASK1), a MAP3 K family member, is upregulated at the wound site to recruit and activate macrophages [86]. Failure to recruit macrophages to the wound in ASK1 mutant mice causes hair growth to be delayed at the wound. Furthermore, transplantation of activated macrophages is sufficient to induce hair growth. Macrophages may also contribute to HFSC activation in normal homeostasis. During telogen, macrophages residing around the follicle upregulate Wnt7b and Wnt10a into the perifollicular environment, however the Wnt ligands are not released from the macrophages unless they undergo apoptosis [87]. The signal that triggers macrophage apoptosis is unknown, but at the end of telogen, much of the resident macrophage population undergoes apoptosis, releasing the Wnt ligands that in turn promote HF entry into anagen. Ablation of the macrophages pharmacologically or with genetically inducible diphtheria toxin mice is also sufficient to release the Wnt ligands and initiate entry into anagen [87]. This is seemingly at odds with the previous finding that transplantation of macrophages is also sufficient for anagen induction, although it is possible the transplanted macrophages must also undergo apoptosis to affect hair growth.

Mice lacking $\gamma\delta$ T cells have also been shown to have shortened telogen stages [88], suggesting immune cells in general may have more influence on the hair than just protection from pathogens. Perhaps loss of immune cells to establish immune privilege is necessary for anagen entry, and the apoptotic release of Wnt ligands seen in macrophages is a signal to the HF that this has been achieved. Conversely, it is unclear whether return of immune cells after anagen is necessary. During anagen, the HF actively inhibits immune cells by the local production of immunosuppressants such as macrophage migration inhibitory factor [85], however it would be interesting to see if the bulge expresses factors to attract macrophages and other immune cells back following anagen.

6.7 *Arrector Pili Muscle*

The arrector pili muscle (APM) attaches to the HF at the bulge throughout the hair cycle and is responsible for piloerection, which aids in thermoregulation. The attachment of the APM at the bulge depends on the extracellular matrix protein nephronectin, which is secreted by bulge cells into the underlying basement membrane, [89]. In cell culture, dermal cells expressing the nephronectin receptor α8-integrin, which were used as surrogate for muscle cells, attached to extracellular nephronectin and in turn stimulated the expression of smooth muscle markers. These data suggested that bulge epithelial cells utilize nephronectin to promote muscle differentiation and perhaps dictate where the APM attaches to the follicle. Full nephronectin knockout caused the APM attachment site to shift upwards to the isthmus, with α8-integrin instead interacting with another nephronectin family member EGFL6. Full knockout of α8-integrin causes the APM to randomly attach either to the bulge or the EGFL6+ isthmus. Despite the loss of APM contact at the bulge, there appeared to be no change in the bulge or HF morphology [89]. However, it is unknown if augmenting differentiation of APM at the bulge has any effect.

In summary, the bulge SCs seem to be involved with their neighboring cells in distinct manners, following different rules of interaction. First, the bulge SCs stay intimately connected with their direct progeny in the inner layer and depend on them for their survival, quiescence, and to inhibit differentiation. Second, the bulge SCs seem to receive instructions from the DP regarding their activation and differentiation, but we have yet to learn about any instructive responses from epithelial cells towards the DP. With that said, the bulge is capable, in the right setting to build its own DP from neighboring dermal cells. Similarly, this is seen with the APM, blood vessels, and nerves, demonstrating the bulge is equipped with the mechanisms necessary to recruit its neighbors for proper HF physiology. In the case of the APM and nerves, this recruitment occurs without receiving feedback signals and without any clear effect on HFSC status or hair growth, while for dermal cells, blood vessels, and fat cells a meaningful cross-talk is apparent.

7 Maintenance and Aging

Throughout their lifetime, SCs are prone to acquiring mutations that could lead to disease and cancer. Niches are commonly viewed as special microenvironments, protecting SCs from assaults such as genomic damage or proliferating and differentiating signals, so that they can survive long-term and maintain their potential to support the organism throughout its lifetime (Fig. 3). A low oxygen environment helps the SCs avoid oxidative stress and promotes quiescence. Also, high oxygen and ROS are known signals for differentiation [75, 76], which SCs must avoid if they are to maintain their potential long-term. As suggested above (see above section on blood vessels) the bulge SCs might not only be equipped with the right environment for maintenance but may in fact actively contribute to creating that environment around their niche.

Bulge quiescence is not only an essential mechanism for ensuring sufficient SCs persist for long-term hair growth. Not surprisingly, quiescence also proves to be a valuable method of tumor suppression. While bulge cells may serve as cancer initiating cells (discussed further in the next section), it has recently been shown that normal activation of the HFSCs is in fact required for tumorigenesis [90]. Using an inducible oncogenic Kras mutation ($Kras^{G12D}$), White et al. demonstrated that bulge hyperplasia and tumorigenesis only occurred when $Kras^{G12D}$ was induced during the telogen-to-anagen transition, the period when bulge cells are activated to divide for self-renewal. During both telogen and late anagen/catagen, the bulge cells are quiescent and unable to initiate tumorigenesis. This quiescence-associated resistance to $Kras^{G12D}$ is not simply a matter of not

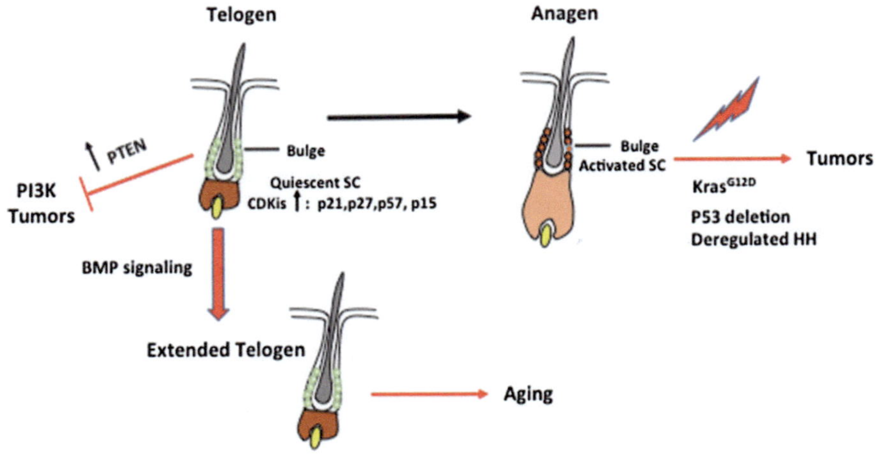

Fig. 3 Quiescence maintains the stem cell pool by preventing their premature exhaustion. Expression of cyclin-dependent kinase inhibitors (CDKIs) such as p21, p27, p57, and p15 promote quiescence. The extended telogen appears to be a result of increased sensitivity of HFSCs to Bmp signaling which leads to aging. Quiescence also acts as a method of tumor suppression, quiescent bulge cells are resistant to KrasG12D dependent tumors *via* tumor suppressor Pten, which inhibit the PI (3) K pathway. Bulge tumorigenesis only occurred when KrasG12D was induced during the telogen-to-anagen transition i.e. upon stem cell activation. Activated HFSCs are the origin of different cancers caused by genetic alterations such as deregulated HH signaling, conditional activation of oncogenic Kras and deletion of p53 in K15+ bulge cells but not from matrix progenitor cells

proliferating. Rather, quiescent SCs depend on Pten to inhibit the phosphatidylinositol-3-OH kinase (PI(3)K) pathway downstream of Ras.

Since the bulge is located in the proximity of the skin surface, HFSCs are susceptible to damage from UV radiation. Therefore, the HFSCs must have been endowed with their own intrinsic properties allowing them to maintain their integrity. In contrast with embryonic SCs, which readily undergo apoptosis in response to DNA damage, it has been shown that HFSCs resist apoptosis in favor of repairing the damage [91]. Bulge cells express higher levels of the anti-apoptotic gene Bcl-2 compared to the rest of the epidermis, which helps HFSCs resist apoptosis following exposure to ionizing radiation [92]. They also express high levels of the key non-homologous end joining repair pathway protein DNA-PK, allowing HFSCs to repair the damage more rapidly than the rest of the epidermal populations.

In addition to specialized repair, the bulge also expresses a number of transcription factors necessary to maintain stemness [27, 93, 94]. Two transcription factors, Sox9 and Lhx2 have been found to maintain the stemness of bulge cells by promoting quiescence and inhibiting differentiation [93–95]. Sox9 is expressed early on in embryonic HF morphogenesis and is necessary for SC specification [96, 97]. When Sox9 is ablated from the adult HF, the bulge cells are unable to return to quiescence after activation in anagen and begin terminally differentiating into epidermal cells [94]. Similarly, epithelial Lhx2 knockout mice display enhanced bulge proliferation as well as abnormal differentiation into sebaceous gland cells [93, 95]. On the other hand, transcription factor Tbx1 acts to allow HFSCs to exit quiescence and self renew by suppressing Bmp signaling during anagen [27]. As such, epithelial Tbx1 knockout follicles have prolonged quiescence and challenging the SCs with depilation induced activation results in depletion of the bulge and dormant HFs resembling those found during aging.

Aging is associated with a decline in maintenance for various tissues. In hair, aging is associated with greying, extended quiescence, hair shaft thinning, and decreased density. Age related hair graying is caused by depletion of melanocyte SCs due to apoptosis and premature differentiation [49, 98]. One of the hallmarks of aging that leads to a decline in SC functionality and senescence is shortened telomeres [99]. Since telomere shortening is a consequence of DNA replication, maintaining quiescence is likely important for delaying telomere shortening. Additionally, bulge cells express telomerase, a specialized DNA polymerase that extends telomeres [100, 101], and therefore maintain longer telomeres than the rest of the HF [102]. However, telomerase is insufficiently expressed in adult SCs to maintain telomere length, and indeed bulge cells display significantly shorter telomeres in aged mice than in young mice [102]. Bulge cells that reach critically short telomere length lose their ability to activate and leave the niche to contribute to hair growth [100].

The hair cycle also displays age-related changes, characterized by extended periods of telogen [103]. These extended telogens appear to be a result of increased sensitivity of HFSCs to Bmp signaling. Not only do aged HFs resist homeostatic cycling, but also, even forced entry into anagen by depilation is significantly delayed in older mice. Further, following depilation, transcription profiles of aged anagen HFSCs more closely resemble HFSCs from young telogen than young anagen. This age-related change in behavior and shortening of telomeres may reflect a shift in priorities for HFSCs, as hair growth and regeneration is compromised in favor of the tumor suppressing benefits of an extended quiescence.

8 Bulge as the Origin of Skin Cancers

While the long-term maintenance of SCs is necessary for tissue regeneration, their extended lifespan also puts them at an increased risk of acquiring oncogenic mutations to initiate tumors and cancers [104]. Genetic lineage tracing has placed the bulge as the cell of origin for the two most common forms of cancer: skin basal cell carcinoma (BCC) and squamous cell carcinoma (SCC). BCCs are characterized by deregulation of Hedgehog (HH) signaling. Using a conditional Smoothened mutant (SmoM2) to constitutively activate HH signaling and induce BCC formation, genetic lineage tracing showed that BCCs originate from the infundibulum and interfollicular epidermis [105]. Conditionally activating HH in either the bulge or matrix populations with lineage specific Cre reporter expression was unable to initiate BCCs. However, in a different BCC model utilizing a deletion of the HH inhibitor Ptch1 and exposure to ionizing radiation gave rise to BCCs predominantly originating from K15+ bulge SCs [106]. These opposing models likely represent two different subtypes of BCCs. A further complication in the identification of the cell of origin for skin cancer may lie in the recruitment of bulge SCs to the interfollicular epidermis in response to injury [13, 43]. Indeed, in HH activation by Ptch1 deletion or Gli1 overexpression, wounding promotes the contribution of bulge cells to tumor formation [107].

Bulge SCs have also been shown to act as the cell of origin for SCCs [108, 109]. Lineage tracing has shown that SCCs could be initiated from the conditional activation of oncogenic Kras and deletion of p53 in K15+ bulge cells but not from matrix progenitor cells [108]. A bulge origin for SCCs has also been demonstrated by lineage tracing in tumors induced chemically with DMBA/TPA [110]. In addition to originating from HFSCs, SCCs have also been demonstrated to utilize the intrinsic SC machinery to promote tumorigenicity, such as with the HFSC transcription factor Runx1 [31, 109]. Intriguingly, while Runx1 is dispensable for HFSC survival and long-term function [30], it is strictly required for both tumor initiation and maintenance in DMBA/TPA and Kras induced mice [109]. Identification of the bulge as cancer initiating cells provides an important starting point for determining the genetic lesions involved in tumor initiation. Importantly, beyond extending our understanding of cancer development, the gene signature of the bulge may help to identify key pathways or targets, such as Runx1, for new therapeutic approaches.

9 Conclusion

As our knowledge of the HF progresses, our conception of the bulge has expanded far beyond simply a storehouse of SCs awaiting activation. Regulation of HFSC behavior is an intricate process that involves the bulge acting as a central signaling hub, integrating signals from the rest of the HF and its non-epithelial neighbors. Normal hair growth not only depends on the DP, but also on signaling from adipocytes and macrophages. Additionally, the bulge does not simply receive signals, but also transmits signals to attract blood vessels, neurons, and muscle cells to form a complete follicular unit. As our tools to manipulate and examine the bulge and its neighbors expand, it will be exciting to discover what new interactions can be unveiled.

Examination of SC self-renewal and differentiation has been and will continue to be an area of intense interest, bringing us ever closer to the goal of tissue regeneration. However, of equal importance is SC quiescence. A main function of the bulge niche environment is concerned with enforcing HFSC quiescence, which is as an important mechanism for genomic maintenance and tumor suppression. Quiescent HFSCs are not just waiting idly in their niche. Quiescence is likely not just a “resting” phase but rather a more dynamic process than previously believed, with HFSCs preparing for the next stage of activation. Further investigation of quiescence is needed to better understand its role in SC maintenance and disease states.

References

1. Muller-Rover S, Handjiski B, van der Veen C, Eichmuller S, Foitzik K, McKay IA, et al. A comprehensive guide for the accurate classification of murine hair follicles in distinct hair cycle stages. J Invest Dermatol. 2001;117(1):3–15. doi:10.1046/j.0022-202x.2001.01377.x.
2. Waghmare SK, Bansal R, Lee J, Zhang YV, McDermitt DJ, Tumbar T. Quantitative proliferation dynamics and random chromosome segregation of hair follicle stem cells. EMBO J. 2008;27(9):1309–20. doi:10.1038/emboj.2008.72.
3. Zhang YV, Cheong J, Ciapurin N, McDermitt DJ, Tumbar T. Distinct self-renewal and differentiation phases in the niche of infrequently dividing hair follicle stem cells. Cell Stem Cell. 2009;5(3):267–78. doi:10.1016/j.stem.2009.06.004.
4. Cotsarelis G. Epithelial stem cells: a folliculocentric view. J Invest Dermatol. 2006;126 (7):1459–68. doi:10.1038/sj.jid.5700376.
5. Fuchs E. Scratching the surface of skin development. Nature. 2007;445(7130):834–42. doi:10.1038/nature05659.
6. Rompolas P, Greco V. Stem cell dynamics in the hair follicle niche. Semin Cell Dev Biol. 2014;25–26:34–42. doi:10.1016/j.semcdb.2013.12.005.
7. Yang CC, Cotsarelis G. Review of hair follicle dermal cells. J Dermatol Sci. 2010;57(1):2–11. doi:10.1016/j.jdermsci.2009.11.005.
8. Sennett R, Rendl M. Mesenchymal-epithelial interactions during hair follicle morphogenesis and cycling. Semin Cell Dev Biol. 2012;23(8):917–27. doi:10.1016/j.semcdb.2012.08.011.
9. Goldstein J, Horsley V. Home sweet home: skin stem cell niches. Cell Mol Life Sci. 2012;69 (15):2573–82. doi:10.1007/s00018-012-0943-3.
10. Cotsarelis G, Sun TT, Lavker RM. Label-retaining cells reside in the bulge area of pilosebaceous unit: implications for follicular stem cells, hair cycle, and skin carcinogenesis. Cell. 1990;61(7):1329–37.
11. Oshima H, Rochat A, Kedzia C, Kobayashi K, Barrandon Y. Morphogenesis and renewal of hair follicles from adult multipotent stem cells. Cell. 2001;104(2):233–45.
12. Claudinot S, Nicolas M, Oshima H, Rochat A, Barrandon Y. Long-term renewal of hair follicles from clonogenic multipotent stem cells. Proc Natl Acad Sci USA. 2005;102 (41):14677–82. doi:10.1073/pnas.0507250102.
13. Tumbar T, Guasch G, Greco V, Blanpain C, Lowry WE, Rendl M, et al. Defining the epithelial stem cell niche in skin. Science. 2004;303(5656):359–63. doi:10.1126/science.1092436.
14. Trempus CS, Morris RJ, Bortner CD, Cotsarelis G, Faircloth RS, Reece JM, et al. Enrichment for living murine keratinocytes from the hair follicle bulge with the cell surface marker CD34. J Invest Dermatol. 2003;120(4):501–11. doi:10.1046/j.1523-1747.2003.12088.x.
15. Lyle S, Christofidou-Solomidou M, Liu Y, Elder DE, Albelda S, Cotsarelis G. The C8/144B monoclonal antibody recognizes cytokeratin 15 and defines the location of human hair follicle stem cells. J Cell Sci. 1998;111(Pt 21):3179–88.
16. Morris RJ, Liu Y, Marles L, Yang Z, Trempus C, Li S, et al. Capturing and profiling adult hair follicle stem cells. Nat Biotechnol. 2004;22(4):411–7. doi:10.1038/nbt950.
17. Lee J, Tumbar T. Hairy tale of signaling in hair follicle development and cycling. Semin Cell Dev Biol. 2012;23(8):906–16. doi:10.1016/j.semcdb.2012.08.003.

18. Hsu YC, Pasolli HA, Fuchs E. Dynamics between stem cells, niche, and progeny in the hair follicle. Cell. 2011;144(1):92–105. doi:10.1016/j.cell.2010.11.049.
19. Fuchs E. The tortoise and the hair: slow-cycling cells in the stem cell race. Cell. 2009;137 (5):811–9. doi:10.1016/j.cell.2009.05.002.
20. Li L, Clevers H. Coexistence of quiescent and active adult stem cells in mammals. Science. 2010;327(5965):542–5. doi:10.1126/science.1180794.
21. Plikus MV, Mayer JA, de la Cruz D, Baker RE, Maini PK, Maxson R, et al. Cyclic dermal BMP signalling regulates stem cell activation during hair regeneration. Nature. 2008;451 (7176):340–4. doi:10.1038/nature06457.
22. Kobielak K, Stokes N, de la Cruz J, Polak L, Fuchs E. Loss of a quiescent niche but not follicle stem cells in the absence of bone morphogenetic protein signaling. Proc Natl Acad Sci USA. 2007;104(24):10063–8. doi:10.1073/pnas.0703004104.
23. Horsley V, Aliprantis AO, Polak L, Glimcher LH, Fuchs E. NFATc1 balances quiescence and proliferation of skin stem cells. Cell. 2008;132(2):299–310. doi:10.1016/j.cell.2007.11.047.
24. Takeda N, Jain R, Leboeuf MR, Padmanabhan A, Wang Q, Li L, et al. Hopx expression defines a subset of multipotent hair follicle stem cells and a progenitor population primed to give rise to K6+ niche cells. Development. 2013;140(8):1655–64. doi:10.1242/dev.093005.
25. Castilho RM, Squarize CH, Chodosh LA, Williams BO, Gutkind JS. mTOR mediates Wnt-induced epidermal stem cell exhaustion and aging. Cell Stem Cell. 2009;5(3):279–89. doi:10.1016/j.stem.2009.06.017.
26. Nakajima T, Inui S, Fushimi T, Noguchi F, Kitagawa Y, Reddy JK, et al. Roles of MED1 in quiescence of hair follicle stem cells and maintenance of normal hair cycling. J Invest Dermatol. 2013;133(2):354–60. doi:10.1038/jid.2012.293.
27. Chen T, Heller E, Beronja S, Oshimori N, Stokes N, Fuchs E. An RNA interference screen uncovers a new molecule in stem cell self-renewal and long-term regeneration. Nature. 2012;485(7396):104–8. doi:10.1038/nature10940.
28. Blanpain C, Lowry WE, Geoghegan A, Polak L, Fuchs E. Self-renewal, multipotency, and the existence of two cell populations within an epithelial stem cell niche. Cell. 2004;118 (5):635–48. doi:10.1016/j.cell.2004.08.012.
29. Lee J, Hoi CS, Lilja KC, White BS, Lee SE, Shalloway D, et al. Runx1 and p21 synergistically limit the extent of hair follicle stem cell quiescence in vivo. Proc Natl Acad Sci USA. 2013;110(12):4634–9. doi:10.1073/pnas.1213015110.
30. Osorio KM, Lee SE, McDermitt DJ, Waghmare SK, Zhang YV, Woo HN, et al. Runx1 modulates developmental, but not injury-driven, hair follicle stem cell activation. Development. 2008;135(6):1059–68. doi:10.1242/dev.012799.
31. Hoi CS, Lee SE, Lu SY, McDermitt DJ, Osorio KM, Piskun CM, et al. Runx1 directly promotes proliferation of hair follicle stem cells and epithelial tumor formation in mouse skin. Mol Cell Biol. 2010;30(10):2518–36. doi:10.1128/MCB.01308-09.
32. Devgan V, Mammucari C, Millar SE, Brisken C, Dotto GP. p21WAF1/Cip1 is a negative transcriptional regulator of Wnt4 expression downstream of Notch1 activation. Genes Dev. 2005;19(12):1485–95. doi:10.1101/gad.341405.
33. Pippa R, Espinosa L, Gundem G, Garcia-Escudero R, Dominguez A, Orlando S, et al. p27Kip1 represses transcription by direct interaction with p130/E2F4 at the promoters of target genes. Oncogene. 2012;31(38):4207–20. doi:10.1038/onc.2011.582.
34. Greco V, Chen T, Rendl M, Schober M, Pasolli HA, Stokes N, et al. A two-step mechanism for stem cell activation during hair regeneration. Cell Stem Cell. 2009;4(2):155–69. doi:10.1016/j.stem.2008.12.009.
35. Ito M, Kizawa K, Hamada K, Cotsarelis G. Hair follicle stem cells in the lower bulge form the secondary germ, a biochemically distinct but functionally equivalent progenitor cell population, at the termination of catagen. Differentiation. 2004;72(9–10):548–57. doi:10.1111/j.1432-0436.2004.07209008.x.

36. Lee SE, Sada A, Zhang M, McDermitt DJ, Lu SY, Kemphues KJ, et al. High Runx1 levels promote a reversible, more-differentiated cell state in hair-follicle stem cells during quiescence. Cell Rep. 2014;6(3):499–513. doi:10.1016/j.celrep.2013.12.039.
37. Zhang YV, White BS, Shalloway DI, Tumbar T. Stem cell dynamics in mouse hair follicles: a story from cell division counting and single cell lineage tracing. Cell Cycle. 2010;9 (8):1504–10.
38. Rompolas P, Mesa KR, Greco V. Spatial organization within a niche as a determinant of stem-cell fate. Nature. 2013;502(7472):513–8. doi:10.1038/nature12602.
39. Osorio KM, Lilja KC, Tumbar T. Runx1 modulates adult hair follicle stem cell emergence and maintenance from distinct embryonic skin compartments. J Cell Biol. 2011;193(1):235–50. doi:10.1083/jcb.201006068.
40. Hsu YC, Li L, Fuchs E. Transit-amplifying cells orchestrate stem cell activity and tissue regeneration. Cell. 2014;157(4):935–49. doi:10.1016/j.cell.2014.02.057.
41. Barry ER, Camargo FD. The Hippo superhighway: signaling crossroads converging on the Hippo/Yap pathway in stem cells and development. Curr Opin Cell Biol. 2013;25(2):247–53. doi:10.1016/j.ceb.2012.12.006.
42. Taylor G, Lehrer MS, Jensen PJ, Sun TT, Lavker RM. Involvement of follicular stem cells in forming not only the follicle but also the epidermis. Cell. 2000;102(4):451–61.
43. Ito M, Liu Y, Yang Z, Nguyen J, Liang F, Morris RJ, et al. Stem cells in the hair follicle bulge contribute to wound repair but not to homeostasis of the epidermis. Nat Med. 2005;11 (12):1351–4. doi:10.1038/nm1328.
44. Horsley V, O'Carroll D, Tooze R, Ohinata Y, Saitou M, Obukhanych T, et al. Blimp1 defines a progenitor population that governs cellular input to the sebaceous gland. Cell. 2006;126 (3):597–609. doi:10.1016/j.cell.2006.06.048.
45. Snippert HJ, Haegebarth A, Kasper M, Jaks V, van Es JH, Barker N, et al. Lgr6 marks stem cells in the hair follicle that generate all cell lineages of the skin. Science. 2010;327 (5971):1385–9. doi:10.1126/science.1184733.
46. Jensen KB, Collins CA, Nascimento E, Tan DW, Frye M, Itami S, et al. Lrig1 expression defines a distinct multipotent stem cell population in mammalian epidermis. Cell Stem Cell. 2009;4(5):427–39. doi:10.1016/j.stem.2009.04.014.
47. Jaks V, Barker N, Kasper M, van Es JH, Snippert HJ, Clevers H, et al. Lgr5 marks cycling, yet long-lived, hair follicle stem cells. Nat Genet. 2008;40(11):1291–9. doi:10.1038/ng.239.
48. Nishimura EK, Jordan SA, Oshima H, Yoshida H, Osawa M, Moriyama M, et al. Dominant role of the niche in melanocyte stem-cell fate determination. Nature. 2002;416(6883):854–60. doi:10.1038/416854a.
49. Nishimura EK. Melanocyte stem cells: a melanocyte reservoir in hair follicles for hair and skin pigmentation. Pigment Cell Melanoma Res. 2011;24(3):401–10. doi:10.1111/j.1755-148X.2011.00855.x.
50. Tanimura S, Tadokoro Y, Inomata K, Binh NT, Nishie W, Yamazaki S, et al. Hair follicle stem cells provide a functional niche for melanocyte stem cells. Cell Stem Cell. 2011;8 (2):177–87. doi:10.1016/j.stem.2010.11.029.
51. Chang CY, Pasolli HA, Giannopoulou EG, Guasch G, Gronostajski RM, Elemento O, et al. NFIB is a governor of epithelial-melanocyte stem cell behaviour in a shared niche. Nature. 2013;495(7439):98–102. doi:10.1038/nature11847.
52. Rabbani P, Takeo M, Chou W, Myung P, Bosenberg M, Chin L, et al. Coordinated activation of Wnt in epithelial and melanocyte stem cells initiates pigmented hair regeneration. Cell. 2011;145(6):941–55. doi:10.1016/j.cell.2011.05.004.
53. Rompolas P, Deschene ER, Zito G, Gonzalez DG, Saotome I, Haberman AM, et al. Live imaging of stem cell and progeny behaviour in physiological hair-follicle regeneration. Nature. 2012;487(7408):496–9. doi:10.1038/nature11218.
54. Chi W, Wu E, Morgan BA. Dermal papilla cell number specifies hair size, shape and cycling and its reduction causes follicular decline. Development. 2013;140(8):1676–83. doi:10.1242/dev.090662.

55. Rendl M, Lewis L, Fuchs E. Molecular dissection of mesenchymal-epithelial interactions in the hair follicle. PLoS Biol. 2005;3(11):e331. doi:10.1371/journal.pbio.0030331.
56. Botchkarev VA, Botchkareva NV, Nakamura M, Huber O, Funa K, Lauster R, et al. Noggin is required for induction of the hair follicle growth phase in postnatal skin. FASEB J: official publication of the Federation of American Societies for Experimental Biology. 2001;15 (12):2205–14. doi:10.1096/fj.01-0207com.
57. Clavel C, Grisanti L, Zemla R, Rezza A, Barros R, Sennett R, et al. Sox2 in the dermal papilla niche controls hair growth by fine-tuning BMP signaling in differentiating hair shaft progenitors. Dev Cell. 2012;23(5):981–94. doi:10.1016/j.devcel.2012.10.013.
58. Woo WM, Zhen HH, Oro AE. Shh maintains dermal papilla identity and hair morphogenesis via a Noggin-Shh regulatory loop. Genes Dev. 2012;26(11):1235–46. doi:10.1101/gad.187401.112.
59. Hu B, Lefort K, Qiu W, Nguyen BC, Rajaram RD, Castillo E, et al. Control of hair follicle cell fate by underlying mesenchyme through a CSL-Wnt5a-FoxN1 regulatory axis. Genes Dev. 2010;24(14):1519–32. doi:10.1101/gad.1886910.
60. Rendl M, Polak L, Fuchs E. BMP signaling in dermal papilla cells is required for their hair follicle-inductive properties. Genes Dev. 2008;22(4):543–57. doi:10.1101/gad.1614408.
61. Enshell-Seijffers D, Lindon C, Kashiwagi M, Morgan BA. beta-catenin activity in the dermal papilla regulates morphogenesis and regeneration of hair. Dev Cell. 2010;18(4):633–42. doi:10.1016/j.devcel.2010.01.016.
62. Gat U, DasGupta R, Degenstein L, Fuchs E. De Novo hair follicle morphogenesis and hair tumors in mice expressing a truncated beta-catenin in skin. Cell. 1998;95(5):605–14.
63. Millar SE. Molecular mechanisms regulating hair follicle development. J Invest Dermatol. 2002;118(2):216–25. doi:10.1046/j.0022-202x.2001.01670.x.
64. Barrandon Y, Li V, Green H. New techniques for the grafting of cultured human epidermal cells onto athymic animals. J Invest Dermatol. 1988;91(4):315–8.
65. Driskell RR, Watt FM. Understanding fibroblast heterogeneity in the skin. Trends Cell Biol. 2015;25(2):92–9. doi:10.1016/j.tcb.2014.10.001.
66. Collins CA, Kretzschmar K, Watt FM. Reprogramming adult dermis to a neonatal state through epidermal activation of beta-catenin. Development. 2011;138(23):5189–99. doi:10.1242/dev.064592.
67. Biernaskie J, Paris M, Morozova O, Fagan BM, Marra M, Pevny L, et al. SKPs derive from hair follicle precursors and exhibit properties of adult dermal stem cells. Cell Stem Cell. 2009;5(6):610–23. doi:10.1016/j.stem.2009.10.019.
68. Rahmani W, Abbasi S, Hagner A, Raharjo E, Kumar R, Hotta A, et al. Hair follicle dermal stem cells regenerate the dermal sheath, repopulate the dermal papilla, and modulate hair type. Dev Cell. 2014;31(5):543–58. doi:10.1016/j.devcel.2014.10.022.
69. Festa E, Fretz J, Berry R, Schmidt B, Rodeheffer M, Horowitz M, et al. Adipocyte lineage cells contribute to the skin stem cell niche to drive hair cycling. Cell. 2011;146(5):761–71. doi:10.1016/j.cell.2011.07.019.
70. Mecklenburg L, Tobin DJ, Muller-Rover S, Handjiski B, Wendt G, Peters EM, et al. Active hair growth (anagen) is associated with angiogenesis. J Invest Dermatol. 2000;114(5):909–16. doi:10.1046/j.1523-1747.2000.00954.x.
71. Yano K, Brown LF, Detmar M. Control of hair growth and follicle size by VEGF-mediated angiogenesis. J Clin Investig. 2001;107(4):409–17. doi:10.1172/JCI11317.
72. Stenn KS, Fernandez LA, Tirrell SJ. The angiogenic properties of the rat vibrissa hair follicle associate with the bulb. J Invest Dermatol. 1988;90(3):409–11.
73. Amoh Y, Li L, Yang M, Moossa AR, Katsuoka K, Penman S, et al. Nascent blood vessels in the skin arise from nestin-expressing hair-follicle cells. Proc Natl Acad Sci USA. 2004;101 (36):13291–5. doi:10.1073/pnas.0405250101.
74. Xiao Y, Woo WM, Nagao K, Li W, Terunuma A, Mukouyama YS, et al. Perivascular hair follicle stem cells associate with a venule annulus. J Invest Dermatol. 2013;133(10):2324–31. doi:10.1038/jid.2013.167.

75. Bigarella CL, Liang R, Ghaffari S. Stem cells and the impact of ROS signaling. Development. 2014;141(22):4206–18. doi:10.1242/dev.107086.
76. Hamanaka RB, Glasauer A, Hoover P, Yang S, Blatt H, Mullen AR et al. Mitochondrial reactive oxygen species promote epidermal differentiation and hair follicle development. Sci Signal. 2013;6(261):ra8. doi:10.1126/scisignal.2003638.
77. Mohyeldin A, Garzon-Muvdi T, Quinones-Hinojosa A. Oxygen in stem cell biology: a critical component of the stem cell niche. Cell Stem Cell. 2010;7(2):150–61. doi:10.1016/j.stem.2010.07.007.
78. Rathman-Josserand M, Genty G, Lecardonnel J, Chabane S, Cousson A, Francois Michelet J, et al. Human hair follicle stem/progenitor cells express hypoxia markers. J Invest Dermatol. 2013;133(8):2094–7. doi:10.1038/jid.2013.113.
79. Chen X, Tian Y, Yao L, Zhang J, Liu Y. Hypoxia stimulates proliferation of rat neural stem cells with influence on the expression of cyclin D1 and c-Jun N-terminal protein kinase signaling pathway in vitro. Neuroscience. 2010;165(3):705–14. doi:10.1016/j.neuroscience.2009.11.007.
80. Gustafsson MV, Zheng X, Pereira T, Gradin K, Jin S, Lundkvist J, et al. Hypoxia requires notch signaling to maintain the undifferentiated cell state. Dev Cell. 2005;9(5):617–28. doi:10.1016/j.devcel.2005.09.010.
81. Botchkarev VA, Eichmuller S, Johansson O, Paus R. Hair cycle-dependent plasticity of skin and hair follicle innervation in normal murine skin. J Comp Neurol. 1997;386(3):379–95.
82. Brownell I, Guevara E, Bai CB, Loomis CA, Joyner AL. Nerve-derived sonic hedgehog defines a niche for hair follicle stem cells capable of becoming epidermal stem cells. Cell Stem Cell. 2011;8(5):552–65. doi:10.1016/j.stem.2011.02.021.
83. Christoph T, Muller-Rover S, Audring H, Tobin DJ, Hermes B, Cotsarelis G, et al. The human hair follicle immune system: cellular composition and immune privilege. Br J Dermatol. 2000;142(5):862–73.
84. Meyer KC, Klatte JE, Dinh HV, Harries MJ, Reithmayer K, Meyer W, et al. Evidence that the bulge region is a site of relative immune privilege in human hair follicles. Br J Dermatol. 2008;159(5):1077–85. doi:10.1111/j.1365-2133.2008.08818.x.
85. Paus R, Nickoloff BJ, Ito T. A 'hairy' privilege. Trends Immunol. 2005;26(1):32–40. doi:10.1016/j.it.2004.09.014.
86. Osaka N, Takahashi T, Murakami S, Matsuzawa A, Noguchi T, Fujiwara T, et al. ASK1-dependent recruitment and activation of macrophages induce hair growth in skin wounds. J Cell Biol. 2007;176(7):903–9. doi:10.1083/jcb.200611015.
87. Castellana D, Paus R, Perez-Moreno M. Macrophages contribute to the cyclic activation of adult hair follicle stem cells. PLoS Biol. 2014;12(12):e1002002. doi:10.1371/journal.pbio.1002002.
88. Kloepper JE, Kawai K, Bertolini M, Kanekura T, Paus R. Loss of gammadelta T cells results in hair cycling defects. J Invest Dermatol. 2013;133(6):1666–9. doi:10.1038/jid.2013.17.
89. Fujiwara H, Ferreira M, Donati G, Marciano DK, Linton JM, Sato Y, et al. The basement membrane of hair follicle stem cells is a muscle cell niche. Cell. 2011;144(4):577–89. doi:10.1016/j.cell.2011.01.014.
90. White AC, Khuu JK, Dang CY, Hu J, Tran KV, Liu A, et al. Stem cell quiescence acts as a tumour suppressor in squamous tumours. Nat Cell Biol. 2014;16(1):99–107. doi:10.1038/ncb2889.
91. Liu JC, Lerou PH, Lahav G. Stem cells: balancing resistance and sensitivity to DNA damage. Trends Cell Biol. 2014;24(5):268–74. doi:10.1016/j.tcb.2014.03.002.
92. Sotiropoulou PA, Candi A, Mascre G, De Clercq S, Youssef KK, Lapouge G, et al. Bcl-2 and accelerated DNA repair mediates resistance of hair follicle bulge stem cells to DNA-damage-induced cell death. Nat Cell Biol. 2010;12(6):572–82. doi:10.1038/ncb2059.
93. Rhee H, Polak L, Fuchs E. Lhx2 maintains stem cell character in hair follicles. Science. 2006;312(5782):1946–9. doi:10.1126/science.1128004.

94. Kadaja M, Keyes BE, Lin M, Pasolli HA, Genander M, Polak L, et al. SOX9: a stem cell transcriptional regulator of secreted niche signaling factors. Genes Dev. 2014;28(4):328–41. doi:10.1101/gad.233247.113.
95. Folgueras AR, Guo X, Pasolli HA, Stokes N, Polak L, Zheng D, et al. Architectural niche organization by LHX2 is linked to hair follicle stem cell function. Cell Stem Cell. 2013;13 (3):314–27. doi:10.1016/j.stem.2013.06.018.
96. Vidal VP, Chaboissier MC, Lutzkendorf S, Cotsarelis G, Mill P, Hui CC, et al. Sox9 is essential for outer root sheath differentiation and the formation of the hair stem cell compartment. Curr Biol: CB. 2005;15(15):1340–51. doi:10.1016/j.cub.2005.06.064.
97. Nowak JA, Polak L, Pasolli HA, Fuchs E. Hair follicle stem cells are specified and function in early skin morphogenesis. Cell Stem Cell. 2008;3(1):33–43. doi:10.1016/j.stem.2008.05.009.
98. Nishimura EK, Granter SR, Fisher DE. Mechanisms of hair graying: incomplete melanocyte stem cell maintenance in the niche. Science. 2005;307(5710):720–4. doi:10.1126/science.1099593.
99. Lopez-Otin C, Blasco MA, Partridge L, Serrano M, Kroemer G. The hallmarks of aging. Cell. 2013;153(6):1194–217. doi:10.1016/j.cell.2013.05.039.
100. Flores I, Cayuela ML, Blasco MA. Effects of telomerase and telomere length on epidermal stem cell behavior. Science. 2005;309(5738):1253–6. doi:10.1126/science.1115025.
101. Flores I, Blasco MA. The role of telomeres and telomerase in stem cell aging. FEBS Lett. 2010;584(17):3826–30. doi:10.1016/j.febslet.2010.07.042.
102. Flores I, Canela A, Vera E, Tejera A, Cotsarelis G, Blasco MA. The longest telomeres: a general signature of adult stem cell compartments. Genes Dev. 2008;22(5):654–67. doi:10.1101/gad.451008.
103. Keyes BE, Segal JP, Heller E, Lien WH, Chang CY, Guo X, et al. Nfatc1 orchestrates aging in hair follicle stem cells. Proc Natl Acad Sci USA. 2013;110(51):E4950–9. doi:10.1073/pnas.1320301110.
104. White AC, Lowry WE. Refining the role for adult stem cells as cancer cells of origin. Trends Cell Biol. 2015;25(1):11–20. doi:10.1016/j.tcb.2014.08.008.
105. Youssef KK, Van Keymeulen A, Lapouge G, Beck B, Michaux C, Achouri Y, et al. Identification of the cell lineage at the origin of basal cell carcinoma. Nat Cell Biol. 2010;12 (3):299–305. doi:10.1038/ncb2031.
106. Wang GY, Wang J, Mancianti ML, Epstein EH Jr. Basal cell carcinomas arise from hair follicle stem cells in Ptch1(±) mice. Cancer Cell. 2011;19(1):114–24. doi:10.1016/j.ccr.2010.11.007.
107. Kasper M, Jaks V, Are A, Bergstrom A, Schwager A, Svard J, et al. Wounding enhances epidermal tumorigenesis by recruiting hair follicle keratinocytes. Proc Natl Acad Sci USA. 2011;108(10):4099–104. doi:10.1073/pnas.1014489108.
108. White AC, Tran K, Khuu J, Dang C, Cui Y, Binder SW, et al. Defining the origins of Ras/p53-mediated squamous cell carcinoma. Proc Natl Acad Sci USA. 2011;108(18):7425–30. doi:10.1073/pnas.1012670108.
109. Scheitz CJ, Lee TS, McDermitt DJ, Tumbar T. Defining a tissue stem cell-driven Runx1/Stat3 signalling axis in epithelial cancer. EMBO J. 2012;31(21):4124–39. doi:10.1038/emboj.2012.270.
110. Li S, Park H, Trempus CS, Gordon D, Liu Y, Cotsarelis G, et al. A keratin 15 containing stem cell population from the hair follicle contributes to squamous papilloma development in the mouse. Mol Carcinog. 2013;52(10):751–9. doi:10.1002/mc.21896.

The Sebaceous Gland Stem Cell Niche

Karen Reuter and Catherin Niemann

Abbreviations

AP-1	Activator protein 1
B	Bulge
BCC	Basal cell carcinoma
BLIMP1	B-lymphocyte-induced maturation protein 1
BMP	Bone Morphogenetic Protein
BrdU	5-bromo-2`desoxyuridine
CD34	Hematopoietic progenitor cell antigen CD34
DAPI	4′,6-Diamidin-2-phenylindol
DMBA	Dimethylbenz-[a]-anthracene
EMT	Epithelial-mesenchymal transition
EDAR	Ectodysplasin A Receptor
FGF	Fibroblast growth factors
Hh	Hedgehog
HF	Hair follicle
hr	Hairless
HS	Heparin sulfate
IFE	Interfollicular epidermis
Itga6	α6-integrin
JZ	Junctional zone
K14	Keratin14
K15	Keratin15
Kras	V-Ki-ras2 Kirsten rat sarcoma viral oncogene homolog
Lef1	Lymphoid enhancer factor 1
Lgr5/6	Leucine-rich repeat-containing G-protein coupled receptor 5/6

K. Reuter · C. Niemann (✉)
Center for Biochemistry, Medical Faculty and Center for Molecular Medicine Cologne, University of Cologne, Cologne, Germany
e-mail: cnieman1@uni-koeln.de

C. Niemann
Institute for Biochemistry, Center for Molecular Medicine Cologne, University of Cologne, Robert-Koch-Str. 21, D-50931 Cologne, Germany

K. Turksen (ed.), *Tissue-Specific Stem Cell Niche*,
Stem Cell Biology and Regenerative Medicine,
DOI 10.1007/978-3-319-21705-5_2

Lrig1	Leucine rich immunoglobuline-like factor 1
Nfatc1	Nuclear factor of activated T-cell c1
NMSC	Nonmelanoma skin cancer
Plet1	Placenta-expressed transcript 1
Ptch1	Patched1
PI	Propidium iodide
Rac1	Ras-related C3 botulinum toxin substrate 1
Ras	Rat sarcoma
SC	Stem cell
SCC	Squamous cell carcinoma
SCD1	Stearoyl-Coenzym A Desaturase 1
SCLT	Sebaceous carcinoma-like tumors
SG	Sebaceous gland
Smad7	Mothers against decapentaplegic homolog 7
Smurf2	SMAD Specific E3 Ubiquitin Protein Ligase 2
Snail	Snail family zinc finger
Sox 9	SRY (sex determining region Y)-box 9
Str	Stroma
T	Tumor
TCF	T-cell factor
Tfam	Transcription factor A, mitochondrial
TPA	12-O-tetradecanoyl-phorbol-13-acetate
UI	Upper isthmus

1 Physiology and Function of the Sebaceous Gland

The skin within its cellular complexity accomplishes many vital functions. Eminently, it forms the protective barrier against harmful insults of the environment and prevents water loss of our body. The essential functions of the skin are assured by the multi-layered epithelium of the interfollicular epidermis (IFE). In addition, over the last years a variety of elegant studies demonstrated an important role of epidermal appendages, including the hair follicle (HF) and sebaceous gland (SG) (Fig. 1) for skin physiology and principal functions of the organ. Although, the SG has not attracted the same intense scientific attention as for instance the HF structure, remarkable progress has been made in our understanding of basic cellular mechanisms that govern SG function and physiology. It is well established that normal development and maintenance of the SG are crucial for homeostasis of mammalian skin [47]. In particular, atrophic SGs and defects in sebaceous lipid production lead to a disturbed barrier function and relevant skin diseases.

The main task of the SGs is to continuously produce mature, differentiated sebocytes and to release sebum to lubricate and waterproof the skin [47, 65, 80].

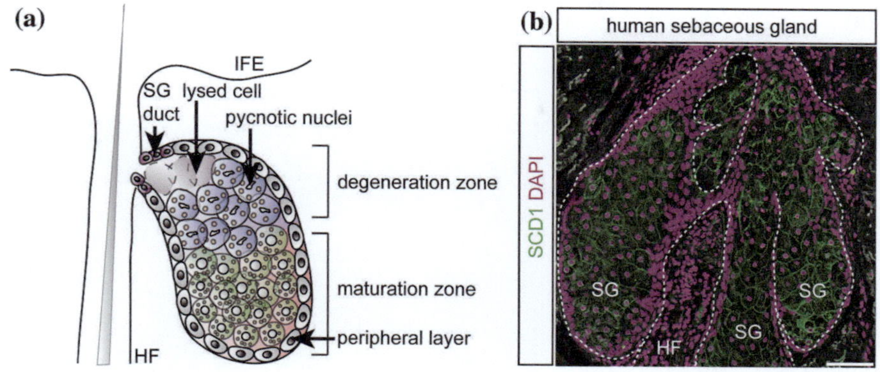

Fig. 1 Morphology and cellular composition of the sebaceous gland. **a** Schematic illustration of the sebaceous gland, presenting the different compartments of sebocyte differentiation in mice. Cells leaving the peripheral layer undergo a defined program of cellular differentiation, including enlargement and production of lipid droplets (maturation zone), followed by progressive maturation and accumulation of lipid droplets until cell lysis of terminal differentiated sebocytes to release sebum (degeneration zone). **b** Staining of human sebaceous glands for the sebocyte marker SCD1 (*green*) and nuclei stained by DAPI (*magenta*). IFE, interfollicular epidermis; SG, sebaceous gland; HF, hair follicle; SCD1, stearoyl-Coenzyme A desaturase 1; DAPI, 4′,6-Diamidin-2-phenylindol. Scale bar: 50 μm

Therefore, mature sebocytes need to be replaced by proliferating cells localizing to the basal layer at the peripheral zone of the gland. These basal undifferentiated cells are attached to a basement membrane separating the gland structure from the surrounding dermal tissue. SG cells leaving the basal layer undergo a well-defined program of cellular differentiation (Fig. 1a, b). The underlying molecular and cellular signals steering this important process are not well understood yet but most likely involve intrinsic cellular signaling pathways as well as regulation by hormones, the extracellular matrix and surrounding stromal cells. Adjacent to the single layer of basal cells, sebocytes localize to the maturation zone of the gland. In this compartment, sebocytes enlarge and produce lipid droplets [74]. As sebocytes are squeezed towards the center of the gland they accumulate more lipid droplets and progressively mature. Finally, mature sebocytes reach the necrosis zone of the gland. Pyknotic nuclei are indicative that sebocytes are about to degenerate and to release the lipid-containing sebum into the specialized secretory duct of the gland and the follicular canal to eventually reach the surface of the skin (Fig. 1).

As essential part of the pilosebaceous unit, the SG is associated with the junctional zone of the HF structure (Figs. 1 and 2). Linking the SG to the HF structure, SG duct cells thought to play an important role balancing the crosstalk between HF junctional zone and the SG, which is essential for normal gland homeostasis. In addition, SGs are formed independently from the HF and can be found as specialized glands in distinct regions of mammalian skin, e.g. as Meibomian glands of the eyelid or as Fordyce's spots of the oral epithelium [65].

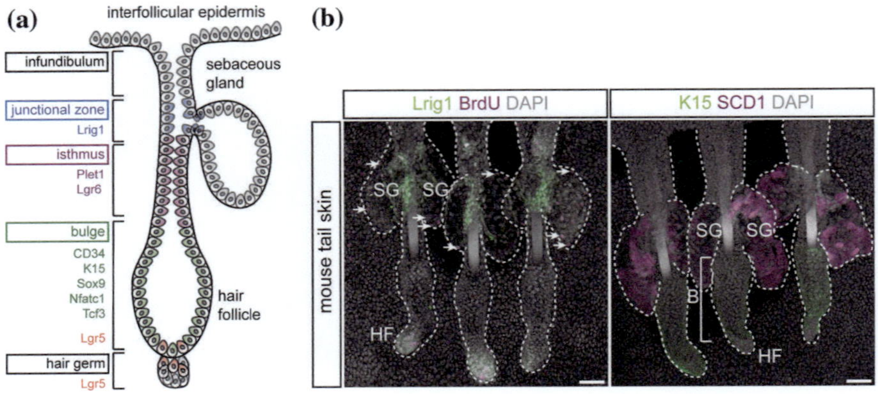

Fig. 2 Stem and progenitor cell populations of the adult hair follicle and the sebaceous gland. **a** Schematic presentation of distinct stem cell compartments localizing to the junctional zone (*blue*), isthmus (*purple*), bulge (*green*) and the hair germ (*red*). Marker for stem and progenitor cells are listed next to the corresponding compartment. **b** Epidermal whole mounts from mouse tail skin stained for the junctional zone marker (*left*) Lrig1 (*green*) and proliferation marker BrdU (*magenta*). Arrows indicate proliferating cells within the peripheral layer of the sebaceous gland. Staining of bulge stem cell marker (*right*) Keratin15 (*green*) and the sebocyte marker SCD1 (*magenta*). Nuclei were stained with DAPI (*grey*). SG, sebaceous gland; HF, hair follicle; B, bulge; Lrig1, leucine rich immunoglobuline-like factor 1; BrdU, 5-bromo-2'desoxyuridine; K15, Keratin15; SCD1, stearoyl-coenzyme A desaturase 1. Scale bar: 50 μm

2 Development of the Sebaceous Gland and Establishment of Progenitor Pools

The development of the SG is coupled to HF morphogenesis thereby forming the pilo-sebaceous unit of mammalian skin. HF morphogenesis is initiated during embryogenesis as a result from intensive neuroectodermal-mesodermal interactions and extensive molecular signaling between the different cell types, epithelial and fibroblast cells [16, 41, 67]. Based on defined morphological criteria, the process of HF development has been divided into eight distinct stages. Changes in morphology and progressive maturation results in a complex miniorgan that is the product of a sequence of tightly regulated signaling events, including the Wnt, EDAR, Bmp, Hedgehog and FGF pathways, among others [15, 43]. In the course of formation of the HF structure, keratinocytes start to form a placode (stage 1), a thickening of the epidermis, which is progressing into a hair germ (stage 2) and subsequently a bulbous peg (stage 3 to 5). First sebocytes of the future SG are seen in the upper part of the developing follicular structure at stage 5 (Fig. 3) [19, 56, 64]. As the follicle goes through stages 6 to 8 finally forming a mature HF, emerging sebocytes form a glandular structure that remains associated with the HF via the sebaceous duct. The complex cellular and molecular processes of shaping the SG and defining the correct size and cellular organization (Fig. 1) are not understood yet and require more detailed investigations.

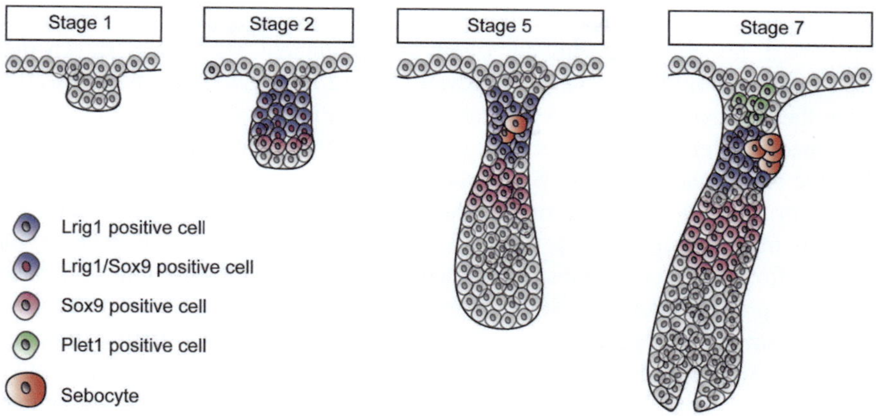

Fig. 3 Establishment of stem cell compartments during morphogenesis of the sebaceous gland. Schematic illustration presenting dynamic pattern of stem cell marker expression during the development of the pilosebaceous unit, starting from placode formation (stage 1), progressing into a hair germ (stage 2), forming the bulbous germ (stage 5) and subsequently bulbous peg (stage 7). Of note Lrig1 and Sox9 positive cells show overlapping localizations at stage 2 and are separating at stage 5, when first sebocytes develop. Lrig1, leucine rich immunoglobuline-like factor 1; Sox9, SRY sex determining region Y-box 9; Plet1, Placenta-expressed transcript 1

Recent studies have been looking into the potential role of epidermal stem and progenitor cells in the process of sebocyte differentiation and SG development. In general, various epidermal SC marker molecules are expressed early during HF formation, including Sox9, Keratin15 and Lrig1 [19, 32, 53]. Initially, Sox9, a marker of the future HF bulge and Lrig1, which is decorating stem cells of the junctional zone in adult skin (Fig. 2) are both expressed by the same progenitor cells contributing to hair germ formation. As HF formation proceeds and lineage specification takes place, Sox9 and Lrig1 expressing cells disperse and ultimately localize to the respective regions within the pilosebaceous unit (Fig. 3) [19]. Sox9-positive SCs that populate the future HF bulge region are involved in SG formation. It has been demonstrated that depletion of Sox9 from embryonic mouse epidermis leads to a block in SG morphogenesis [53]. Thus, Sox9 expression during early stages of HF formation, perhaps by precursor cells that are also positive for Lrig1, is required for proper SG development. Alternatively, Sox9 expression by early bulge SCs could promote differentiation of the sebocyte cell lineage by currently unknown mechanisms.

Recently, it was shown that sebocytes are generated by Lrig1-positive SCs. Detailed analysis of this process revealed that subsequent SG cells are establishing from asymmetric cell fate decision of Lrig1 SCs [19]. Sebocytes emerging during development are enclosed by Lrig1-positive SCs that undergo cell division. In contrast, cells positive for the sebocyte marker SCD1 do not express Lrig1 and do not proliferate. However, the molecular mechanisms regulating the positioning of the Lrig1 precursor pool require further investigations. Furthermore, it is currently not known how cell fate specification within the Lrig1 compartment is controlled. Of interest, for pilo-sebaceous units with two prominent SGs it was shown that these emerge form one cluster of sebocytes derived from Lrig1-positive cells [19]. Once the sebocyte cell lineage is established, the Plet1-positive precursor pool is generated. Therefore, these precursor cells of the HF isthmus seem not to play a decisive role in the process of SG formation (Fig. 3) [19, 52].

SG formation is depending on normal function of mitochondria, particularly the electron transport chain [34]. Ablation of Tfam, a key maintenance factor for mtDNA from mouse epidermis and subsequent loss of the electron transport chain leads to a profound defect in SG morphogenesis. Astonishingly, the Lrig1-positive SC compartment is generated in Tfam-deficient mice demonstrating that the establishment of the Lrig1 SC compartment is not sufficient to guarantee formation of SGs. Thus, it will be important to identify the additional instructive signals steering SG morphogenesis.

3 Stem Cells in Sebaceous Gland Homeostasis and Renewal

Work by many laboratories has suggested that different SC pools maintain SG homeostasis [47]. As presented in Fig. 4, one can envision different scenarios how SGs are maintained on a cellular level:

(1) Renewal of the gland occurs independent of HF SCs by unipotent progenitor cells localizing to the SG duct or the outermost proliferative cell layer of the gland (Fig. 4e, f). Although such a SG SC has not been identified yet, a previous study investigating retrovirus-mediated gene transfer to genetically mark cutaneous epithelial stem cells in mouse skin revealed that individually labeled cells within the SG seem to at least partially contribute to SG renewal [22]. Furthermore, Blimp1 (B-lymphocyte-induced nuclear maturation protein1) was described as a marker for sebocyte precursor cells governing SG homeostasis [28]. However, lineage tracing experiments following the fate of genetically marked cells in vivo demonstrated that Blimp1-positive cells do not give rise to proliferative and differentiating sebocytes [36]. Additionally, several reports have shown that Blimp1 is expressed by differentiated cells of the IFE, the SG and inner root sheath of HFs [8, 10, 36, 39, 66].

(2) SGs are maintained by progeny of HF SCs repopulating the basal proliferating compartment of the SG (Fig. 4a–d). Indeed, there is evidence that stem and progenitor cells of the HF junctional zone and the isthmus contribute to SG renewal [32, 52]. Lineage tracing of Lrig1- and Lgr6-positive keratinocytes revealed that these cells are capable to renew SG tissue in vivo [54, 68]. In addition, HF bulge SCs can also contribute to SG maintenance in vivo. This has been documented by genetic lineage tracing and fate mapping of Keratin15-positive bulge SCs [42, 57, 58]. Interestingly, analyzing progeny of bulge cells expressing Keratin19 revealed that these keratinocytes are not involved in SG maintenance, at least under

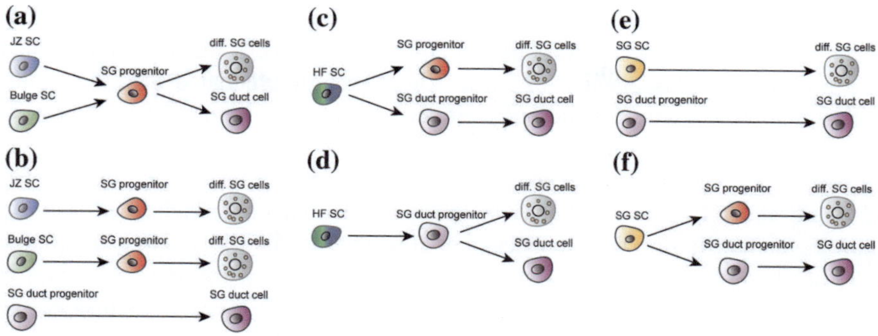

Fig. 4 Potential scenarios of SC contribution to sebaceous gland homeostasis. **a** Junctional zone stem cells and bulge stem cells give rise to one common sebaceous gland progenitor pool. Common sebaceous gland progenitor cells can subsequently differentiate into differentiated sebaceous gland cells and sebaceous gland duct cells. **b** Junctional zone stem cells and bulge stem cells contribute to different sebaceous gland progenitors, which finally differentiate into sebocytes. Sebaceous gland duct progenitor cells differentiate directly into sebocytes. **c** Hair follicle stem cells contribute to sebaceous gland progenitor cells which give rise to differentiated duct cells. Hair follicle stem cells also form sebaceous gland duct progenitor which are precursor cells for sebaceous duct cells. **d** Hair follicle stem cells give rise to sebaceous gland duct progenitors, which than generate both, differentiated sebocytes and sebaceous gland duct cells. **e** Sebaceous gland stem cells and sebaceous gland duct progenitor can differentiate into differentiated sebaceous gland cells and sebaceous gland duct cells respectively. **f** Sebaceous gland stem cells produce sebaceous gland progenitors differentiating into mature gland cells. Sebaceous gland stem cells also generate duct progenitor give rise to sebaceous gland duct cells. The different options of generating a functional sebaceous gland are not mutually exclusive. JZ, junctional zone; SC, stem cell; SG, sebaceous gland

homeostatic conditions [54]. The heterogeneous composition of the HF SC bulge, differences within the activation state and the localization of a particular labeled SC within the niche could all impact on the result of individual lineage tracing studies. In fact, expression of multiple bulge SC marker molecules only partially overlap in their expression pattern, including Sox9, Keratin15, CD34, Nfatc1, Keratin19, Lgr5 and Gli1 [29, 59, 71]. HF bulge SCs present a heterogeneous population comprising a quiescent SC pool and SCs that are rapidly activated for HF regeneration and wound response [69, 76, 77]. Moreover, there is functional evidence that bulge SCs exist in different states of activation and are differentially prone to respond to homeostatic clues, e.g. regulation by the circadian molecular clock mechanisms [30]. In the future, innovative techniques like the intravital live cell imaging will allow visualizing individual labeled SCs and following their responses and lineage contribution within their native environment [6].

Clearly, there is intense cross-regulation between bulge and sebaceous gland [59]. The intimate relationship between the HF bulge SCs and the SG has been observed in the Rhino mutant mouse where the bulge SC compartment is impaired. In this mutant, the *hairless* gene activity is lost leading to the disintegration of bulge SCs and defects in hair growth. Remarkably, shortly after hair loss the SGs disappear. In contrast, another mutant mice with recessive mutation of the *hairless* locus (*hr*), HFs also disintegrate but retain some bulge cells that can be activated and proliferate. Consequently, *hr* mice exhibit well-differentiated SGs, thus indicating that the presence of the bulge SC compartment is required for proper SG maintenance [55].

Next, there is good evidence that SGs have significant impact on the HF structure and HF SCs. Loss of SGs and sebocyte function can lead to scarring alopecia demonstrating the dependence of HFs on SGs [5, 70]. However, until now it is not known how the crosstalk between SG tissue and the HF bulge SCs is controlled on a molecular level and further studies are needed to dissect the potential contribution of individual HF and SG SCs and the cellular environment.

Until now, little is known about the cellular origin and maintenance of the sebaceous duct connecting SG and hair canal (Fig. 4e, f). Previous results analyzing expression of marker molecules suggest that SG duct cells are different from SG cells. Particularly, Lrig1 is detected within sebaceous duct cells but not in differentiating sebocytes [36, 54]. Furthermore, activation of the Hedgehog signaling pathway stimulates enlargement of specifically the SG duct cell lineage in Gli2 transgenic mice [24]. Obviously, more research is required to unravel if the SG duct is maintained by its own pool of progenitors and if such progenitor cells contribute to the homeostasis and regeneration of the SG.

4 Stem Cell Regulatory Mechanisms

Over the last years, important work by many laboratories has demonstrated that a variety of different factors can modulate SG morphogenesis and SG function, among them hormones, cytokines, signaling pathways as well as cell-cell and cell-matrix interactions [47, 65, 73, 75]. However, much less is known about molecular mechanisms operating directly on stem and progenitor cells of the SG.

One pathway that is essential in directing cell fate decisions in HF bulge SCs is canonical Wnt/β-catenin signaling [9, 51]. It is well established that transcription factors of the TCF/Lef1 family, which are mediating canonical Wnt/β-catenin activity, control bulge SC activation for cyclic HF regeneration and HF differentiation [25, 38]. In addition, previous data demonstrated that blocking TCF/Lef1 activity promotes SG lineage selection by HF stem and progenitor cells [40, 48]. Importantly, inhibition of TCF/Lef1 signaling in bulge SCs results in expansion of the Lrig1 SC pool and induces de novo SG formation [57]. In contrast, expression of TCF3 in mouse epidermis leads to a block in sebocyte specification and the lack of SG development [45]. One important mechanism of repressing canonical Wnt/β-catenin signaling in sebocytes involves Smad7 activity. It has been shown that Smad7 directly binds to β-catenin and recruits Smurf2, an ubiquitin E3 ligase that leads to the degradation of cytoplasmic β-catenin [26]. Taken together, these results reveal that suppression of β-catenin/TCF/Lef1 activity in epidermal stem and progenitor cells is driving SG cell specification and proper SG differentiation (Fig. 5).

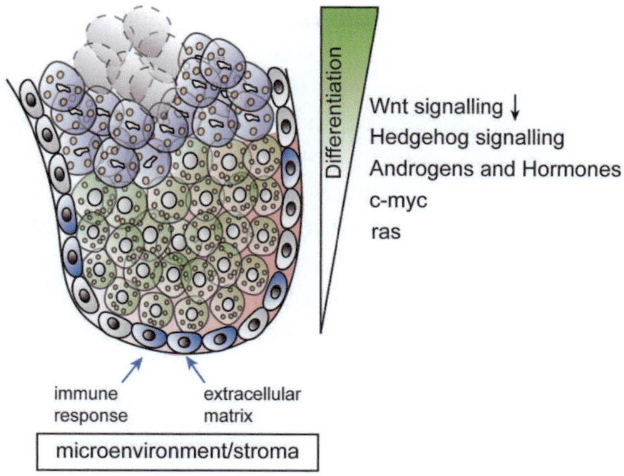

Fig. 5 Regulators of sebaceous gland homeostasis. Sebaceous gland cell proliferation (*blue arrows*) and differentiation are influenced by different signaling pathways, hormones, signal molecules and the microenvironment

Another pathway governing sebocyte proliferation and differentiation is Hedgehog signaling. The binding of the hedgehog ligand to its receptor patched-1 leads to activation of the co-receptor smoothened thereby initiating a sequence of intracellular signaling events. Subsequently, a cascade of signaling reactions in primary cilia is induced culminating in the Gli transcription factor activation and hedgehog target gene transcription [61]. There is good evidence that stimulating Hedgehog signaling activity promotes SG differentiation whereas inhibition of the canonical Hedgehog pathway by overexpressing a dominant-negative Gli mutant blocks sebocyte differentiation (Fig. 5) [2, 24]. Additionally, it has been shown that Indian hedgehog expression can increase sebocyte proliferation and maturation in vitro and drives differentiation of sebaceous gland tumors in vivo [33, 50]. However, more detailed studies are required to identify the underlying molecular and cellular mechanisms and to demonstrate which stem and progenitor cell is mediating these effects.

Recently, it was shown that increasing KRas signaling by expressing an oncogenic constitutive active mutant KRas G12D in bulge SCs, promotes SG cell fate and leads to enlarged SGs (Fig. 5) [37]. This highlights again that modulating the HF bulge SC compartment can induce tremendous responses of the SG tissue indicating an intimate crosstalk between the different cellular compartments.

Epithelial-mesenchymal interactions trigger the development of the pilosebaceous unit and cyclic renewal of the HF [67]. This important cellular interrelationship is controlled by the composition of the extracellular matrix that can also modulate SC function and activation [78]. There is strong evidence that the extracellular matrix and cellular environment impact on SG morphology and function (Fig. 5). A recent study unraveled an important role of heparin sulfate (HS), a proteoglycan found in the extracellular matrix and on the cell surface for HF and SG physiology. In particular, HF ablation from mouse epidermis results in an increase in the number of HFs and SGs and induced SG hyperplasia with excessive sebum production [11]. Based on the observation that HFs and SGs are strongly affected, HS is suggested to regulate SC compartments involved in HF and SG maintenance and regeneration. However, the underlying molecular mechanisms and the type of SC pool involved have not been discovered yet.

Cyclic HF regeneration and HF SC activation is modulated by other cell types that most likely also affect SG physiology and SCs maintaining SG homeostasis [23]. These include adipocytes and neurons that provide important signaling molecules, including hedgehog ligands and BMP2 [7, 18]. Furthermore, inflammatory and immune cells impact on epidermal SC function as demonstrated by the regulation of hair loss and wound repair [17, 35, 62]. In the future, it will be important to study their role in normal function of sebocytes as well as their involvement in defective SG tissue, including the formation and progression of sebaceous tumors.

It is well established that hormones regulate SG activity [73, 75]. In particular, androgens can modulate sebocyte proliferation in vitro and in vivo [1, 13, 79]. Interestingly, the androgen receptor has been identified as a Myc target gene in mouse epidermis [44] suggesting that Myc activity on SG function is mediated by androgen signaling [10]. However it is not known if androgens directly affect SC pools that are driving SG maintenance.

5 Stem Cells and Sebaceous Gland Pathologies

The SG is implicated in the pathogenesis of relevant skin diseases including forms of severe acne and some skin tumors. Several different types of sebaceous tumors are found in patients, ranging from benign well-differentiated sebaceous adenoma to malignant and aggressive sebaceous carcinoma (Fig. 6a, b) [47].

In recent years, some important studies have shed light on molecular mechanisms involved in the generation of sebaceous lesions. Although there is a constant increase in the number of mutant mice strains displaying defective SG morphology as well as abnormal sebocyte function, only a few genetic mouse models have been described displaying sebaceous tumors [47].

Transgenic mice overexpressing c-myc, an oncogene that has been revealed to affect epidermal SCs, show an increase in number and size of SGs and generate sebaceous adenomas in response to a two-step carcinogenesis protocol involving topical application of the carcinogen DMBA and a tumor-promoting agent TPA [3, 27]. Generally, DMBA/TPA treatment of mouse skin results in benign squamous papilloma with the ability to progress into malignant squamous skin carcinoma (SCC) [14] demonstrating that c-myc is specifically promoting a program of sebocyte differentiation in nonmelanoma skin cancer (NMSC).

Importantly, mutations within the N-terminus of the transcription factor Lef1 have been identified in human sebaceous adenoma and eyelid sebaceous carcinoma [31, 72]. These mutations prevent the binding of Lef1 to β-catenin and result in a block of β-catenin-dependent transcription of Wnt target genes. Transgenic mice

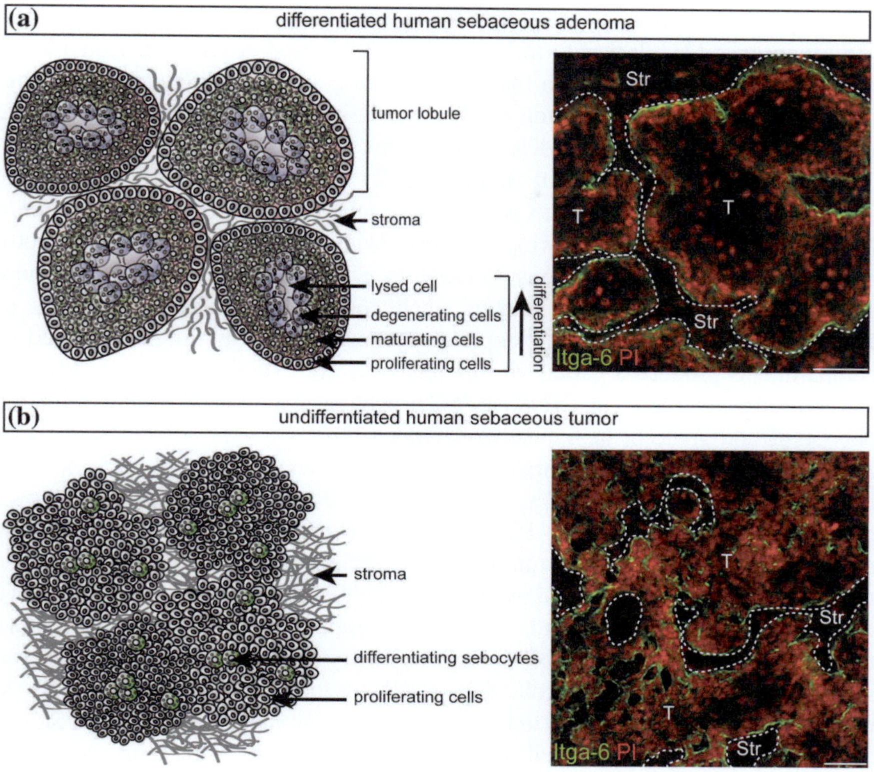

Fig. 6 Morphology of different types of sebaceous tumors. **a** Illustration depicting the morphology of well-differentiated sebaceous adenoma (*left*). Note the lobular architecture of sebaceous adenomas and the presence of mature sebocytes in the center of tumor lobules. Proliferating cells leaving the peripheral layer undergo a defined program of sebocyte maturation. Immunofluorescence staining of Itga-6 (*green*) labeling the periphery of tumor lobules in differentiated human sebaceous adenoma. Nuclei were stained using PI (*red*). **b** Schematic presentation of undifferentiated human sebaceous tumors/carcinomas, illustrating the loss of well-defined tumor architecture and a decrease of sebocyte differentiation (*left*). Arrows indicate proliferating cells (*grey*) and differentiated sebocytes (*green*). Itga-6 staining in undifferentiated human sebaceous tumors throughout the tumor mass. Itga-6, α6-Integrin; PI, propidium iodide; T, tumor tissue; Str, stroma. Scale bar: 100 μm

expressing a similar mutant form of Lef1 under control of the Keratin14 promoter (K14ΔNLef1 mice) display enlarged and de novo SGs and develop spontaneous sebaceous tumors [48]. This demonstrates that mutant Lef1 activity indeed is a highly relevant mechanism driving sebaceous tumor formation in mammalian skin. In addition, treatment of K14ΔNLef1 mice with the carcinogen DMBA results in sebaceous tumors in high frequency within a short period of time [49].

Another example how sebaceous gland differentiation is regulated in the process of skin tumourigenesis came from studies manipulating the AP-1 transcription factor. In particular, the typical response to chemical carcinogenesis was changed from squamous tumors towards sebaceous adenomas following blocking AP-1 activity in mice [21]. It was further shown that inhibition of AP-1 induced a block in β-catenin/TCF/Lef1 function thereby promoting sebocyte differentiation within the skin tumors. Thus, repression of Wnt/β-catenin signaling is a common feature within the different mouse models of sebaceous tumor formation [46].

Interestingly, expression of mutant Lef1 specifically in HF bulge SCs (K15ΔNLef1 mice) also lead to spontaneous sebaceous skin tumors. Skin tumor formation is accelerated following treatment with a single dose of DMBA [60]. Astonishingly, sebaceous lesions forming in K15ΔNLef1 mice display a more aggressive and malignant phenotype when compared to sebaceous tumors of K14ΔNLef1 mice. Thus, mice expressing mutant Lef1 in different epidermal compartments produce a variety of different sebaceous skin lesions demonstrating that the cell population, e.g. SC versus SC progeny has a tremendous impact on the phenotype and grade of malignancy of tumors [60].

Recently, a number of elaborate and elegant in vivo studies have unraveled that many different types of NMSC arise from multipotent epidermal SCs [4]. Our own experiments have shown that HF bulge SCs constitute one cell of origin for mutantLef1-driven sebaceous tumors. More specifically, lineage tracing experiments of bulge SCs show clonal expansion of individual labeled Keratin15-positive cells and their contribution to sebaceous adenoma formation in K14ΔNLef1 transgenic mice [60]. However, sebaceous tumors are not monoclonal derived suggesting that non-labeled cells (other SCs or non-SCs) could contribute to tumors. In the future, it will be important to investigate the potential role of SCs of the upper isthmus and junctional zone as well as cells of the sebaceous duct in the process of sebaceous tumor development. Of note, Lrig1-positive SCs of the junctional zone do not constitute a cell of origin for squamous skin tumors [54] but have the potential to contribute to the initiation of basal cell carcinoma (BCC) following inactivation of the hedgehog co-receptor ptch1 [60].

Analyzing the underlying molecular mechanism of sebaceous tumourigenesis revealed that mutant Lef1 interferes with SC-specific surveillance mechanisms of the HF bulge, including the control of DNA damage response and proliferation [60]. These data support the observation that bulge SCs are important for the SG lineage in normal skin and in epidermal tumors.

It has been shown that markers of different HF SC compartments are expressed in sebaceous tumors, including bulge SC marker Keratin15, CD34 and Sox9 (Fig. 7) [60, 63]. Remarkably, the expression level of SC marker for the junctional zone and upper isthmus, Lrig1 and Plet1 respectively, correlate with the malignant progression of sebaceous skin tumors [60]. Here, the SC-regulatory small GTPase Rac1 was shown to promote Lrig1 production by tumor cells and to induce progression of benign sebaceous adenoma to malignant sebaceous carcinoma-like tumors (SCLT) [20]. It was also shown that epidermis-specific overexpression of the EMT-inducing transcription factor Snail induces various types of skin carcinoma, including SG carcinoma [12]. This was accompanied by an increase in

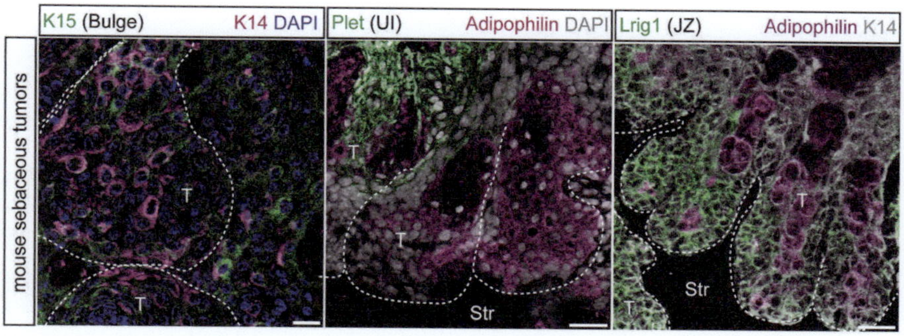

Fig. 7 Stem cell marker expression in mouse sebaceous tumors. Immunofluorescence staining of the bulge stem cell marker Keratin15 (*green*) together with Keratin14 (*magenta*) (*left*), the upper isthmus marker Plet1 (*green*) stained with Adipophilin (*magenta*) (*middle*) and the junctional zone marker Lrig1 (*green*) together with Adipophilin (*red*) and Keratin14 (*grey*). Nuclei were stained using DAPI. K15, Keratin15; K14, Keratin14; Lrig1, leucine rich immunoglobuline-like factor 1; Plet1, Placenta-expressed transcript 1. Scale bar: 25 μm

progenitor cells of the upper isthmus expressing Plet1. However, the specific role of different types HF SCs in sebaceous carcinomas needs to be investigated in more detail.

Taken together, work of recent years has just begun to unravel the molecular and cellular mechanisms driving the process of SG disease, particularly the formation of sebaceous tumors in vivo. Further studies are needed to better understand how the individual pathways are interconnected and which SC is targeted and drives the disease in patients.

6 Concluding Remarks

Over the last years several distinct stem cell compartment have been identified within mammalian epidermis contributing to SG morphogenesis and maintenance of its normal function. Good progress has been made to develop elaborate in vivo mouse models, which allow deciphering the molecular mechanisms steering SG differentiation and we have begun to understand that interfering with normal SC function impacts on SG physiology. Clearly, more research is required to unravel the complexity and the specific contribution of diverse SC pools to SG homeostasis and patho-physiologies of the gland. In addition, more detailed studies will lead to the identification of cellular and molecular signals determining the SG cell fate.

Acknowledgement The authors like to thank all the members of the Niemann lab for their valuable contribution and the helpful comments and discussions. The work was supported by Köln Fortune to KR and by the DFG (SFB 829, project A3) to CN.

References

1. Akamatsu H, Zouboulis CC, Orfanos CE. Control of human sebocyte proliferation in vitro by testosterone and 5-alpha-dihydrotestosterone is dependent on the localization of the sebaceous glands. J Invest Dermatol. 1992;99:509–11. doi:10.1111/1523-1747.ep12616181.
2. Allen M, Grachtchouk M, Sheng H, et al. Hedgehog signaling regulates sebaceous gland development. Am J Pathol. 2003;163:2173–8. doi:10.1016/S0002-9440(10)63574-2.
3. Arnold I, Watt FM. c-Myc activation in transgenic mouse epidermis results in mobilization of stem cells and differentiation of their progeny. Curr Biol. 2001;11:558–68.
4. Beck B, Blanpain C. Unravelling cancer stem cell potential. Nat Rev Cancer. 2013;13:727–38. doi:10.1038/nrc3597.
5. Binczek E, Jenke B, Holz B, et al. Obesity resistance of the stearoyl-CoA desaturase-deficient (scd1-/-) mouse results from disruption of the epidermal lipid barrier and adaptive thermoregulation. Biol Chem. 2007;388:405–18. doi:10.1515/BC.2007.046.
6. Brown S, Greco V. Stem Cells in the Wild: Understanding the World of Stem Cells through Intravital Imaging. Cell Stem Cell. 2014;15:683–6. doi:10.1016/j.stem.2014.11.006.
7. Brownell I, Guevara E, Bai CB, et al. Nerve-derived sonic hedgehog defines a niche for hair follicle stem cells capable of becoming epidermal stem cells. Cell Stem Cell. 2011;8:552–65. doi:10.1016/j.stem.2011.02.021.
8. Chang DH, Calame KL. The dynamic expression pattern of B lymphocyte induced maturation protein-1 (Blimp-1) during mouse embryonic development. Mech Dev. 2002;117:305–9. doi:10.1016/S0925-4773(02)00189-2.
9. Clevers H, Loh KM, Nusse R. An integral program for tissue renewal and regeneration: wnt signaling and stem cell control. Science (80-). 2014;346:1248012–1248012. doi:10.1126/science.1248012.
10. Cottle DL, Kretzschmar K, Schweiger PJ, et al. c-MYC-induced sebaceous gland differentiation is controlled by an androgen receptor/p53 axis. Cell Rep. 2013;3:427–41. doi:10.1016/j.celrep.2013.01.013.
11. Coulson-Thomas VJ, Gesteira TF, Esko J, Kao W. Heparan sulfate regulates hair follicle and sebaceous gland morphogenesis and homeostasis. J Biol Chem. 2014;289:25211–26. doi:10.1074/jbc.M114.572511.
12. De Craene B, Denecker G, Vermassen P, et al. Epidermal Snail expression drives skin cancer initiation and progression through enhanced cytoprotection, epidermal stem/progenitor cell expansion and enhanced metastatic potential. Cell Death Differ. 2014;21:310–20. doi:10.1038/cdd.2013.148.
13. Deplewski D, Rosenfield RL. Growth hormone and insulin-like growth factors have different effects on sebaceous cell growth and differentiation. Endocrinology. 1999;140:4089–94. doi:10.1210/endo.140.9.6957.
14. DiGiovanni J. Multistage carcinogenesis in mouse skin. Pharmacol Ther. 1992;54:63–128.
15. Duverger O, Morasso MI. To grow or not to grow: hair morphogenesis and human genetic hair disorders. Semin Cell Dev Biol. 2014;25–26:22–33. doi:10.1016/j.semcdb.2013.12.006.
16. Duverger O, Morasso MI. Epidermal patterning and induction of different hair types during mouse embryonic development. Birth Defects Res Part C Embryo Today Rev. 2009;87:263–72. doi:10.1002/bdrc.20158.
17. Eming SA, Krieg T, Davidson JM. Inflammation in wound repair: molecular and cellular mechanisms. J Invest Dermatol. 2007;127:514–25. doi:10.1038/sj.jid.5700701.
18. Festa E, Fretz J, Berry R, et al. Adipocyte Lineage Cells Contribute to the Skin Stem Cell Niche to Drive Hair Cycling. Cell. 2011;146:761–71. doi:10.1016/j.cell.2011.07.019.
19. Frances D, Niemann C. Stem cell dynamics in sebaceous gland morphogenesis in mouse skin. Dev Biol. 2012;363:138–46. doi:10.1016/j.ydbio.2011.12.028.
20. Frances D, Sharma N, Pofahl R, et al. A role for Rac1 activity in malignant progression of sebaceous skin tumors. Oncogene. 2015;. doi:10.1038/onc.2014.471.

21. Gerdes MJ, Myakishev M, Frost NA, et al. Activator protein-1 activity regulates epithelial tumor cell identity. Cancer Res. 2006;66:7578–88. doi:10.1158/0008-5472.CAN-06-1247.
22. Ghazizadeh S, Taichman LB. Multiple classes of stem cells in cutaneous epithelium: a lineage analysis of adult mouse skin. EMBO J. 2001;20:1215–22. doi:10.1093/emboj/20.6.1215.
23. Goldstein J, Horsley V. Home sweet home: skin stem cell niches. Cell Mol Life Sci. 2012;69:2573–82. doi:10.1007/s00018-012-0943-3.
24. Gu L-H, Coulombe PA. Hedgehog signaling, keratin 6 induction, and sebaceous gland morphogenesis: implications for pachyonychia congenita and related conditions. Am J Pathol. 2008;173:752–61. doi:10.2353/ajpath.2008.071089.
25. Haegebarth A, Clevers H. Wnt signaling, lgr5, and stem cells in the intestine and skin. Am J Pathol. 2009;174:715–21. doi:10.2353/ajpath.2009.080758.
26. Han G, Li AG, Liang Y-Y, et al. Smad7-induced beta-catenin degradation alters epidermal appendage development. Dev Cell. 2006;11:301–12. doi:10.1016/j.devcel.2006.06.014.
27. Honeycutt KA, Waikel RL, Koster MI, et al. The effect of c-myc on stem cell fate influences skin tumor phenotype. Mol Carcinog. 2010;49:315–9. doi:10.1002/mc.20617.
28. Horsley V, O'Carroll D, Tooze R, et al. Blimp1 defines a progenitor population that governs cellular input to the sebaceous gland. Cell. 2006;126:597–609. doi:10.1016/j.cell.2006.06.048.
29. Jaks V, Kasper M, Toftgård R. The hair follicle-a stem cell zoo. Exp Cell Res. 2010;316:1422–8. doi:10.1016/j.yexcr.2010.03.014.
30. Janich P, Pascual G, Merlos-Suárez A, et al. The circadian molecular clock creates epidermal stem cell heterogeneity. Nature. 2011;480:209–14. doi:10.1038/nature10649.
31. Jayaraj P, Sen S, Sharma A, et al. Eyelid sebaceous carcinoma: a novel mutation in lymphoid enhancer-binding factor 1 (LEF1). Br J Dermatol. 2015;. doi:10.1111/bjd.13706.
32. Jensen KB, Collins CA, Nascimento E, et al. Lrig1 expression defines a distinct multipotent stem cell population in mammalian epidermis. Cell Stem Cell. 2009;4:427–39. doi:10.1016/j.stem.2009.04.014.
33. Kakanj P, Reuter K, Séquaris G, et al. Indian hedgehog controls proliferation and differentiation in skin tumorigenesis and protects against malignant progression. Cell Rep. 2013;4:340–51. doi:10.1016/j.celrep.2013.06.037.
34. Kloepper JE, Baris OR, Reuter K, et al. Mitochondrial function in murine skin epithelium is crucial for hair follicle morphogenesis and epithelial-mesenchymal interactions. J Invest Dermatol. 2015;. doi:10.1038/jid.2014.475.
35. Kloepper JE, Kawai K, Bertolini M, et al. Loss of γδ T cells results in hair cycling defects. J Invest Dermatol. 2013;133:1666–9. doi:10.1038/jid.2013.17.
36. Kretzschmar K, Cottle DL, Donati G, et al. BLIMP1 is required for postnatal epidermal homeostasis but does not define a sebaceous gland progenitor under steady-state conditions. Stem cell reports. 2014;3:620–33. doi:10.1016/j.stemcr.2014.08.007.
37. Lapouge G, Youssef KK, Vokaer B, et al. Identifying the cellular origin of squamous skin tumors. Proc Natl Acad Sci U S A. 2011;108:7431–6. doi:10.1073/pnas.1012720108.
38. Lowry WE, Blanpain C, Nowak JA, et al. Defining the impact of beta-catenin/Tcf transactivation on epithelial stem cells. Genes Dev. 2005;19:1596–611. doi:10.1101/gad.1324905.
39. Magnúsdóttir E, Kalachikov S, Mizukoshi K, et al. Epidermal terminal differentiation depends on B lymphocyte-induced maturation protein-1. Proc Natl Acad Sci U S A. 2007;104:14988–93. doi:10.1073/pnas.0707323104.
40. Merrill BJ, Gat U, DasGupta R, Fuchs E. Tcf3 and Lef1 regulate lineage differentiation of multipotent stem cells in skin. Genes Dev. 2001;15:1688–705. doi:10.1101/gad.891401.
41. Millar SE. Molecular mechanisms regulating hair follicle development. J Invest Dermatol. 2002;118:216–25. doi:10.1046/j.0022-202x.2001.01670.x.
42. Morris RJ, Liu Y, Marles L, et al. Capturing and profiling adult hair follicle stem cells. Nat Biotechnol. 2004;22:411–7. doi:10.1038/nbt950.
43. Nakamura M, Schneider MR, Schmidt-Ullrich R, Paus R. Mutant laboratory mice with abnormalities in hair follicle morphogenesis, cycling, and/or structure: an update. J Dermatol Sci. 2013;69:6–29. doi:10.1016/j.jdermsci.2012.10.001.

44. Nascimento EM, Cox CL, MacArthur S, et al. The opposing transcriptional functions of Sin3a and c-Myc are required to maintain tissue homeostasis. Nat Cell Biol. 2011;13:1395–405. doi:10.1038/ncb2385.
45. Nguyen H, Rendl M, Fuchs E. Tcf3 governs stem cell features and represses cell fate determination in skin. Cell. 2006;127:171–83. doi:10.1016/j.cell.2006.07.036.
46. Niemann C. Differentiation of the sebaceous gland. Dermatoendocrinol. 2009;1:64–7.
47. Niemann C, Horsley V. Development and homeostasis of the sebaceous gland. Semin Cell Dev Biol. 2012;23:928–36. doi:10.1016/j.semcdb.2012.08.010.
48. Niemann C, Owens DM, Hülsken J, et al. Expression of DeltaNLef1 in mouse epidermis results in differentiation of hair follicles into squamous epidermal cysts and formation of skin tumours. Development. 2002;129:95–109.
49. Niemann C, Owens DM, Schettina P, Watt FM. Dual role of inactivating Lef1 mutations in epidermis: tumor promotion and specification of tumor type. Cancer Res. 2007;67:2916–21. doi:10.1158/0008-5472.CAN-06-3427.
50. Niemann C, Unden AB, Lyle S, et al. Indian hedgehog and beta-catenin signaling: role in the sebaceous lineage of normal and neoplastic mammalian epidermis. Proc Natl Acad Sci U.S.A 2003;100 Suppl :11873–11880. doi:10.1073/pnas.1834202100.
51. Niemann C, Watt FM. Designer skin: lineage commitment in postnatal epidermis. Trends Cell Biol. 2002;12:185–92.
52. Nijhof JGW, Braun KM, Giangreco A, et al. The cell-surface marker MTS24 identifies a novel population of follicular keratinocytes with characteristics of progenitor cells. Development. 2006;133:3027–37. doi:10.1242/dev.02443.
53. Nowak JA, Polak L, Pasolli HA, Fuchs E. Hair follicle stem cells are specified and function in early skin morphogenesis. Cell Stem Cell. 2008;3:33–43. doi:10.1016/j.stem.2008.05.009.
54. Page ME, Lombard P, Ng F, et al. The epidermis comprises autonomous compartments maintained by distinct stem cell populations. Cell Stem Cell. 2013;13:471–82. doi:10.1016/j.stem.2013.07.010.
55. Panteleyev AA, Rosenbach T, Paus R, Christiano AM. The bulge is the source of cellular renewal in the sebaceous gland of mouse skin. Arch Dermatol Res. 2000;292:573–6.
56. Paus R, Müller-Röver S, Van Der Veen C, et al. A comprehensive guide for the recognition and classification of distinct stages of hair follicle morphogenesis. J Invest Dermatol. 1999;113:523–32. doi:10.1046/j.1523-1747.1999.00740.x.
57. Petersson M, Brylka H, Kraus A, et al. TCF/Lef1 activity controls establishment of diverse stem and progenitor cell compartments in mouse epidermis. EMBO J. 2011;30:3004–18. doi:10.1038/emboj.2011.199.
58. Petersson M, Frances D, Niemann C. Lineage tracing of hair follicle stem cells in epidermal whole mounts. Methods Mol Biol. 2013;989:45–60. doi:10.1007/978-1-62703-330-5_5.
59. Petersson M, Niemann C. Stem cell dynamics and heterogeneity: implications for epidermal regeneration and skin cancer. Curr Med Chem. 2012;19:5984–5992.
60. Petersson M, Reuter K, Brylka H, et al. Interfering with stem cell-specific gatekeeper functions controls tumour initiation and malignant progression of skin tumours. Nat Commun. 2015;6:5874. doi:10.1038/ncomms6874.
61. Petrova R, Joyner AL. Roles for Hedgehog signaling in adult organ homeostasis and repair. Development. 2014;141:3445–57. doi:10.1242/dev.083691.
62. Petukhova L, Duvic M, Hordinsky M, et al. Genome-wide association study in alopecia areata implicates both innate and adaptive immunity. Nature. 2010;466:113–7. doi:10.1038/nature09114.
63. Qiu W, Lei M, Li J, et al. Activated hair follicle stem cells and Wnt/β-catenin signaling involve in pathnogenesis of sebaceous neoplasms. Int J Med Sci. 2014;11:1022–8. doi:10.7150/ijms.8383.
64. Schmidt-Ullrich R, Paus R. Molecular principles of hair follicle induction and morphogenesis. BioEssays. 2005;27:247–61. doi:10.1002/bies.20184.
65. Schneider MR, Paus R. Sebocytes, multifaceted epithelial cells: lipid production and holocrine secretion. Int J Biochem Cell Biol. 2010;42:181–5. doi:10.1016/j.biocel.2009.11.017.

66. Sellheyer K, Krahl D. Blimp-1: a marker of terminal differentiation but not of sebocytic progenitor cells. J Cutan Pathol. 2010;37:362–70. doi:10.1111/j.1600-0560.2009.01434.x.
67. Sennett R, Rendl M. Mesenchymal-epithelial interactions during hair follicle morphogenesis and cycling. Semin Cell Dev Biol. 2012;23:917–27. doi:10.1016/j.semcdb.2012.08.011.
68. Snippert HJ, Haegebarth A, Kasper M, et al. Lgr6 marks stem cells in the hair follicle that generate all cell lineages of the skin. Science. 2010;327:1385–9. doi:10.1126/science.1184733.
69. Sotiropoulou PA, Candi A, Blanpain C. The majority of multipotent epidermal stem cells do not protect their genome by asymmetrical chromosome segregation. Stem Cells. 2008;26:2964–73. doi:10.1634/stemcells.2008-0634.
70. Sundberg JP, Boggess D, Sundberg BA, et al. Asebia-2 J (Scd1(ab2 J)): a new allele and a model for scarring alopecia. Am J Pathol. 2000;156:2067–75. doi:10.1016/S0002-9440(10)65078-X.
71. Tadeu AMB, Horsley V. Epithelial stem cells in adult skin. Curr Top Dev Biol. 2014;107:109–31. doi:10.1016/B978-0-12-416022-4.00004-4.
72. Takeda H, Lyle S, Lazar AJF, et al. Human sebaceous tumors harbor inactivating mutations in LEF1. Nat Med. 2006;12:395–7. doi:10.1038/nm1386.
73. Thiboutot D. Regulation of human sebaceous glands. J Invest Dermatol. 2004;123:1–12. doi:10.1111/j.1523-1747.2004.t01-2-.x.
74. Thody AJ, Shuster S. Control and function of sebaceous glands. Physiol Rev. 1989;69:383–416.
75. Tóth BI, Oláh A, Szöllosi AG, et al. "Sebocytes' makeup": novel mechanisms and concepts in the physiology of the human sebaceous glands. Pflugers Arch. 2011;461:593–606. doi:10.1007/s00424-011-0941-6.
76. Tumbar T, Guasch G, Greco V, et al. Defining the epithelial stem cell niche in skin. Science. 2004;303:359–63. doi:10.1126/science.1092436.
77. Waghmare SK, Bansal R, Lee J, et al. Quantitative proliferation dynamics and random chromosome segregation of hair follicle stem cells. EMBO J. 2008;27:1309–20. doi:10.1038/emboj.2008.72.
78. Watt FM, Huck WTS. Role of the extracellular matrix in regulating stem cell fate. Nat Rev Mol Cell Biol. 2013;14:467–73. doi:10.1038/nrm3620.
79. Zouboulis CC. [The sebaceous gland]. Hautarzt 2010;61:467–8, 4704, 476–7. doi:10.1007/s00105-009-1894-y.
80. Zouboulis CC, Adjaye J, Akamatsu H, et al. Human skin stem cells and the ageing process. Exp Gerontol. 2008;43:986–97. doi:10.1016/j.exger.2008.09.001.
81. Zouboulis CC, Baron JM, Böhm M, et al. Frontiers in sebaceous gland biology and pathology. Exp Dermatol. 2008;17:542–51. doi:10.1111/j.1600-0625.2008.00725.x.

Inner Ear Stem Cell Niche

Zhengqing Hu

Abbreviations

ABR	Auditory brainstem response
Atoh1	Atonal Homolog 1
BDNF	Brain-derived neurotrophic factor
Bmp	Bone morphogenetic protein
BrdU	Bromodeoxyuridine
Dll1	Delta-like 1
Dlx3	Distal-Less Homeobox 3
Dlx5	Distal-Less Homeobox 5
EGF	Epidermal growth factor
EMT	Epithelial to mesenchymal transition
ES cell	Embryonic stem cell
Eya1	Eyes Absent Homolog 1
FACS	Fluorescence activated cell sorting
FGFs	Fibroblast Growth Factors
Foxi1	Forkhead Box I1
Gata3	GATA binding protein 3
Gbx2	Gastrulation Brain Homeobox 2
GFAP	Glial fibrillary acidic protein
GFP	Green fluorescent protein
Hes1	Hes family bHLH transcription factor 1
Hes5	Hes family bHLH transcription factor 5
IGF	Insulin-like growth factor
iPSC	Induced pluripotent stem cells
Islet1	Insulin gene enhancer protein 1
Jag1	Jagged 1
Jag2	Jagged 2
Klf4	Kruppel-like factor 4

Z. Hu (✉)
Department of Otolaryngology-Head and Neck Surgery, Wayne State University
School of Medicine, 550 E Canfield St. 258 Lande, Detroit, MI 48201, USA
e-mail: zh@med.wayne.edu

K. Turksen (ed.), *Tissue-Specific Stem Cell Niche*,
Stem Cell Biology and Regenerative Medicine,
DOI 10.1007/978-3-319-21705-5_3

Lfng	LFNG O-fucosylpeptide 3-beta-N-acetylglucosaminyltransferase
Lgr5	Leucine-rich repeat-containing G-protein coupled receptor 5
LIF	Leukemia inhibitory factor
MUCs	Mouse utricle sensory epithelium-derived prosensory-like cells
NGF	Nerve growth factor
NT-3	Neurotrophin-3
Oct4,	Octamer-binding transcription factor 4
Pax	Paired box protein
P-zero	Myelin protein zero
RT-PCR	Reverse transcription polymerase chain reaction
SGN	Spiral ganglion neurons
Six1	Sine Oculis Homeobox 1
Smad3	Sma, mothers against decapentaplegic 3
Snai1	Snail Family Zinc Finger 1
Snai2	Snail Family Zinc Finger 2
Sox1	SRY-related HMG-box 1
Sox2	SRY-related HMG-box 2
Sox3	SRY-related HMG-box 3
Sox9	SRY-related HMG-box 9
TGF-α	Transforming growth factor alpha
TGF-β	The transforming growth factor beta
TUJ1	Beta III Tubulin
Zeb1	Zinc finger E-box binding homeobox 1
Zeb2	Zinc finger E-box binding homeobox 2
ZO-1	Zona occuldens protein 1

1 Inner Ear and Hearing Loss

In the inner ear, hair cell is the primary receptor for sound in the auditory system and for movement and sense of balance in the vestibular system. Sensory hair cell possesses a distinct morphological structure called hair bundle locating above the apex of the columnar-shaped cell body. In response to sound or motion signals, hair bundle is deflected to stimulate hair cells, which release glutamine to fire inner ear neurons, including spiral ganglion and vestibular ganglion neurons. The inner ear neurons in turn transfer signals to the cochlear and vestibular nucleus in the central nerve system.

Hair cells and inner ear neurons are vulnerable to a number of insults, including sound overstimulation, ototoxic drugs, gene disorders, and aging. In non-mammalian vertebrates including birds and fish, hair cells can be regenerated throughout their life [1, 2]. However, damage to mammalian hair cells is usually irreversible and causes permanent hearing loss and other inner ear disorders, such as tinnitus and balance impairment. These inner ear disorders affect the daily activity of millions of people in the world. In 2012, World Health Organization estimates that approximately 360 million persons (5.3 % of the world's population) in the world suffer from hearing loss [3]. In these hearing impaired persons worldwide, 32 million are children. The reported hearing loss population seems to be higher in developed countries. For example, the National Institute on Deafness and Other Communication Disorders (NIDCD) of the United States of America (USA) states that approximately 13 % (30 million) Americans aged 12 years or older have hearing loss in both ears in the USA, which is based on standard hearing examinations (NIDCD Quick Statistics). In the USA, approximately 2–3 out of 1000 children are born with a detectable level of hearing loss in one or both ears.

Hearing prostheses, including hearing aids, cochlear implants, and brainstem implants have been used clinically to treat hearing loss and have obtained promising results [4–7]. However, these prostheses are not able to biologically replace damaged cellular components, including hair cells and neurons. Additionally, their efficacy are limited in noisy environment, music detection, and patients with disrupt auditory pathway including cochlear nerve damage in traumatic brain injury [4–7]. With advances in stem cell biology and regeneration medicine, stem/progenitor cell populations have been identified in the mammalian inner ear and have been investigated in vitro and in vivo [7]. However, the stem cell niches that are critical for the maintenance and differentiation of inner ear stem cells have been rarely studied. In this chapter, I will discuss stem cell niches that are related to hair cell and neural stem/progenitor cells in the inner ear.

2 Hair Cell Stem Cell Niche

In the developing mammalian inner ear, stem/progenitor cells are able to generate new hair cells. However, this generation ability is significantly decreased after birth, which indicates that either hair cell stem/progenitors lose their regeneration ability, or the microenvironment in the inner ear changes after birth, therefore does not allow hair cell generation in the postnatal and adult inner ear. Currently, postnatal and adult mouse hair cell stem/progenitors have been identified and investigated in vitro, whereas research in vivo has been rarely reported. Understanding the microenvironment that is fundamental to the maintenance of hair cell stem/progenitor cells in vitro will potentially provide cues and help in vivo identify hair cell stem/progenitor cells and guide them to differentiate into new hair cells. In this section, I will discuss the microenvironment for generation of hair cell stem/progenitor cells in the development, followed by identification of in vitro culture conditions for main types of hair cell stem/progenitor cells.

2.1 Generation of Hair Cells in Vertebrate Embryos

During vertebrate development, a patch of ectoderm adjacent to the hindbrain thickens and becomes the otic placode, which begins to invaginate to form a pit, deepens into a cup, and finally closing over to form otic vesicle as differentiation proceeds [8]. In the past decades, many studies have attempted to address what are the molecular markers of otic cells. During gastrulation, otic placode precursors appear to be mixed with cells destined to give rise to neural, neural crest, and other ectodermal derivatives [9]. The earliest specific molecular markers of the otic placode seem to be the transcription factor *Pax8*, *Pax2*, *Foxi1*, and *Sox9* [10–15]. Many studies show that other transcription factors are also involved in invagination, including *Eya1*, *Gata3*, *Nkx5.1*, *Gbx2*, *Sox3*, *Dlx5*, and *Bmps* [15–22]. The microenvironment that is critical for otic placode formation has been examined and it was suggested that signals from the hindbrain, the notochord, the surrounding cranial paraxial mesoderm, and the neural crest provide inductive and patterning information for otic placode formation. Hindbrain is adjacent to the otic placode, and the cranial paraxial mesoderm lies under the otic placode. The role of the cranial paraxial mesoderm in otic placode formation has been explored and it was found that removal of paraxial mesoderm precursors blocked otic placode formation even in the presence of the hindbrain, indicating that the paraxial mesoderm may be involved in inducing the otic placode [23]. Recent studies of gain- and loss-of-function experiments confirmed that cranial paraxial mesoderm may play a critical role in the induction of the otic placode [24–27]. The role of the hindbrain is still controversial, as it is still hard to rule out the role of host mesoderm where the donor hindbrain transplants. The major candidates for signaling molecules that are involved in otic placode formation include Fgf family (i.e. Fgf3, Fgf8, and Fgf10) [27–33]. A recent study using zebrafish shows that Fgf and hedgehog act on a symmetric otic pre-pattern to determine anterior and posterior otic formation, respectively [34]. Wnt signaling was suggested to regulate timely expression of Fgf3 and Fgf8 in the hindbrain, which in turn induce otic generation [35]. Additionally, Wnt seems to play a role in directing Pax2 (+) cells to otic placodal fate and promoting dorsal otocysts [36].

As otic placode forms, a patch of cells in the otic placode start to express prosensory genes, including *Bmp4*, *Jag1/Notch*, *Lfng*, and *Sox2* [37]. A proportion of cell in prosensory patch will express transcriptional factor *Atoh1* (in mouse also called *Math1*) and further develop into hair cells for the vestibular by embroynic days 12.5 (E12.5) and auditory systems as early as E13.5–14.5 [38]. In vertebrate inner ear sensory epithelium, each hair cell is separated from the next by an interceding supporting cell, which forms an invariant and alternating mosaic pattern. It is important to understand how prosensory progenitor cells give rise to a hair cell and its immediate neighbors as supporting cells. Notch-mediated lateral inhibition is hypothesized to play a major role in the determination of mosaic pattern of hair cells and supporting cells. The Notch receptor is comprised of three major domains, extracellular, transmembrane, and intracellular domains. During early development in the mouse (E9–10), *Notch* and *Jag1* are broadly expressed in the otocyst [39, 40], proneural gene *Lfng* and *Dll1* are expressed in the anteroventral portion of the otocyst [41, 42], which will develop into neuroblasts and delaminate to form auditory and vestibular ganglion. By E12, *Jag1* is expressed in the prosensory area which will develop into sensory epithelium containing hair cells and supporting cells in the auditory and vestibular organs [43]. As development proceeds and immediately after (or approximately at) E12, *Atoh1* expressed in a specific proportion of cells in the prosensory domain [38]. *Atoh1* positive cells will become hair cells whereas *Atoh1* negative cells will become supporting cells. At E13–14, *Notch1*, *Jag1*, *Lfng*, and *Atoh1* are expressed in all of the sensory epithelia in the inner ear, whereas *Hes5*, *Dll1*, and *Jag2* are solely expressed in the vestibular system. At E14–15, *Dll1* and *Atoh1* are restricted to hair cells in the auditory and vestibular organs [38, 41]. By E17.5, *Jag1* expression is restricted to the supporting cells of each of the sensory patches in the inner ear [41]. Before hair cell generation, Notch signaling levels are equal among individual cells in the prosensory domain in the otocyst. As one cell in the presensory patch starts to express higher levels of Notch ligand *Dll1*, more Notch receptors are engaged in neighboring cells that elicit a strong Notch signal compared with the ligand-expressing cells. Activation of Notch in any one cell will reduce the expression of Notch ligand in that cell, and this change results in a feedback loop that a cell with decreased *Dll1* expression will less likely differentiate into hair cells. Accordingly, the Notch signaling receiving cells are inhibited from developing into hair cells, and therefore become supporting cells. In contrast, the Notch signaling initiating cell, which expresses high levels of Notch ligand *Dll1*, will adopt a hair cell fate. This hypothesis has been tested in mutant mouse models and hair cell organ cultures with the application of Notch signaling inhibitors. In mouse knockout mutant deletion studies, hair cell number was observe to be increased in mutants with deletion of *Jag2*, *Hes1*, or *Hes5*, which suggested that deletion of transcription factors in Notch signaling was able to stimulate hair cell generation in mutants [43–45]. In organ cultures and in vivo studies, inhibition of Notch signaling using γ-secretase inhibitors resulted in the generation of new hair cells [46–49], indicating that inhibition of Notch signaling is capable of generating new hair cells in organ culture and in vivo. Moreover, *Dll1/Jag2* double mutant or *Dll1* conditional knockout study suggests that Notch ligand *Dll1* and *Jag2* seem to act synergistically to regulate hair cell and supporting cell differentiation [50, 51]. In general, newly-formed hair cells produce inhibitory signals that repress the generation of hair cells in their immediate neighbors via Notch signaling. It is unclear, however, what is the molecule/signaling that determines the initialization of Notch signaling in a specific proportion of cells in the sensory patch.

2.2 Generation of Hair Cell Progenitors and Hair Cells Using Pluripotent Stem Cells

Pluripotent embryonic stem (ES) cells are able to differentiate into cell types of three germ layers, including otic placode-derived hair cell progenitors and hair cells. It was reported that it is possible to generate hair cell stem/progenitor cells from murine ES cells via a stepwise culture method. When exposed to 10 days of EGF and IGF-1 followed by 8 days of FGF2, ES cell-originated embryonic body-derived cells started to express *Otx2*, *Nestin*, and otic placode and otic vesicle markers *Pax2*, *BMP7*, and *Jag1* that are normally shown in prosensory domain in the developing inner ear [52]. Additionally, these hair cell progenitor cells differentiated into cells expressing hair cell markers in the absence of EGF, IGF-1 and FGF2, which indicates that these growth factors were not only important for the induction of prosensory-like hair cell stem/progenitor cells from pluripotent ES cells, but also play essential roles in the maintenance of the self-renewal of hair cell stem/progenitor cells.

In another study using mouse pluripotent ES cells, Koehler and his colleagues induced the generation of hair cell progenitor cells via stepwise differentiation in three-dimensional culture [53]. In their study, the authors applied a precise temporal control of signaling pathway to ES cells when ES cell aggregates transformed sequentially into non-neural, preplacodal, and otic-placode-like sensory epithelial cells. First (culture days 1–5), it was found that BMP treatment resulted in the down-regulation of neuroectoderm marker *Sox1* whereas up-regulate the expression non-neural ectoderm marker *Dlx3*, which suggested that BMP signaling was required for non-neural ectoderm induction of ES cell aggregates. However, BMP treatment also led to the expression of mesendodermal marker; therefore, co-application of TGF-β inhibitor was used to suppress aberrant mesendoderm induction. Second (culture days 5–8), a subsequent BMP inhibition together with FGF signaling activation were required for non-neural ectoderm to become a preplacodal fate, and further developed into otic placode. Third (culture days 9–12), BMP/TGF-β inhibition/BMP inhibition/FGF-treated cell aggregates were transferred to a serum-free floating culture condition approximately 3–4 days after BMP inhibition/FGF treatment. It was proposed that BMP/TGF-β inhibition/BMP inhibition/FGF-treated cells were able to be self-guided and differentiate into default cell fate and become sensory hair cell automatically. Endogenous Wnt signaling seems to play a role in inducing otic vesicle formation, as treatment of Wnt signaling inhibitor significantly decreased the number of prosensory vesicles. Finally (culture days 12–24), BMP/TGF-β inhibition/BMP inhibition/FGF-treated cells developed into cells expressing hair cell genes, proteins, and hair bundles when they were maintained in the serum-free floating culture. In general, changes in the culture condition of ES cells lead to the induction of hair cell stem/progenitor cells. This includes that BMP activation and TGF-β inhibition were involved in non-neural ectoderm induction, and subsequent inhibition of BMP together with FGF activation resulted in preplacodal induction, which triggered self-guided differentiation of hair cells. However, the mechanism that is involved in the self-guided hair cell differentiation remains to be elucidated.

A possible source for pluripotent stem cells is induced pluripotent stem (iPS) cells, which can be generated from mature somatic cells by overexpressing pluripotent genes [54–56]. Oshima and his colleagues generated iPS cells by infecting Math1/nGFP embryonic fibroblasts with retrovirus overexpressing *Oct4*, *Sox2*, *Klf4*, and *C-myc* [57]. The authors used several steps to generate hair cell progenitors and hair cells. First, induction of ectoderm from pluripotent iPS and ES cells. The authors generated embryoid bodies from either iPS cells or ES cells. It was found that embryoid bodies-derived cells differentiated into cells expressing ectodermal, mesoderm and endoderm linages. The application of Wnt signaling inhibitor, the Smad3 inhibitor, and IGF-1 significantly reduce the expression of mesodermal and endodermal markers, whereas majority of the cells were guided into an ectodermal lineage with these treatment. Second, induction of otic cell fate. FGF2 was added to cell culture and it was found approximately 20–30 % of iPS or ES derivatives express *Pax2*, a gene expressed during inner ear development. The authors also found that application of FGF receptor inhibitor SU5402 abolished *Pax2* induction. Third, generation of hair cell progenitor and hair cells. The authors plated the induced iPS or ES cells on fibronectin, gelatin, mouse embryonic fibroblast (MEF) feeders, and mitotically-inactived chicken utricle stromal cells, and the authors found the expression of *Atoh1*, which is a hair cell progenitor and early stage marker.

It is observed that FGF signaling was used to induce ectoderm to become otic linage in Oshima and Koehler's studies. However, there are several discrepancies in these two observations. First, Wnt signaling inhibitor, TGF-β signaling inhibitor, IGF, and FGF2 were involved in the generation of otic progenitors in Oshima's study. In Koehler's observation, BMP was used for 4–5 days to induce non-neural ectoderm differentiation, followed by preplacodal and otic placode induction using BMP signaling inhibitor and FGF signaling activation. Second, both studies used Wnt signaling inhibitors. However, Oshima et al. used Wnt signaling inhibitors to guide ES cells to an ectodermal linage, whereas Koehler applied Wnt signaling inhibitor to treat induced-otic progenitor cell aggregates and found that Wnt signaling inhibitor was able to decrease the number of Pax2 positive prosensory vesicles. Finally, chicken utricle stromal cells were used for generation of hair cell progenitors and hair cells in Oshima's study. It appears that chicken utricle stromal cells released cytokines and/or growth factors that played major roles in the generation and differentiation of mouse hair cell progenitors. In Koehler's study, a serum-free floating culture was used to activate self-guided hair cell progenitor induction and differentiation.

2.3 Identification of Sphere-Forming Cells from Inner Ear Sensory Epithelia

In the mammalian auditory system, damage to hair cells is usually permanent because the cochlea is unable to replace lost hair cells. In contrast, injured vestibular organs can generate a limited number of hair cells. It is important to understand the cell source for hair cell progenitor which is able to give rise to new hair cell following damage. In order to address this question, pure sensory epithelial cells were harvested from adult mouse utricle of the vestibular system [58], and it was reported that adult mouse utricles contained cells that may display the features of stem/progenitor cells. These utricle-derived cells were able to proliferate and form spherical structures in culture. In addition, these cells expressed genes of the embryonic inner ear, which indicates that these cells may possess the features of hair cell progenitors. Following this discovery, cells harvested from the other inner ear structures, including the organ of Corti, stria vascularis, greater epithelial ridge, lesser epithelial ridge, and cristae of semicircular canals, were reported to have the ability to form spheres in cultures [59]. These sphere-forming cells were able to express genes expressed in developing inner ear and hair cell progenitors. Additionally, Notch signaling seems to play a role in the self-renewal of these sphere-forming cells. In a study using postnatal mouse cochlear sensory epithelium, it was found that Notch signaling inhibitor resulted in a significant decrease in sphere formation, whereas Jag1 treatment seems to be able to maintain cochlear stem/progenitor cells [60].

In the study of sphere-forming cells, dissociated cells were cultured in serum-free culture medium that contained growth factors, including EGF, FGF2, IGF-1, and LIF. It was found that a combination of EGF and IGF-1 had a partially additive effect, whereas LIF did not enhanced sphere formation [58]. This study indicates that growth factors EGF and IGF-1 seem to play a vital role in promoting dissociated cells to form spherical structures. It is important to understand whether cell in a dissociated condition is required for cell proliferation in the presence of growth factors, and whether cells in organ culture still possess this proliferation ability with the supplement of growth factors. One direct approach to address this question is to apply growth factors to organ cultures where the cells are not dissociated. In a study using rat sensory epithelial sheet, it was found that FGFs, IGF-1, IGF-2, TGF-α, and EGF stimulated the proliferation of utricular epithelial cells. These observations indicate that growths factors are able to promote cell growth in both dissociated cells and organ cultures. The next question is whether these hair cell progenitors can only proliferate in vitro milieu, and whether the in vivo microenvironment allows growth factors to stimulate cell growth. In an in vivo study, guinea pigs were treated with ototoxic gentamycin followed by the infusion of growth factors. It was found that in the presence of growth factors including TGF-α, IGF-1, and retinoic acid, gentamicin-treated guinea pig utricles exhibited significantly enhanced hair cell renewal, which was indicated by scanning microscopy examination of stereocilia bundles [61]. Therefore, both in vitro and in vivo evidences show that hair cell progenitors in the mammalian vestibular organ can proliferate with the supplement of growth factors, which indicated that growth factors may be necessary for the self-renewal of hair cell progenitors in the inner ear stem cell niche.

2.4 Generation of Hair Cell Progenitor via the Induction of a Proportion of Supporting Cells

Whether sphere-formation is required and a critical step in hair cell progenitor identification in vitro? In organ culture and in vivo, sphere formation was less likely during cell proliferation and/or hair cell renewal. It is noted that the microenvironment in dissociated cell culture and organ culture is different: floating culture in dissociated cell culture whereas adherent culture in organ culture. Therefore, it is necessary to explore whether hair cell stem/progenitor cells can proliferate in adherent culture without sphere formation. In a study using transgenic mice in which supporting cells were marked by expression of green fluorescence protein (GFP) under the control of $P27^{Kip1}$, postnatal cochleae were dissociated and fluorescence activated cell sorting (FACS) was used to obtain purified supporting cells expressing $P27^{Kip1}$-GFP [62]. Purified supporting cells from wild-type or $P27^{Kip1}$ mutant mice were seeded with feeder layers containing embryonic periotic mesenchymal cells in serum-containing medium supplemented with EGF and FGF2. It was found that $P27^{Kip1}$-GFP supporting cells down-regulated the expression of $P27^{Kip1}$ and incorporated 5-bromo-2′-deoxyuridine (BrdU) after 2 days in culture, which indicates that a proportion of post-mitotic supporting cells were able to re-entered the S-phase in vitro in adherent culture in the presence of EGF, FGF2, and feeder layers without sphere formation.

In culture of *Lgr5* positive supporting cells isolated from mouse cochleae, mitomycin-inactivated embryonic chicken utricle mesenchymal tissues were used as a feeder layer [63]. After 10 days in serum-free medium containing DMEM/F12, B27, and N2, $Lgr5^+$ cells formed epithelial colonies that were labeled by anti-pan-cytokeratin antibodies. This study indicates that embryonic chicken utricle mesenchymal tissues may be involved in inducing $Lgr5^+$ cells to re-enter the cell cycle and become hair cell progenitors. However, it remains unclear whether $Lgr5^+$ cells are able to form epithelial colonies in the absence of chick utricle mesenchyme. Another research group isolated *Lgr5* and *Sox2* positive cells from postnatal mouse organ of Corti using FACS sorting and cultured $Lgr5^+$ and $Sox2^+$ cells in serum free medium containing DMEM/F12, N2, B27, EGF, FGF2, and IGF-1 [64]. It was found that both $Sox2^+$ and $Lgr5^+$ cells were able to form colonies/spheres and expressed hair cell protein myosin VIIa after 10 days of culture, which suggests that growth factors including FGF2, EGF, and IGF-1 may be sufficient for $Lgr5^+$ or $Sox2^+$ supporting cells to re-enter the cells cycle and form hair cell progenitor colonies then further become hair cells. Based on current data, both chicken inner ear mesenchyme and growth factors seem to be necessary and sufficient in induction of cell cycle re-entry of hair cell progenitors. However, a number of questions remain to be answered. For example, what is the role of chick inner ear mesenchyme in hair cell progenitor identification and self-renewal? Does chick inner ear mesenchyme only provide a matrix or also release growth factors and cytokines to assist hair cell progenitor self-renewal?

2.5 *Generation of Hair Cell Progenitors via Epithelial-to-Mesenchymal Transition (EMT)*

In adult mammalian inner ear, sensory epithelial cells are mainly composed of two types: hair cells and supporting cells. These hair cells and supporting cells are differentiated columnar-shaped epithelial cells that have exited cell cycle. In other words, adult hair cells and supporting cells are mature epithelial cells that are not able to divide into new cells. Unlike skin and intestine epithelia, regeneration of sensory hair cell is severely limited in the inner ear following insults to the inner ear sensory epithelia. This poor regeneration ability indicates that (a) stem/progenitor cells in the adult inner ear sensory epithelium have lost the ability to re-enter the cell cycle; and/or (b) microenvironment in the damaged inner ear is not sufficient to stimulate resident stem/progenitor cells, if there is any, to re-enter the cell cycle and robustly regenerate hair cells. In general, adult mammalian inner ear stem/progenitor cells are not observed to re-enter cell cycle following damage to the inner ear because of either intrinsic and/or extrinsic deficiencies. Previous studies have shown that stem/progenitor cells in the adult vestibular and auditory systems are able to re-enter the cell cycle in vitro [58, 65]. One of the major challenges here is how to induce inner ear stem/progenitor cells to re-enter cell cycle and what microenvironment cues are required.

Cell reprogramming has been applied in stem cell and regeneration research and a dedifferentiation program seems to be involved in inducing somatic cells to become pluripotent stem cells [54–56, 66]. This dedifferentiation program is essential for dedifferentiating mature epithelial cells into stem/progenitor cells [67]. When epithelial cells from pancreatic islets are cultured in vitro, they dedifferentiate into mesenchymal-like stem/progenitor cells that are able to re-enter the cell cycle via epithelial to mesenchymal transition (EMT) [68, 69]. During development, EMT is fundamental to body plan formation and organ generation [70, 71]. In adult, EMT is reported to be related to tissue recovery and cancer migration. Recently, EMT is found to be involved in stem cell biology and prosensory cell generation [7, 72–74].

EMT seems to be essential for in vitro epithelial stem cell generation based on following observations. First, induction of EMT in immortalized human mammary epithelial cells results in the acquisition of mesenchymal phenotype and expression of stem cell markers [75]. Second, EMT is found to give rise to multipotency of immortalized oral mucosal epithelial cells [76]. Third, the pluripotent transcription factor, *Oct4*, activates early EMT, which subsequently enhances reprogramming of pluripotent stem cells [74]. In the mammalian inner ear, pure sensory epithelial cells were harvested from adult mouse utricles and were dissociated into singular cells, which were plated to two-dimensional cultures. The columnar-shaped hair cells with hair bundles usually died with a week of culture, whereas supporting cells survived for longer periods. In the two-dimensional culture, solitary supporting cell gradually lost their columnar shape and started to expand on 2D substrates during passage 1 to passage 2 (Fig. 1a–d). Phalloidin-labeled F-actin, which normally forms an "actin ring" in epithelial cells [77], assembled into irregular "stress fibers" that are usually found in mesenchymal cells [78] (Fig. 1e–f). Immunofluorescence of passage 1 cells showed that they did not express the epithelial markers E-cadherin and zonula occludens (ZO-1) (Fig. 1g–h), hair cell marker myosin VIIa [79], or cytokeratin, which is usually expressed in epithelial supporting cells [80, 81] (Fig. 1j–m), indicating that supporting cells had dedifferentiated. It was found that approximately 85 % of the cell clones proliferated for 3–4 passages and then stopped growing, while the remaining (15 %) proliferated and could be maintained in culture for at least 30–40 passages (unpublished data).

Complementary cell proliferation assays, including BrdU incorporation and Cyquant NF cell proliferation assays, were used to characterize the proliferation of mouse utricle sensory epithelium-derived prosensory-like cells (MUCs) [73]. Reverse transcription PCR (RT-PCR) and immunofluorescence showed that MUCs expressed the mesenchymal markers *Snai1*, *Snai2*, *Zeb1*, *Zeb2*, *vimentin*, and *fibronectin* (Fig. 2), indicating supporting cells likely converted into mesenchymal-like cells. In addition, RT-PCR, western blotting, and immunofluorescence showed that MUCs expressed the prosensory markers *Bmp4*, *Sox2*, *Lfng*, *Islet1*, *Eya1*, *Dlx5*, and *Six1* [37, 82], and the stem cell markers *nestin*, *Sox2*, *Oct4*, *Nanog*, and *GFAP* (Fig. 3). These data suggest that MUCs might possess some of the prosensory cell features. It should be noted that MUCs also expressed markers found in both prosensory and supporting cells, including *Sox2*, *Jag1*, *Notch1*, and $P27^{kip1}$ (Fig. 3). Generally, pure sensory epithelium derived from adult mouse utricle seems to undergo EMT to become proliferative stem/progenitor cells expressing prosensory markers when they were cultured on adherent cultures, which indicates that EMT might be a possible mechanism that is able to convert mature supporting cells to re-enter the cell cycle to adopt a stem/progenitor cell fate.

3 Inner Ear Neural Stem Cell and Stem Cell Niche

In the mammalian inner ear, two types of afferent nerve fibers connect the ear to the brainstem. In the auditory system, spiral ganglion neurons (auditory neurons) are responsible for transferring sound signals from cochlear hair cells to the cochlear nucleus in the brainstem. In the vestibular system, the peripheral projections of vestibular ganglion neurons connect vestibular hair cells, whereas the central

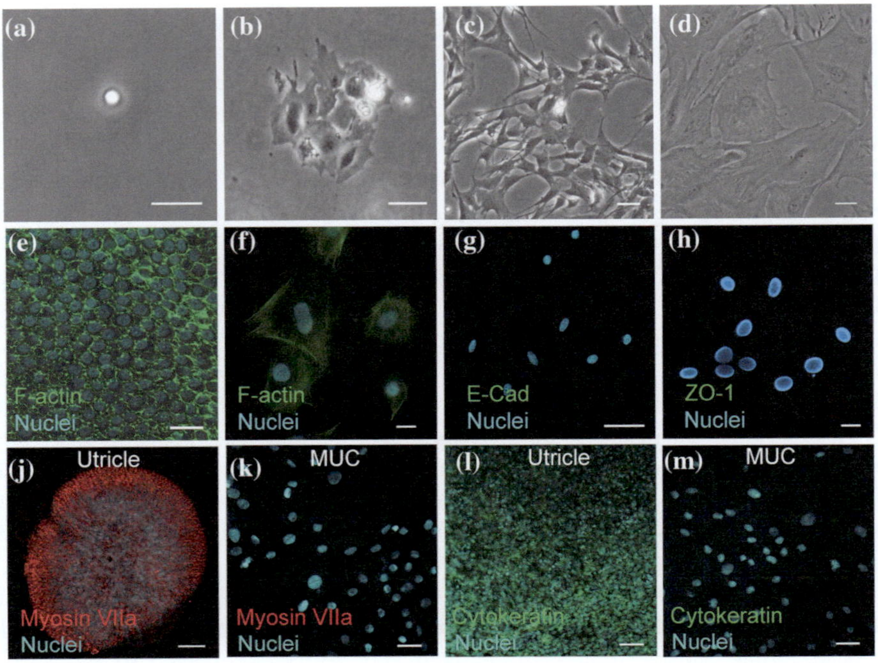

Fig. 1 The mouse utricle supporting cells dedifferentiated into MUCs in 2D cultures. **a–d** The solitary supporting cell was obtained from dissociation of pure sensory epithelia of adult mouse utricle (**a**). The cell grew to a small island in 3–4 days (**b**), gradually lost columnar shape and became flat during passage 1 (**c**) and passage 2 (**d**). **e–f** Phalloidin-labeled F-actin, which normally forms an "actin ring" in utricle sensory epithelium (**e**), was changed into irregular mesenchymal-like "stress fibers" in passage 1 MUCs (**f**). **g–h** Epithelial markers, such as E-cadherin (E-cad) and ZO-1 were not detected in passage 1 MUCs. **j–k** Hair cell marker Myosin VIIa was detected in postnatal day 5 normal mouse utricle (J), but was not observed in passage 1 MUCs (K). **l–m** Cytokeratin that was found in postnatal day 5 normal mouse utricle sensory epithelium (**l**) was not observed in passage 1 MUCs (**m**). *Scale* 20 μm in A, E, F, G, H, K, M; 50 μm in B, C, D, J, and L

processes form contacts with the nuclei in the brainstem. During development, neuroblasts are able to generate new auditory and vestibular ganglion neurons. However, adult inner ear ganglion neurons are not able to regenerate after damage or loss, indicating that (a) the neural stem/progenitor cells do not exist in the adult inner ear, (b) adult neural stem/progenitor cells may have lost the ability to re-enter the cell cycle to generate new neurons, or (c) the microenvironment in the adult inner ear does not permit neural stem/progenitor cells to self-renew and generate new neurons. In this section, I will discuss generation of inner ear ganglia during development, followed by in vitro and in vivo identification and differentiation of neural stem/progenitor cells in the inner ear.

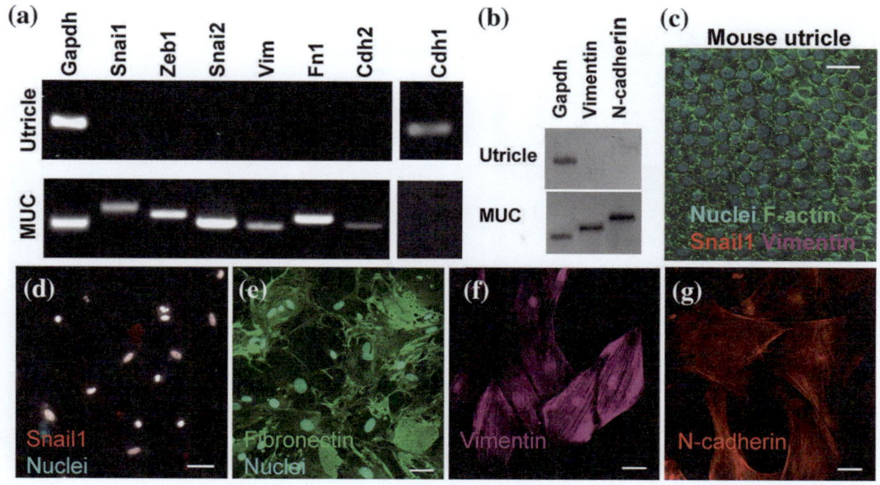

Fig. 2 Passage 10 MUCs cultured on 2D substrates expressed mesenchymal features. **a** RT-PCR data showed that MUCs expressed mesenchymal but not epithelial markers, while mouse utricle sensory epithelial cells did not express mesenchymal genes. **b** Western blotting revealed that mesenchymal markers Vimentin and N-cadherin were detected in MUCs but not in mouse utricle sensory epithelia. **c–g** Immunofluorescence showed that normal mouse utricle sensory epithelial cells did not express Snail1 and Vimentin (**c**), while MUCs expressed mesenchymal cell markers Snail1, Fibronectin, Vimentin, and N-cadherin (**d–g**). *Scale* 20 μm in **c–g**

3.1 Generation of Inner Ear Ganglion in the Inner Ear During Development

During mammalian development (around embryonic day 9 in mouse), the peripheral neural progenitors for both auditory and vestibular system migrate from the otic vesicle as neuroblasts, which will develop as cochleovestibular ganglion (also called statoacoustic ganglion) and differentiate into spiral and vestibular ganglion neurons. Almost as soon as these cells leave, they begin to extend neurites back into the otocyst in order to form neural connections with cells that will develop into hair cells. However, the molecular cues that are responsible for guiding the delamination and migration of neuroblasts have not been thoroughly understood. Cell source outside of otic vesicle also seems to contribute to cochleovestibular ganglion formation. It is reported that cells derived from neural crest participate in the formation of cochleovestibular ganglion [83, 84]. Neuroblasts in the mammalian cochleovestibular ganglion differentiate into inner ear neurons in the auditory and vestibular ganglia. Proneural basic helix-loop-helix transcriptional factors including neurogenin 1 and neuroD1 have been shown to be critical for the development of inner ear sensory neurons during development [85, 86]. It was further found that overexpression of exogenous neurogenin 1 and neuroD1 in the nonsensory epithelial region in the inner ear led to the formation of neuronal cells [87]. In addition, spiral ganglion neurons were absent in cochlea from Sox2 (Lcc/Lcc) mice, which

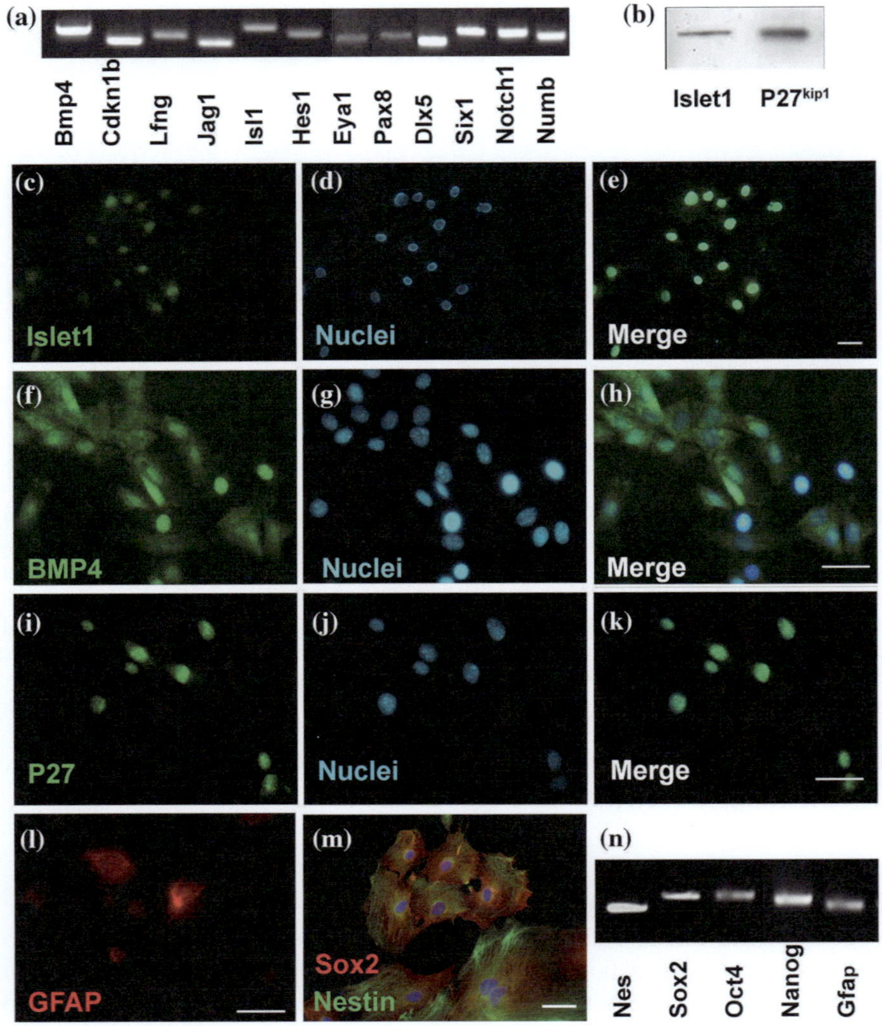

Fig. 3 Passage 10 MUCs cultured on 2D substrates showed prosensory features. **a** RT-PCR results revealed the expression of a number of prosensory genes. **b** Western blotting data showed that MUCs expressed prosensory proteins Islet1 and P27^{kip1}. **c–k** Immunofluorescence demonstrated the expression of transcription factor Islet1 and P27^{kip1} in the nucleus and Bmp4 located in the cytoplasm. **l–n** MUCs were assayed for general stem cell markers and it was found that MUCs expressed *Oct4*, *Nanog*, *Gfap*, *Nes* (Nestin), and *Sox2*, as shown in immunofluorescence (L-M) and RT-PCR (N). *Scale* 50 μm in E, H, K, and L; 20 μm in M

indicates that Sox2 is critical for auditory ganglion formation in the developing inner ear. These studies suggest that intrinsic transcriptional factors including Sox2, neurogenins and neuroD1 are important for the generation of new inner ear ganglion neurons. As for extrinsic factors, BDNF and or NT-3 mutant mice showed fewer neurons in the inner ear [88, 89], indicating that neurotrophins may be important to the generation and/or maintenance of inner ear neurons. IGF-1 mutant mice showed decreased size of spiral ganglion neurons, decreased levels of neurofilament and myelin P-zero, abnormal synaptic proteins in the spiral ganglion neurons, which suggests that lack of IGF-1 may affect the survival, differentiation, and maturation of spiral ganglion neurons [90]. It remains unclear whether neurotrophins are required for the generation of inner ear neurons, or neurotrophins are solely essential for the maintenance of naive and/or mature inner ear neurons. The microenvironment cue that is critical for the initiation of inner ear neuron differentiation is still an understudied research field.

3.2 In Vitro *Conditions Required for Generation and Maintenance of Inner Ear Ganglion Stem/Progenitor Cells*

Identification of spiral ganglion stem/progenitor cells have been studied in vitro using the culture methods that are similar to culturing neural stem cells from the brain, i.e. subventricular zone and subgranular zone. It was found that cells derived from mouse embryonic cochleovestibular ganglion were able to proliferate and form spheres in serum-free culture medium containing DMEM/F12, B27, N2, FGF2, and EGF [91]. Further, these sphere-forming cells expressed proteins that are expressed in neural stem cells, including nestin and Sox2, which indicates that neural/progenitor cells may exist in these neural sphere-forming cells. This study also suggests that cells derived from embryonic cochleovestibular ganglion are able to expand and develop into spherical structures in serum-free medium.

In addition to the culture of inner ear neural stem/progenitor cells derived from mouse embryos, cells derived from adult mouse spiral ganglion have been studied. It was reported that cells harvested from adult spiral ganglion tissues were able to proliferate and form spherical structures in serum-free culture medium containing neural supplement, EGF, FGF2, and IGF-1 [92]. In some of the studies, IGF-1 was not included in the culture medium [93, 94]. One study explored the effect of growth factors on the neurosphere formation ability of spiral ganglion-derived cells [92]. It was found that either EGF alone or IGF-1 alone has very limited effects on sphere formation, whereas FGF2 only or combination of growth factors in the presence of FGF2 seems to be able to promote sphere formation, including FGF2 + EGF, FGF2 + IGF-1, and FGF2 + EGF + IGF-1. Supplementation of cell surface receptor and co-receptor heparan sulfate was found to be able to augment the effect of growth factors on sphere formation. Further, it was observed that the number of sphere-forming cells derived from spiral ganglion decreased from birth to adult. For example, the number of spheres per 10,000 cells were $\sim$100 at postnatal day 1, <10 at postnatal day 14, <5 at postnatal day 21, and none at 6 weeks after birth.

3.3 *Differentiation of Inner Ear Neural Stem/Progenitor Cells* in Vitro

Generally, inner ear ganglion-derived neural stem/progenitor cells are able to differentiate when they are plated on substrate in the culture medium containing serum and/or growth factors. In the presence of neurotrophins, embryonic cochleovestibular ganglion-derived neural stem/progenitor cells differentiated into cells expressing neuronal intermediate filament and tubulin, including neurofilament and β-III tubulin (TUJ1) [91]. Further, these neural stem/progenitor cells showed the ability to become glutamatergic neurons because vesicle glutamate transporter 1 was detected in differentiated neurons using specific antibodies. Generation of glutamatergic neuronal-like cells indicates that neuroblasts cultured in vitro have the capacity to develop along default program to become glutamatergic spiral ganglion neuron-like cells. In the culture of adult spiral ganglion-derived stem/progenitor cells, postnatal spiral ganglion-derived stem/progenitor cells were cultured on gelatin-coated substrate and exposed to neurotrophins including BDNF and NT-3, and they were able to differentiate into cells expressing the neuronal proteins neurofilament and TUJ1, and the glial cell marker GFAP [92]. Scanning electron microscope showed neuronal-like cells were bi-polar. However, the function of these in vitro-generated neurons has been rarely reported using electrophysiology.

To study the effect of growth factor on cell fate determination of spiral ganglion stem/progenitor cells, NGF was applied to neural stem/progenitor cell cultures. It was found that NGF affected the cell fate of spiral ganglion stem/progenitor cells in a concentration-dependent manner. Low concentration of NGF (2–5 ng/mL) promoted cell proliferation. Medium concentration of NGF (20–40 ng/mL) stimulated cells to differentiate into bi-polar SGN-like cells expressing glutamatergic proteins. High concentration of NGF (100 ng/mL) could rescue cells from induced apoptosis [91].

3.4 *Progress in the* in Vivo *Studies*

It is still unclear whether adult mammalian inner ear ganglion neural stem/progenitor cells, if there is any, are able to re-enter the cell cycle and become new neurons in a mature inner ear microenvironment. And it has not been determined what kind of local microenvironment is required and/or necessary for the self-renewal of neural stem/progenitor cells in the adult inner ear. Therefore, it is important to address these questions in an in vivo model. In an ouabain-induced inner ear damage model in which spiral ganglion neurons were selectively damaged, it was found that Sox2 expression is significantly increased in ouabain-treated inner ears [95]. Immunofluorescence showed that the number of Sox2 positive glial cells increased in 3 days after ouabain treatment. Similarly, the number of BrdU positive cells increased in 3 and 7 days after treatment. Additionally, immunostaining showed that approximately 70 % of BrdU positive cells in the cochlear nerve were Sox2 positive. These results showed that following cochlear nerve damage, a proportion of glial cells along the cochlear nerve may re-enter the cell cycle and simultaneously increase the expression of Sox2. However, it is unclear whether these proliferative Sox2 cells are the candidates for resident neural stem/progenitor cells, which are able to differentiate into new neuronal phenotype. It is also important to identify what molecular signaling is activated in response to ouabain damage, and how the local microenvironment in the inner ear contribute to cell cycle re-entry and Sox2 up-regulation in those glial cells.

In addition to identifying resident stem/progenitor cells in the inner ear, exogenous neurons and stem/progenitor cells have been transplanted into the inner ear in order to replace damaged spiral ganglion neurons [96–108]. One of the critical challenges for cell transplantation is the survival of transplanted cells. The survival of implanted cell is related, at least in part, to the donor cell types. Brain-derived neural stem cells usually have a low survival rate. Approximately 0.05 % of embryonic mouse brain-derived neural stem cells survived when they were allografted implanting into adult mouse inner ear [109]. It was found that ~0.04–0.07 % implanted adult mouse brain-derived neural stem cells survived following xenografted implantation into the inner ear of adult guinea pigs [97]. In studies using standard stereological methods [110–112], samples were collected from every other sections (~12 μm interval among the sections) all through the cochlea and evaluated the number of surviving cells in each cochlea implanted with ES cells [113]. Although it was found that >10 cells/section in over 40 % of collected sections, the overall survival rate of ES cells was still <1 %. Similarly, in the adult mouse inner ear transplanted with embryonic cochleovestibular ganglion-derived neural stem cells, it was found that poor survival rate of transplanted cells seems to be the major challenge, indicating that microenvironment in the mature inner ear may not be capable of supporting the survival of transplanted cells [91]. To increase the survival of transplanted cells, NGF was infused into the inner ear that has been transplanted with inner ear neural stem/progenitor cells. It was observed that exogenous neurotrophins were able to promote the survival of transplanted cells [91]. Transcription factors such as neurogenin seem to be involved in guiding transplanted cells into bi-polar glutamatergic cell fate [97, 114]. Functional evaluation of transplanted cells has been performed. One of the studies indicated no significant changes on electrically-evoked auditory brainstem responses (eABR) [115], whereas the other showed modest hearing threshold changes in acoustically-evoked ABR (aABR) [116]. It is notable that transplanted cell types and host animals were different in these two studies: transplantation of mouse embryonic ganglion into guinea pigs in one study, whereas human ES-derived otic progenitors into gerbils in the other one. Generally, remarkable progresses have been made in the inner ear transplantation; however, some of the major challenges remain. For example, the local milieu that is necessary and important for the survival and differentiation of transplanted cells has not been fully understood.

4 Summary and Perspective

In the past decade, inner ear stem/progenitor cells have been identified and investigated both in vitro and in vivo. It seems that cells harvested from majority of the inner ear structures including sensory epithelium and ganglion have demonstrated the ability of proliferation and differentiation into several inner ear cell types. Induction of other cell types, including multipotent neural stem cells and pluripotent ES cells and iPS cells, into inner ear cell types has shown promising results. The local microenvironments, including the cell types that maintain stem cell niche and the molecular signaling pathways that are able to activate resident stem/progenitor cells to re-enter the cell cycle, remain to be elucidated. In the in vitro cultures, stem cells rely on certain cytokines and/or growth factors to maintain their cell identity. For instance, mouse ES cells depend on the existence of LIF to maintain their self-renewal in the culture. The LIF receptors and intracellular pathways that are critical for mouse ES cell self-renewal have been well studied. In the inner ear stem cell research, culture conditions that are critical for maintaining inner ear stem/progenitor cells have been identified; however, the corresponding receptors and/or intracellular pathways have not been thoroughly understood. In the in vivo studies, exogenous cells including pluripotent ES cells, multipotent neural stem cells, and inner ear-derived stem/progenitor cells have been transplanted into the mature inner ear. It is noted that the microenvironment in the adult inner ear might not been ready to adopt exogenous cells. Therefore, identifying and providing necessary microenvironment support including growth factors/cytokines/matrix that are critical for the survival, self-renewal, and/or differentiation of transplanted stem/progenitor cells remain the major challenges in the field.

Acknowledgments This work is supported by R01 (R01DC013275) from NIH. The author thank Ms. Yang Zhou, Xiaoyang Li, and Yiyun Jiang for their valuable comments to the manuscript.

References

1. Corwin JT, Cotanche DA. Regeneration of sensory hair cells after acoustic trauma. Science. 1988;240:1772–4.
2. Ryals BM, Rubel EW. Hair cell regeneration after acoustic trauma in adult Coturnix quail. Science. 1988;240:1774–6.
3. WHO. WHO global estimates on prevalence of hearing loss. Mortality and burden of diseases and prevention of blindness and deafness WHO, 2012.
4. Lim HH, Lenarz T, Joseph G, et al. Electrical stimulation of the midbrain for hearing restoration: insight into the functional organization of the human central auditory system. J Neurosci. 2007;27:13541–51.
5. Mowry SE, Woodson E, Gantz BJ. New frontiers in cochlear implantation: acoustic plus electric hearing, hearing preservation, and more. Otolaryngol Clin North Am. 2012;45:187–203.

6. Colletti L, Shannon R, Colletti V. Auditory brainstem implants for neurofibromatosis type 2. Curr Opin Otolaryngol Head Neck Surg. 2012;20:353–7.
7. Hu Z, Ulfendahl M. The potential of stem cells for the restoration of auditory function in humans. Regen Med. 2013;8:309–18.
8. Noramly S, Grainger RM. Determination of the embryonic inner ear. J Neurobiol. 2002;53:100–28.
9. Streit A. Extensive cell movements accompany formation of the otic placode. Dev Biol. 2002;249:237–54.
10. Wright E, Hargrave MR, Christiansen J, et al. The Sry-related gene Sox9 is expressed during chondrogenesis in mouse embryos. Nat Genet. 1995;9:15–20.
11. Solomon KS, Kudoh T, Dawid IB, et al. Zebrafish foxi1 mediates otic placode formation and jaw development. Development. 2003;130:929–40.
12. Saint-Germain N, Lee YH, Zhang Y, et al. Specification of the otic placode depends on Sox9 function in Xenopus. Development. 2004;131:1755–63.
13. Pfeffer PL, Gerster T, Lun K, et al. Characterization of three novel members of the zebrafish Pax2/5/8 family: dependency of Pax5 and Pax8 expression on the Pax2.1 (noi) function. Development. 1998;125:3063–74.
14. Heller N, Brandli AW. Xenopus Pax-2 displays multiple splice forms during embryogenesis and pronephric kidney development. Mech Dev. 1997;69:83–104.
15. Groves AK, Bronner-Fraser M. Competence, specification and commitment in otic placode induction. Development. 2000;127:3489–99.
16. Solomon KS, Fritz A. Concerted action of two dlx paralogs in sensory placode formation. Development. 2002;129:3127–36.
17. Sahly I, Andermann P, Petit C. The zebrafish eya1 gene and its expression pattern during embryogenesis. Dev Genes Evol. 1999;209:399–410.
18. Mizuseki K, Kishi M, Matsui M, et al. Xenopus Zic-related-1 and Sox-2, two factors induced by chordin, have distinct activities in the initiation of neural induction. Development. 1998;125:579–87.
19. Liu D, Chu H, Maves L, et al. Fgf3 and Fgf8 dependent and independent transcription factors are required for otic placode specification. Development. 2003;130:2213–24.
20. Liu A, Joyner AL. EN and GBX2 play essential roles downstream of FGF8 in patterning the mouse mid/hindbrain region. Development. 2001;128:181–91.
21. George KM, Leonard MW, Roth ME, et al. Embryonic expression and cloning of the murine GATA-3 gene. Development. 1994;120:2673–86.
22. Esterberg R, Fritz A. dlx3b/4b are required for the formation of the preplacodal region and otic placode through local modulation of BMP activity. Dev Biol. 2009;325:189–99.
23. Orts F, Jimenez-Collado L, Jimenez-Collado J. Regulation of the embryo after the extirpation of Hensen's node. Consequences on the differentiation of the otic placode. Arch Anat Histol Embryol. 1971;54:1–11.
24. Mendonsa ES, Riley BB. Genetic analysis of tissue interactions required for otic placode induction in the zebrafish. Dev Biol. 1999;206:100–12.
25. Maulding K, Padanad MS, Dong J, et al. Mesodermal Fgf10b cooperates with other fibroblast growth factors during induction of otic and epibranchial placodes in zebrafish. Dev Dyn. 2014;243:1275–85.
26. Kil SH, Streit A, Brown ST, et al. Distinct roles for hindbrain and paraxial mesoderm in the induction and patterning of the inner ear revealed by a study of vitamin-A-deficient quail. Dev Biol. 2005;285:252–71.
27. Phillips BT, Bolding K, Riley BB. Zebrafish fgf3 and fgf8 encode redundant functions required for otic placode induction. Dev Biol. 2001;235:351–65.
28. Zelarayan LC, Vendrell V, Alvarez Y, et al. Differential requirements for FGF3, FGF8 and FGF10 during inner ear development. Dev Biol. 2007;308:379–91.
29. Wright TJ, Mansour SL. Fgf3 and Fgf10 are required for mouse otic placode induction. Development. 2003;130:3379–90.

30. McKay IJ, Lewis J, Lumsden A. The role of FGF-3 in early inner ear development: an analysis in normal and kreisler mutant mice. Dev Biol. 1996;174:370–8.
31. Maroon H, Walshe J, Mahmood R, et al. Fgf3 and Fgf8 are required together for formation of the otic placode and vesicle. Development. 2002;129:2099–108.
32. Mahmood R, Mason IJ, Morriss-Kay GM. Expression of Fgf-3 in relation to hindbrain segmentation, otic pit position and pharyngeal arch morphology in normal and retinoic acid-exposed mouse embryos. Anat Embryol (Berl). 1996;194:13–22.
33. Mansour SL, Goddard JM, Capecchi MR. Mice homozygous for a targeted disruption of the proto-oncogene int-2 have developmental defects in the tail and inner ear. Development. 1993;117:13–28.
34. Hammond KL, Whitfield TT. Fgf and Hh signalling act on a symmetrical pre-pattern to specify anterior and posterior identity in the zebrafish otic placode and vesicle. Development. 2011;138:3977–87.
35. Phillips BT, Storch EM, Lekven AC, et al. A direct role for Fgf but not Wnt in otic placode induction. Development. 2004;131:923–31.
36. Ohyama T, Mohamed OA, Taketo MM, et al. Wnt signals mediate a fate decision between otic placode and epidermis. Development. 2006;133:865–75.
37. Kelley MW. Regulation of cell fate in the sensory epithelia of the inner ear. Nat Rev Neurosci. 2006;7:837–49.
38. Bermingham NA, Hassan BA, Price SD, et al. Math1: an essential gene for the generation of inner ear hair cells. Science. 1999;284:1837–41.
39. Williams R, Lendahl U, Lardelli M. Complementary and combinatorial patterns of Notch gene family expression during early mouse development. Mech Dev. 1995;53:357–68.
40. Lewis AK, Frantz GD, Carpenter DA, et al. Distinct expression patterns of notch family receptors and ligands during development of the mammalian inner ear. Mech Dev. 1998;78:159–63.
41. Morrison A, Hodgetts C, Gossler A, et al. Expression of Delta1 and Serrate1 (Jagged1) in the mouse inner ear. Mech Dev. 1999;84:169–72.
42. Morsli H, Choo D, Ryan A, et al. Development of the mouse inner ear and origin of its sensory organs. J Neurosci. 1998;18:3327–35.
43. Lanford PJ, Lan Y, Jiang R, et al. Notch signalling pathway mediates hair cell development in mammalian cochlea. Nat Genet. 1999;21:289–92.
44. Zine A, Aubert A, Qiu J, et al. Hes1 and Hes5 activities are required for the normal development of the hair cells in the mammalian inner ear. J Neurosci. 2001;21:4712–20.
45. Kiernan A. Notch signaling during cell fate determination in the inner ear. Semin Cell Dev Biol. 2013;24:470–9.
46. Mizutari K, Fujioka M, Hosoya M, et al. Notch inhibition induces cochlear hair cell regeneration and recovery of hearing after acoustic trauma. Neuron. 2013;77:58–69.
47. Lin V, Golub JS, Nguyen TB, et al. Inhibition of Notch activity promotes nonmitotic regeneration of hair cells in the adult mouse utricles. J Neurosci. 2011;31:15329–39.
48. Bramhall NF, Shi F, Arnold K, et al. Lgr5-positive supporting cells generate new hair cells in the postnatal cochlea. Stem Cell Rep. 2:311–322.
49. Korrapati S, Roux I, Glowatzki E, et al. Notch signaling limits supporting cell plasticity in the hair cell-damaged early postnatal murine cochlea. PLoS ONE.8:e73276.
50. Kiernan AE, Cordes R, Kopan R, et al. The Notch ligands DLL1 and JAG2 act synergistically to regulate hair cell development in the mammalian inner ear. Development. 2005;132:4353–62.
51. Brooker R, Hozumi K, Lewis J. Notch ligands with contrasting functions: Jagged1 and Delta1 in the mouse inner ear. Development. 2006;133:1277–86.
52. Li H, Roblin G, Liu H, et al. Generation of hair cells by stepwise differentiation of embryonic stem cells. Proc Natl Acad Sci USA. 2003;100:13495–500.
53. Koehler KR, Mikosz AM, Molosh AI, et al. Generation of inner ear sensory epithelia from pluripotent stem cells in 3D culture. Nature. 2013;500:217–21.

54. Takahashi K, Tanabe K, Ohnuki M, et al. Induction of pluripotent stem cells from adult human fibroblasts by defined factors. Cell. 2007;131:861–72.
55. Takahashi K, Yamanaka S. Induction of pluripotent stem cells from mouse embryonic and adult fibroblast cultures by defined factors. Cell. 2006;126:663–76.
56. Yu J, Vodyanik MA, Smuga-Otto K, et al. Induced pluripotent stem cell lines derived from human somatic cells. Science. 2007;318:1917–20.
57. Oshima K, Shin K, Diensthuber M, et al. Mechanosensitive hair cell-like cells from embryonic and induced pluripotent stem cells. Cell. 2010;141:704–16.
58. Li H, Liu H, Heller S. Pluripotent stem cells from the adult mouse inner ear. Nat Med. 2003;9:1293–9.
59. Oshima K, Grimm CM, Corrales CE, et al. Differential distribution of stem cells in the auditory and vestibular organs of the inner ear. J Assoc Res Otolaryngol. 2007;8:18–31.
60. Savary E, Sabourin JC, Santo J, et al. Cochlear stem/progenitor cells from a postnatal cochlea respond to Jagged1 and demonstrate that notch signaling promotes sphere formation and sensory potential. Mech Dev. 2008;125:674–86.
61. Kopke RD, Jackson RL, Li G, et al. Growth factor treatment enhances vestibular hair cell renewal and results in improved vestibular function. Proc Natl Acad Sci USA. 2001;98:5886–91.
62. White PM, Doetzlhofer A, Lee YS, et al. Mammalian cochlear supporting cells can divide and trans-differentiate into hair cells. Nature. 2006;441:984–7.
63. Chai R, Kuo B, Wang T, et al. Wnt signaling induces proliferation of sensory precursors in the postnatal mouse cochlea. Proc Natl Acad Sci USA. 2012;109:8167–72.
64. Shi F, Kempfle JS, Edge AS. Wnt-responsive Lgr5-expressing stem cells are hair cell progenitors in the cochlea. J Neurosci. 2012;32:9639–48.
65. Shi F, Hu L, Edge AS. Generation of hair cells in neonatal mice by beta-catenin overexpression in Lgr5-positive cochlear progenitors. Proc Natl Acad Sci USA. 2013;110:13851–6.
66. Halley-Stott RP, Pasque V, Gurdon JB. Nuclear reprogramming. Development. 2013;140:2468–71.
67. Tata PR, Mou H, Pardo-Saganta A, et al. Dedifferentiation of committed epithelial cells into stem cells in vivo. Nature. 2013;503:218–23.
68. Gershengorn MC, Hardikar AA, Wei C, et al. Epithelial-to-mesenchymal transition generates proliferative human islet precursor cells. Science. 2004;306:2261–4.
69. Joglekar MV, Hardikar AA. Epithelial-to-mesenchymal transition in pancreatic islet beta cells. Cell Cycle. 2010;9:4077–9.
70. Thiery JP, Acloque H, Huang RY, et al. Epithelial-mesenchymal transitions in development and disease. Cell. 2009;139:871–90.
71. Acloque H, Adams MS, Fishwick K, et al. Epithelial-mesenchymal transitions: the importance of changing cell state in development and disease. J Clin Invest. 2009;119:1438–49.
72. Hu Z, Luo X, Zhang L, et al. Generation of human inner ear prosensory-like cells via epithelial-to-mesenchymal transition. Regen Med. 2012;7:663–73.
73. Zhang L, Hu Z. Sensory epithelial cells acquire features of prosensory cells via epithelial to mesenchymal transition. Stem Cells Dev. 2012;21:1812–21.
74. Liu X, Sun H, Qi J, et al. Sequential introduction of reprogramming factors reveals a time-sensitive requirement for individual factors and a sequential EMT-MET mechanism for optimal reprogramming. Nat Cell Biol. 2013;15:829–38.
75. Mani SA, Guo W, Liao MJ, et al. The epithelial-mesenchymal transition generates cells with properties of stem cells. Cell. 2008;133:704–15.
76. Qiao B, Gopalan V, Chen Z, et al. Epithelial-mesenchymal transition and mesenchymal-epithelial transition are essential for the acquisition of stem cell properties in hTERT-immortalised oral epithelial cells. Biol Cell. 2012;104:476–89.
77. Ivanov AI. Actin motors that drive formation and disassembly of epithelial apical junctions. Front Biosci. 2008;13:6662–81.

78. Pollack V, Scheiber K, Pfaller W, et al. Loss of cytokeratin expression and formation of actin stress fibers in dedifferentiated MDCK-C7 cell lines. Biochem Biophys Res Commun. 1997;241:541–7.
79. Hasson T, Gillespie PG, Garcia JA, et al. Unconventional myosins in inner-ear sensory epithelia. J Cell Biol. 1997;137:1287–307.
80. Oesterle EC, Campbell S. Supporting cell characteristics in long-deafened aged mouse ears. J Assoc Res Otolaryngol. 2009;10:525–44.
81. Lawlor P, Marcotti W, Rivolta MN, et al. Differentiation of mammalian vestibular hair cells from conditionally immortal, postnatal supporting cells. J Neurosci. 1999;19:9445–58.
82. Barald KF, Kelley MW. From placode to polarization: new tunes in inner ear development. Development. 2004;131:4119–30.
83. Sandell LL, Butler Tjaden NE, Barlow AJ, et al. Cochleovestibular nerve development is integrated with migratory neural crest cells. Dev Biol. 2013;385:200–10.
84. Freyer L, Aggarwal V, Morrow BE. Dual embryonic origin of the mammalian otic vesicle forming the inner ear. Development. 2012;138:5403–14.
85. Ma Q, Chen Z, del Barco Barrantes I, et al. neurogenin1 is essential for the determination of neuronal precursors for proximal cranial sensory ganglia. Neuron. 1998;20:469–82.
86. Kim WY, Fritzsch B, Serls A, et al. NeuroD-null mice are deaf due to a severe loss of the inner ear sensory neurons during development. Development. 2001;128:417–26.
87. Puligilla C, Dabdoub A, Brenowitz SD, et al. Sox2 induces neuronal formation in the developing mammalian cochlea. J Neurosci. 30:714–722.
88. Ernfors P, Lee KF, Kucera J, et al. Lack of neurotrophin-3 leads to deficiencies in the peripheral nervous system and loss of limb proprioceptive afferents. Cell. 1994;77:503–12.
89. Ernfors P, Van De Water T, Loring J, et al. Complementary roles of BDNF and NT-3 in vestibular and auditory development. Neuron. 1995;14:1153–64.
90. Camarero G, Avendano C, Fernandez-Moreno C, et al. Delayed inner ear maturation and neuronal loss in postnatal Igf-1-deficient mice. J Neurosci. 2001;21:7630–41.
91. Zhang L, Jiang H, Hu Z. Concentration-dependent effect of NGF on cell fate determination of neural progenitors. Stem Cells Dev. 2011;20:1723–31.
92. Diensthuber M, Zecha V, Wagenblast J, et al. Spiral ganglion stem cells can be propagated and differentiated into neurons and glia. Biores Open Access. 2014;3:88–97.
93. Park KH, Yeo SW, Troy FA 2nd. Expression of polysialylated neural cell adhesion molecules on adult stem cells after neuronal differentiation of inner ear spiral ganglion neurons. Biochem Biophys Res Commun. 2014;453:282–7.
94. Rask-Andersen H, Bostrom M, Gerdin B, et al. Regeneration of human auditory nerve. In vitro/in video demonstration of neural progenitor cells in adult human and guinea pig spiral ganglion. Hear Res. 2005;203:180–91.
95. Lang H, Li M, Kilpatrick LA, et al. Sox2 up-regulation and glial cell proliferation following degeneration of spiral ganglion neurons in the adult mouse inner ear. J Assoc Res Otolaryngol. 2011;12:151–71.
96. Hu Z, Ulfendahl M, Olivius NP. Central migration of neuronal tissue and embryonic stem cells following transplantation along the adult auditory nerve. Brain Res. 2004;1026:68–73.
97. Hu Z, Wei D, Johansson CB, et al. Survival and neural differentiation of adult neural stem cells transplanted into the mature inner ear. Exp Cell Res. 2005;302:40–7.
98. Tateya I, Nakagawa T, Iguchi F, et al. Fate of neural stem cells grafted into injured inner ears of mice. NeuroReport. 2003;14:1677–81.
99. Tamura T, Nakagawa T, Iguchi F, et al. Transplantation of neural stem cells into the modiolus of mouse cochleae injured by cisplatin. Acta Otolaryngol Suppl. 2004:65–68.
100. Iguchi F, Nakagawa T, Tateya I, et al. Surgical techniques for cell transplantation into the mouse cochlea. Acta Otolaryngol Suppl. 2004:43–47.
101. Hildebrand MS, Dahl HH, Hardman J, et al. Survival of partially differentiated mouse embryonic stem cells in the scala media of the guinea pig cochlea. J Assoc Res Otolaryngol. 2005;6:341–54.

102. Coleman B, Hardman J, Coco A, et al. Fate of embryonic stem cells transplanted into the deafened mammalian cochlea. Cell Transplant. 2006;15:369–80.
103. Corrales CE, Pan L, Li H, et al. Engraftment and differentiation of embryonic stem cell-derived neural progenitor cells in the cochlear nerve trunk: growth of processes into the organ of Corti. J Neurobiol. 2006;66:1489–500.
104. Backhouse S, Coleman B, Shepherd R. Surgical access to the mammalian cochlea for cell-based therapies. Exp Neurol. 2008;214:193–200.
105. Bogaerts S, Douglas S, Corlette T, et al. Microsurgical access for cell injection into the mammalian cochlea. J Neurosci Methods. 2008;168:156–63.
106. Praetorius M, Vicario I, Schimmang T. Efficient transfer of embryonic stem cells into the cochlea via a non-invasive vestibular route. Acta Otolaryngol. 2008;128:720–3.
107. Fu Y, Wang S, Liu Y, et al. Study on neural stem cell transplantation into natural rat cochlea via round window. Am J Otolaryngol. 2009;30:8–16.
108. Lang H, Schulte BA, Goddard JC, et al. Transplantation of mouse embryonic stem cells into the cochlea of an auditory-neuropathy animal model: effects of timing after injury. J Assoc Res Otolaryngol. 2008;9:225–40.
109. Iguchi F, Nakagawa T, Tateya I, et al. Trophic support of mouse inner ear by neural stem cell transplantation. NeuroReport. 2003;14:77–80.
110. Eriksson NP, Lindsay RM, Aldskogius H. BDNF and NT-3 rescue sensory but not motoneurones following axotomy in the neonate. NeuroReport. 1994;5:1445–8.
111. Gundersen HJ, Bendtsen TF, Korbo L, et al. Some new, simple and efficient stereological methods and their use in pathological research and diagnosis. APMIS. 1988;96:379–94.
112. Kroustrup JP, Gundersen HJ, Vaeth M. Stereological analysis of three-dimensional structure organization of surfaces in multiphase specimens: statistical methods and model-inferences. J Microsc. 1988;149:135–52.
113. Hu Z, Andang M, Ni D, et al. Neural cograft stimulates the survival and differentiation of embryonic stem cells in the adult mammalian auditory system. Brain Res. 2005;1051:137–44.
114. Reyes JH, O'Shea KS, Wys NL, et al. Glutamatergic neuronal differentiation of mouse embryonic stem cells after transient expression of neurogenin 1 and treatment with BDNF and GDNF: in vitro and in vivo studies. J Neurosci. 2008;28:12622–31.
115. Hu Z, Ulfendahl M, Prieskorn DM, et al. Functional evaluation of a cell replacement therapy in the inner ear. Otol Neurotol. 2009;30:551–8.
116. Chen W, Jongkamonwiwat N, Abbas L, et al. Restoration of auditory evoked responses by human ES-cell-derived otic progenitors. Nature. 2012;490:278–82.

Corneal Limbal Stem Cell Niche

Mercede Majdi, Lisa Wasielewski and Ali R. Djalilian

Abreviations

LSCD	Limbal stem cell deficiency
CESCs	Corneal epithelial stem cells
TA	Transient amplifying
ECM	Extracellular matrix
LSCs	Limbal stem cells
TDCs	Terminally differentiated cells
TACs	Transient amplifying cells
PMCs	Post-mitotic cells

1 Overview

Adult tissue-specific stem cells have the capacity to self-renew and are capable of generating functional differentiated cells that replenish lost cells through an organism's lifetime. Stem cells are thought to share a common set of characteristics, including a high proliferative potential and a long cell cycle, and they have been estimated to make up from 0.5 to less than 10 % of the total cell population [1].

The cornea is a major protective barrier and is the major source of the refractive power of the eye. The epithelial layer of the cornea develops from the surface ectoderm, and the stromal and endothelial layers are derived from neural crest cells

M. Majdi · A.R. Djalilian (✉)
Department of Ophthalmology and Visual Sciences, University of Illinois at Chicago,
1855 W Taylor St, Chicago, IL 60612, USA
e-mail: adjalili@uic.edu

M. Majdi
e-mail: mercedesmajd@gmail.com

L. Wasielewski
Rosalind Franklin University, 3333 Green Bay Rd, North Chicago, IL 60064, USA
e-mail: lisa.wasielewski@my.rfums.org

K. Turksen (ed.), *Tissue-Specific Stem Cell Niche*,
Stem Cell Biology and Regenerative Medicine,
DOI 10.1007/978-3-319-21705-5_4

(mesenchymal tissue); diverse types of stem cells are located in each layer [2]. A natural turnover of corneal epithelial cells takes place when the superficial cells are shed from the corneal surface and are replaced from a population of stem cells. These corneal epithelial stem cells (CESCs) reside in the basal cell layer of the limbus. The limbus is a 1.5–2 cm wide area between the cornea and conjunctiva, also referred to as the corneal-scleral junction. This location of CESCs was first proposed by Davanger and Everson in 1971 [3].

Limbal stem cells (LSCs) are believed to be the primary source of corneal epithelium and have an unlimited proliferative capacity. LSCs give rise to transient amplifying (TA) cells that have a limited proliferative capacity. TA cells migrate centripetally and differentiate into terminally differentiated cells. Terminally differentiated cells are shed from the surface. Limbal stem cells are constantly self-renewing, repairing and regenerating corneal tissues and are responsible for maintaining the corneal epithelium [4]. The limbal niche plays an essential role in maintaining the function of the corneal LSCs and consists of both cellular (i.e. limbal keratocytes) and noncellular (i.e. extracellular matrix) components [5] where SCs are maintained in an undifferentiated state [6]. The stem cell niche and intrinsic genetic programs in the stem cells regulate the delicate balance of self-renewal and differentiation.

2 Experimental Evidence for the Limbal Location of CESCs

There is strong evidence confirming the existence of corneal epithelial stem cells in the limbus. DNA labeling studies have demonstrated that the peripheral cornea undergoes more active DNA synthesis than the central cornea [7]. In vitro studies have demonstrated that limbal basal epithelial cells have a higher proliferation potential in culture than those epithelial cells in the periphery and center of the cornea [8, 9]. Additionally, it has been shown that limbal basal epithelial cells respond to central corneal wounds and to differentiation-inducing agents with much higher proliferative rates compared to the central corneal epithelial cells [10, 11]. Furthermore, limbal and central corneal epithelial cells have different responses to growth factors, retinoic acid, and calcium [12–14].

Evidence for the limbal location of corneal epithelial stem cells is also demonstrated with DNA label-retaining or slow-cycling experiments [15–17]. In these experiments, the DNA of mice cells in the S-phase were labeled over several days by continuous treatment with titrated thymidine and bromodeoxyuridine. Treatment was then stopped, and the mice were allowed to grow and mature. After two months when the epithelial cells had gone through multiple cell divisions, the label had begun to dissipate within the cells. The cells that did not divide thus retained the label for longer periods of time, also known as label-retaining or slow-cycling cells. These label-retaining cells are located at the limbal region in the basal cell population [15].

Keratins are a group of intracellular cytoskeletal proteins that are synthesized by almost all epithelial surfaces. Analysis of these water-soluble proteins gives valuable information about the level of differentiation and lineage of epithelial cells. Keratin 3 (K3) and Keratin 12 (K12) have been thought to be tissue-specific keratins of corneal epithelia [10, 15, 18] that are not expressed in conjunctival epithelium. K3 and K12 are expressed by more differentiated corneal epithelial cells and are synthesized throughout the central corneal and suprabasal limbal epithelium. Limbal basal epithelium lacks these keratins and instead expresses the more primitive keratin 14 (K14), which indicates a less differentiated state.

Yuspa et al. demonstrated that there are two types of proliferative basal cells in terms of their response to phorbol ester tumor promoters [19]. A 'tumor promoter-sensitive' subpopulation are cells that cease mitosis and initiate terminal differentiation in the presence of phorbol ester tumor promoters and are considered to be more differentiated TACs. In contrast, the 'tumor promoter-resistant' subpopulation continues proliferation despite exposure to phorbol ester tumor promoters and may be the target of carcinogenic substances; these can be considered stem cells. Phorbol esters induce proliferation in corneal epithelial cells, subsequently exhausting the proliferative capacity of TACs and causing terminal differentiation. Conversely, limbal stem cells retain their proliferative capacity [11]. A considerably higher population of tumor promoter-resistant cells is located at the limbus than the central corneal epithelium.

Further clinical evidence of the limbal location of corneal epithelial stem cells comes from the evidence that almost all neoplasms arise from stem cells. A vast majority of corneal neoplasms originate from the limbus, and tumors virtually never originate from the central corneal epithelium, which suggests that stem cells reside at the limbus [20]. Additionally, during the re-epithelialization of injured human cornea, the cells have been found to migrate from the limbus toward the central cornea, which supports that corneal epithelial stem cells are located in the limbal basal epithelium [21]. Since then, further studies have confirmed this observation [22–25].

More recently, the theory that the limbal area is the exclusive source of epithelial stem cells has been challenged. It has been shown that stem cells are not limited to the limbal area and are present in the central cornea as well. Majo et al. reported that the entire ocular surface of pig eyes contained oligopotent stem cells (holoclones, described below) with the capacity to generate individual colonies of corneal and conjunctival cells. They also reported that mouse corneal epithelium could be serially transplanted, self-maintained, and contained oligopotent stem cells that can generate goblet cells when provided with a conjunctival environment [26]. Chang et al. showed that within the first 12 h after wounding, the central human corneal epithelial cells are capable of corneal regeneration, even after ablation of the limbus [4]. Even so, the limbal region undoubtedly contains stem cells that are efficient in restoring the corneal surface in very large corneal injuries and is the principle source of stem cells for cell therapy in humans.

3 Characteristics of Limbal Stem Cells

Barrandon and Green classified epidermal keratinocytes into three distinct colonies in terms of size and proliferation capacity [27]. 'Holoclone' colonies have the highest proliferative capacity and are considered to be stem cells with less than 5 % aborted colonies and greater than 100 cell doublings [28]. Keratinocyte cultures containing holoclones can regenerate epidermis that persists for years. In contrast, 'paraclone' colonies have poorest proliferative capacity. They have the lowest proliferative capacity and are considered to be abortive colonies of terminally differentiated cells (TDCs). 'Meroclone' colonies are of mixed composition and are considered to be a reservoir of transient amplifying cells (TACs). Transitions from holoclone to meroclone to paraclone are unidirectional and result in progressively restricted growth potential. Pellegrini et al. demonstrated that only limbal basal cells are able to become holoclone colonies, whereas central corneal epithelia can only give rise to paraclone and meroclone colonies [29].

Adult stem cells have certain intrinsic defining characteristics. Listed below are those properties that CESC have in common with all other stem cells:

(i) Un-differentiated Phenotype with Primitive cytoplasm

Limbal basal stem cells have long been recognized as primitive in terms of differentiation markers [15]. Entering the differentiation pathway implies removal of cell from the stem cell population. Identification of undifferentiated stem cells has relied primarily on the presence or absence of specific phenotypic markers [30]. Limbal basal cell layers preferentially express certain structural proteins, cell adhesion molecules, enzymes, cell cycle regulators, and ABCG2 which is an ATP-binding cassette protein thought to protect LSCs from stress by transporting small regulatory molecules required for their proliferation, differentiation, and apoptosis [31]. The limbal stem cell profile is currently defined as p63, ABCG2, integrin-a9-positive and nestin, E-cadherin, connexin 43, and involucrin-negative [32]. Additionally, the limbal basal cells lack the well-known corneal epithelial differentiation associated keratin pair, CK3 and CK12 discussed above.

As limbal basal cells migrate out of the limbus, their protein expression profile gradually changes. Central corneal epithelium is characterized by loss of α-enolase and melanin pigmentation and the expression of CK3 and CK12, connexin 43 and 50, involucrin, and CLED, which is a calcium linked protein associated with early epithelial differentiation [31].

(ii) High Capacity for Error-free Self-Renewal

Limbal stem cells must have unlimited proliferative capacity to maintain corneal epithelial cell mass under normal and wound healing states. It is universally accepted that stem cells are capable of unlimited self-renewal. Tissue-specific stem cells are defined by their ability to self-renew and give rise to progenitor cells that enter the differentiation pathway. Differentiated cells are generally short-lived; they are produced from a small pool of long-lived stem cells that last throughout life [33, 34]. Cell division within stem cells is asymmetric; there is production of a daughter cell and the remaining parent cell. The daughter cell is destined for the differentiation pathway to become rapidly-dividing TACs, whereas the parent cell serves to replenish the stem cell pool [15, 35]. These TACs constitute the majority of the proliferative cell population in the corneal/limbal epithelium [36]. Integrins, cell-surface Receptors, that settle cell attachment to proteins on Basal Membranes can identify Stem cells as they anchor cells within a tissue. Integrins may prove as the markers which distinguish LSCs from their early TAC progeny. Loss of integrin expression is thought to trigger departure from the niche, initiating a differentiation program [37]. Before leaving the proliferative section, TACs undergo a limited number of cell divisions to become more terminally differentiated, known as TDCs or post-mitotic cells (PMCs).

(iii) Infrequent Proliferation in the Steady State (Slow cycling during homeostatis in vivo)

Although stem cells have high proliferative capacity, under steady state conditions, they exhibit extremely low rates of proliferation [38], indicative of low mitotic activity. Label-retaining, also known as slow-cycling, is one of the intrinsic properties of stem cells [15, 16, 39] and the limbal basal epithelium has been shown to contain slow-cycling cells following pulse-chase labeling of all cells with a DNA precursor (^{3}H-thymidine or bromodeoxyuridine) [38]. This indicates that the limbus is the niche for the stem cells responsible for the long-term renewal of the cornea.

(iv) Pluripotency (Transdifferentiation)

Transdifferentiation or plasticity refers to the differentiation of an adult tissue-specific stem cell into another type of cell or tissue. It has been shown that basal epithelial cells form the adult cornea under an appropriate microenvironment can transdifferentiate to hair follicles [40, 41]. Tissue recombination experiments have shown that adult central corneal cells are able to respond to specific information originating from embryonic dermis; under the right environmental cues, K3/K12 positive corneal epithelial cells can decrease expression of K12 and increase expression keratin 5 and 14 as expressed in the basal cells of the limbus and epidermis and can then proliferate and form hair follicles [40, 41]. It has also been shown in vitro that hair follicle epithelial stem cells are capable of differentiating into corneal epithelial-like cells when exposed to a limbus-specific microenvironment [42]. This indicates the pleuripotent nature of corneal transient amplifying cells.

(v) Discrete Microenvironment or Niche

Niche, a German word meaning ‘nest’, is a specific microenvironment where adult stem cells reside in their tissue of origin without differentiating. Schofield first introduced the concept of niche in 1978 [43]. Niches are three-dimensional SC-sheltering, highly organized interactive structural units which commonly occur at tissue intersections or transition zones (e.g., corneo-limbal, esophago-gastric, endo-ectocervical) [44]. The limbal niche consists of a healthy organized microenvironment containing various factors including secreted cytokines, extracellular matrix (ECM) components such as laminin, collagen type IV, collagen type XII, and tenacin-C and intercellular adhesion factors. These components provide a unique stromal microenvironment for limbal epithelial cells [45]. Niche cells are mesenchymal cells that help maintain the stem phenotype of the LESCs [46] and may provide an environment to protect stem cells from differentiation stimuli, apoptotic stimuli, and any other external insult. Additionally, the niche protects stem cells against excessive proliferation, leading to malignancies.

4 Corneal Limbal Stem Cell Niche

Increasing evidence supports that adult germ and somatic SCs are regulated by their niche [47, 48]. Interactions with the microenvironment are essential for maintaining and activating adult stem cells. Compared to other types of adult somatic stem cells, limbal stem cells are unique in being enriched in an anatomic location that is relatively easy to access. The Palisades of Vogt [49] and limbal crypts are structures proposed as the corneal stem cell niche. Recently, two other structures, the limbal epithelial crypts and focal stromal projections have been reported to contain cells exhibiting limbal stem cell markers. Interestingly, limbal epithelial crypts and focal stromal projections were not found in limbal stem cell deficiency, highlighting the hypothesis that such structures are part of limbal stem cell's niche [50, 51]. These structures are absent in patients with limbal stem cell deficiency (LSCD).

The epithelial-stromal interface in the limbus differs from the interface in the central cornea. In the cornea, the most obvious difference between the stromal matrix and the basement membrane zone is the presence of blood vessels in the limbus, which help form the palisades of Vogt and provides necessary nutrients and growth factors [1, 52]. Bowman's layer is absent in the limbus, meaning corneal stroma directly underlies the epithelial basement membrane. The limbal zone between the Palisades of Vogt and the cornea contains a very rough undulating surface with papillae or 'pegs' of stroma extending upward as shown by ultra-structural and immunohistochemical techniques [53], in contrast to the relatively flat basement membrane in the central cornea. This creates increased basement membrane surface area at the limbus which may also facilitate flow of nutrients and growth factors. This structure offers physical protection in addition to a large surface area that can accommodate increased cell numbers, blood vessels, and other supportive cells such as melanocytes, macrophages, and stromal cells.

The basement membrane beneath the CESCs has a distinct extracellular matrix (ECM) composition in terms of laminin isoforms [54], collagen type 4 alpha chain [54], and the AE27 bone marrow antigen [55]. These limbal basement membrane components might help determine stem cell distribution in the niche as suggested in the intestinal crypt villus [50]. Furthermore, like that of other stem cell niches, the limbal basement membrane might help sequester and hence modulate concentrations of growth factors and cytokines that are released from limbal niche cells for efficient and precise targeting to limbal stem cells [56]. Effectively, the limbal niche plays a crucial role in regulating self-renewal and fate decision of limbal epithelial stem cells [20].

It is believed that limbal basal cells, stromal stem cells and the ECM molecules function as one unit to maintain the reservoir of ocular stem cells. The interactive crosstalk between the ECM, surrounding cells, and soluble signals is critical for stem cell homeostasis or activation. Within the niche, the stromal fibroblasts in particular appear to be one of the important components of the limbal niche given their intimate interactions with the epithelium through the production of cytokines such as hepatocyte growth factor, keratinocyte growth factor, and IL-6 [18, 21, 57]. Limbal fibroblasts appear to play an important role in maintaining the limbal epithelial phenotype by secreting corneal specific factors that are crucial for the corneal epithelial fate [9, 11, 49, 58].

In ex vivo human limbal suspension cultures, it has been shown that the stem cells and niche cells interact with each other, migrate in spiraling patterns, and self-organize to form niche-like compartments that resembal the limbal crypts. These sphere-clusters are enriched with niche cells positive for nestin, vimentin, S100, and p27 and quiescent epithelial stem cells positive for p15, p21, p63α, C/EBPδ, ABCG2, and Pax6 [59]. While the specific features of the limbal niche have not been fully characterized, it likely includes both cellular and extracellular factors [12, 49, 57, 60, 61]. Since there is not one consensus marker for LSCs, a combination of functional, morphological, and immunohistochemical markers is the most useful marker in the present. The neighboring cells in the limbal niche include melanocytes, antigen-presenting Langerhans cells, suppressor T-lymphocytes, and recently identified limbal niche cells [62]. Human limbal niche cells were shown to be a subset of mesenchymal cells immediately subjacent to limbal basal epithelial cells that are as small as 5 μm and heterogeneously express embryonic stem cell markers such as Oct4, Sox2, SSEA4, and Nanog in addition to other stem cell markers such as Nestin, N-Cadherin, and CD34 [63]. In a study by Li et al. using limbal niche cells cultured on plastic in Dulbecco's modified Eagle's medium with fetal bovine serum or coated or three-dimensional Matrigel in embryonic stem cell medium with leukemia inhibitory factor and basic fibroblast growth factor, it was shown that niche cells can generate progenitors with angiogenesis and mesenchymal stem cell potential and may play a role in angiogenesis and regeneration in corneal wound healing [64].

Due to the importance of stem cells within their niche, they are protected by various mechanisms. The undulated basement membrane at the limbus not only increases the surface area for contact with the basement membrane, but also provides protection for the deeply seated stem cells from injury and shearing forces [53]. A rich deposit of pigment on the limbal basal cell offers photoprotection against possible DNA damage via ultraviolet radiation and subsequent generation of oxygen radicals. Additionally, indirect immunohistochemical findings are suggestive of the prominent accumulation of stem cells in the superior and inferior zones of the limbus, where they are naturally protected from by the eyelids. Additionally, the Bell's phenomenon is a defense mechanism where the eye moves upward and outward as the eye closes that serves as protection.

Several studies have indicated that components of the niche can influence the expression of LSC markers. Espanza et al. transplanted rabbit limbal epithelial sheets or central corneal sheets on to either corneal stroma or limbal stroma and examined expression profile of cytokeratin 3 and connexin 43. Regardless of the type of epithelium transplanted, corneal stroma promoted expression of cytokeratin 3 and limbal stroma suppressed it [65], highlighting the influence of components of the niche on the LSCs.

5 Corneal Limbal Stem Cell Deficiency and the Role of the Niche

Major insults to the ocular surface such as chemical injuries or severe autoimmune reactions typically destroy the LSCs, as well as their niche. Any alterations to the cornea can lead to a poor visual outcome and ultimately, visual loss. Pathologies and injuries affecting the limbus lead to limbal stem cell deficiency (LSCD), which can be caused by inherited pathologies or, more commonly, acquired factors such as chemical/thermal injuries, ultraviolet or ionizing radiation, contact lens wear, limbal surgery and conditions such as Stevens-Johnson syndrome [28] (Fig. 1). When the limbal area is partially or completely damaged, the conjunctival epithelium normally prevented from encroaching the corneal surface by LSCs migrates over the stroma, covering the cornea with conjunctival epithelium with goblet cells. This conjunctivalization is usually accompanied by neovascularization and is associated with abnormal fibrovascular tissue covering the corneal surface, leading to chronic inflammation (Fig. 2). Surgical removal of rabbit limbal epithelia results in defective corneal epithelial regeneration with abnormal conjunctival epithelial ingrowths over the cornea [45], which is suggestive that corneal epithelial wounds also have limited capacity to heal in the absence of limbal stem cells.

There is evidence that in certain conditions, LSC functioning may be compromised because of disturbances to the limbal niche, and there are a number of reports in the literature that describe cases of partial LSC deficiency where the disease was reversible with medical therapy [66, 67]; it is likely that these cases may represent

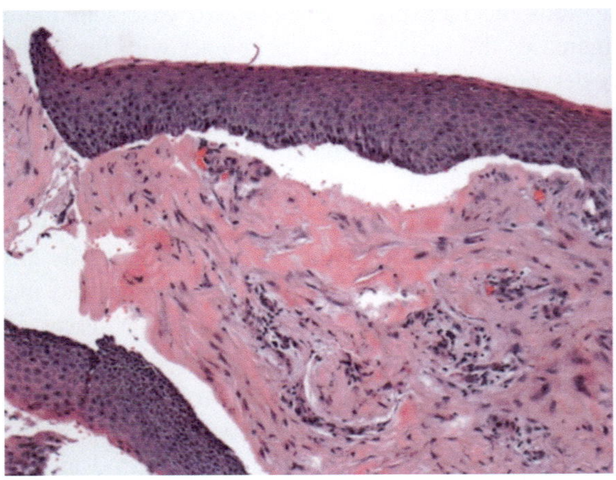

Fig. 1 The histology of Limbal stem cell deficiency. irregular epithelium and variably thick, with surface keratinization. The epithelium is lining fibrocollagenous stroma with dilated, congested blood vessels and patchy chronic inflammation

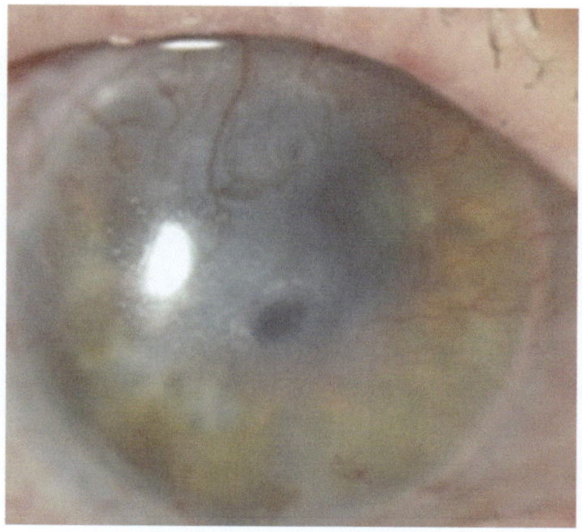

Fig. 2 Limbal Stem Cell Deficiency. Conjunctivalization accompanied by neovascularization and abnormal fibrovascular tissue covering the corneal surface, leading to chronic inflammation due to damage of the limbal area

pathology associated with the dysfunction of the limbal niche [5]. If left untreated, chronic and persistent damage to the limbal niche may lead to permanent loss of the niche, or complete limbal stem cell deficiency.

Successful treatment of LSCD requires repopulation of limbal stem cells. Treatment of LSCD involves surgical intervention to reconstitute the normal corneal epithelium and includes superficial keratectomy and amniotic membrane transplantation, limbal stem cell transplantation, or both [68, 69]. Four types of limbal stem cell transplant procedures have been developed: conjunctival-limbal autologous transplantation, living-related conjunctival-limbal allogenic transplantation, keratolimbal allogenic transplantation (cadaveric donor), and ex vivo expansion of limbal stem cells transplantation [70]. The risk of corneal transplant failure increases with increasing hostility of the local environment in the recipient [71]. Living-related conjunctival limbal allografts may convey an advantage over cadaveric conjunctival limbal allografts and result in increased treatment successes in severe ocular presentations of LSCD. It is much easier to achieve HLA-antigen matches in a living-related donor than a cadaveric one, thus the highly antigenic limbic tissue has a lower risk for rejection in a living related transplant. Surgical restoration of the limbal stem cells must accompany restoration of the niche, or treatment may have limited success. Restoration of the niche involves restoring a healthy ocular surface and optimal environment for the limbal stem cells which can be accomplished by optimal medical therapy, including use of anti-inflammatory therapy, artificial tears, and scleral contact lenses [5, 72].

Amirjamshidi et al. have demonstrated that cell culture media conditioned with human limbal fibroblasts can have a therapeutic benefit in a mouse model of limbal stem cell deficiency. Limbal stem cell deficient mice treated topically with conditioned media derived from human limbal fibroblasts had less conjunctival goblet cells in comparison to the control treated with Dulbecco's serum-free medium. This finding further supports the notion that the essential limbal stem cell niche is provided by limbal fibroblasts and suggests a new, non-invasive option in the treatment of limbal stem cell deficiency [72].

6 Conclusions

Recent advances in stem cell research have introduced a new era of regenerative medicine, where stem cells will enable the regeneration and treatment of diseased tissues. Knowledge about microenvironment of corneal stem cells is helpful for stem cell based regenerative medicine. Restoring a healthy ocular surface in patients requires creating an optimal environment for the limbal stem cells. Replacing factors produced by the limbal niche, in conjunction with restoring a healthy ocular surface, may be a way to restore the niche and the ocular surface as a potential treatment of LSCD.

References

1. Zieske JD. Perpetuation of stem cells in the eye. Eye. 1994;8:163–9.
2. Espandar L, Afshari NA. Adult corneal stem cells and alternative sources for regenerative therapy for the cornea. Ophthalmology. 2013;23(1):1–6.
3. Davanger M, Evensen A. Role of pericorneal papillary structure in renewal of corneal epithelium. Nature. 1971;229:560–1.
4. Chang CY, Green CR, McGhee CN, Sherwin T. Acute wound healing in the humancentral corneal epithelium appears to be independent of limbal stem cell influence. Invest Ophthalmol Vis Sci. 2008;49:5279–86.
5. Kim BY, Riaz KM, Bakhtiani P, Chan CC, Welder JD, Holland EJ, et al. Medically reversible limbal stem cell disease. Ophthalmology. 2014;121:2053–8.
6. Tsai RJ, Tsai RY. From stem cell niche environments to engineering of corneal epithelium tissue. Jpn J Ophthalmol. 2014;58(2):111–9.
7. Hanna C, O'Brien JE. Cell production and migration in the epithelial layer of the cornea. Arch Ophthalmol. 1960;64:536–9.
8. Ebato B, Friend J, Thoft RA. Comparison of limbal and peripheral human corneal epithelium in tissue culture. Invest Ophthalmol Vis Sci. 1988;29:1533–7.
9. Lindberg K, Brown ME, Chaves HV, Kenyan KR, Rheinwald JG. In vitro propagation of human ocular surface epithelial cells for transplantation. Invest Ophthalmol Vis Sci. 1993;34:2672–9.
10. Schermer A, Galvin S, Sun TT. Differetiation-related expression of a major 64 k corneal keratin in vivo and in culture suggests limbal location of corneal epithelial stem cells. J Cell Biol. 1986;103:49–62.
11. Kruse FE, Tseng SC. A tumor promoter-resistant subpopulation of progenitor cells is larger in limbal epithelium than in corneal epithelium. Invest Ophthalmol Vis Sci. 1993;34:2501–11.
12. Tseng SC. Regulation and clinical implications of corneal epithelial stem cells. Mol Biol Rep. 1996;23:47–58.
13. Kruse FE, Tseng SC. Retinoic acid regulates clonal growth and differentiation of cultured limbal and peripheral corneal epithelium. Invest Ophthalmol Vis Sci. 1994;35:2405–20.

14. Kruse FE, Tseng SC. Growth factors modulate clonal growth and differentiation of cultured rabbit limbal and corneal epithelium. Invest Ophthalmol Vis Sci. 1993;34:1963–76.
15. Cotsarelis G, Cheng SZ, Dong G, Sun TT. and Lavker, R.M . Existence of slow- cycling limbal basal epithelial cells that can be preferentially stimulated to proliferate: implications on epithelial stem cells. Cell. 1989;57:201–209.
16. Lavker RM, Sun TT. Epithelial stem cells: the eye provides a vision. Eye. 2003;17:937–40.
17. Lavker RM, Tseng SC, Sun TT. Corneal epithelial stem cells at the limbus: looking at some old problems from a new angle. Exp Eye Res. 2004;78:433–6.
18. Kurpakus MA, Stock EL, Jones JC. Expression of 55-kD/64-kD corneal keratins in ocular surface epithelium. Invest Ophthalmol Vis Sci. 1990;31:448–56.
19. Yuspa SH, Ben T, Hennings H, Lichti U. Divergent responses in epidermal basal cells exposed to the tumor promoter 12-O-tetradecanoylphorbol-13-acetate. Cancer Res. 1982;42:2344–9.
20. Nassiri N, Djalilian AR, Chan CC. Ocular limbal stem cell biology and transplantation. Stem Cell Appl Dis. 2008 Nova Science Publishers.
21. Lemp MA, Mathers WD. Corneal epithelial cell movement in humans. Eye. 1989;3:438–45.
22. Kiritoshi A, SundarRaj N, Thoft RA. Differentiation in cultured limbal epithelium as defined by keratin expression. Invest Ophthalmol Vis Sci. 1991;32:3073–7.
23. Kurpakus MA, Maniaci MT, Esco M. Expression of keratins K12, K4 and K14 during development of ocular surface epithelium. Curr Eye Res. 1994;13:805–14.
24. Liu CY, Zhu G, Westerhausen-Larson A, Converse R, Kao CW. Sun TT KaoWW, Cornea-specific expression of K12 keratin during mouse development. Curr Eye Res. 1993;129110:963–74.
25. Matic M, Petrov IN, Chen S, Wang C, Dimitrijevich SD, Wolosin JM. Stem cells of the corneal epithelium lack connexins and metabolite transfer capacity. Differentiation. 1997;61:251–60.
26. Majo F, Rochat A, Nicolas M, Jaoudé GA, Barrandon Y. Oligopotent stem cells are distributed throughout the mammalian ocular surface. Nature. 2008;456:250–4.
27. Barrandon Y, Green H. Three clonal type of keratinocyte with different capacities for multicaption. Proc Natl Acad Sci USA. 1987;84:2302–6.
28. Barbaro V, Ferrari S, Parekh M, Ponzin D, Parolin C, Di lorio E. Laser scanning confocal microscopy: application in manufacturing and research of corneal stem cells. confocal laser microscopy—Principles and applications in medicine, biology, and the food sciences. 1st edition. Neil Lagali,editor. InTech; 2013.
29. Pellegrini G, Golisano O, Paterna P, Lambiase A, Bonini S, Rama P, et al. Location and clonal analysis of stem cells and their differentiated progency in the human ocular surface. J Cell Biol. 1999;145:769–82.
30. Figueira EC, Girolamo ND, Coroneo MT, Wakefield D. The phenotype of limbal epithelial stem cells. Invest Ophthalmol Vis Sci. 2007;48(1):144–56.
31. Yoon JJ, Ismail S, Sherwin T. Limbal Stem Cells: Central concepts of corneal epithelial homeostasis. World J Stem Cells. 2014;6(4):391–403.
32. Chen Z, de Paiva CS, Luo L, Kretzer FL, Pflugfelder SC, Li DQ. Characterization of putative stem cell phenotype in human limbal epithelia. Stem Cells. 2004;22:355–66.
33. Dexter TM, Spooncer E. Growth and differentiation in the hematopoitic system. Annu Rev Cell Biol. 1989;3:423–41.
34. Jones PH. Epithelial stem cells. BioEssays. 1997;19:683–90.
35. Morrison SJ, Shah NM, Anderson DJ. Regulatory mechanisms in stem cell biology. Cell. 1997;88:287–98.
36. Lehrer MS, Sunn TT, Lavker RM. Strategies of epithelial repair: modulation of stem cell and transit amplifying cell proliferation. J Cell Sci. 1998;111:2867–75.
37. Ordonez P, Di Girolamon N. Concise review: limbal epithelial stem cells: role of the niche microenvironment. Stem Cells. 2012;30:100–7.
38. Dua HS, Azuara-Blanco A. Limbal stem cells of the cornea epithelium. Sur Ophthalmol. 2000;44:415–25.

39. Lavker RM, Wei ZG, Sun TT. Phorbol ester preferentially stimulates mouse fornical conjunctival and limbal epithelial cells to proliferate in vivo. Invest Ophthalmol Vis Sci. 1998;39:301–7.
40. Ferraris C, Chevalier G, Favier B, Jahoda CA, Dhouailly D. Adult corneal epithelium basal cells possess the capacity to activate epidermal, pilosebaceous and sweat gland genetic programs in response to embryonic dermal stimuli. Development. 2000;127(24):5487–95.
41. Pearton DJ, Ferraris C, Dhouailly D. Transdifferentiation of corneal epithelium: evidence for a lineage between the segregation of epidermal stem cells and induction of hair follicles during embryogenesis. Int J Dev Biol. 2004;48:197–201.
42. Blazejewski EA, Schlotzer-schrehardt U, Zenkel M, Bachman B, Chankiewitz E, Jacobi C, Kruse FE. Corneal limbal microenvironment can induce transdifferentiation of hair follicle stem cells into corneal epithelial-like cells. Stem Cells. 2009;27:642–52.
43. Schofield R. The relationship between the spleen colony-forming cell and the haemopoietic stem cell. Blood cells. 1978;4:7–25.
44. McNairn AJ, Guasch G. Epithelial transition zones: Merging microenvironments, niches, and cellular transformation. Eur J Dermatol. 2011;21:21–8.
45. Huang M, Wang B, Wan P, Liang X, Wang X, Liu Y, Zhou Q, Wang Z. Roles of limbal microvascular net and limbal stroma in maintenance of limbal epithelial stem cells. Cell Tissue Res 2014.
46. Polisetty N, Fatima A, Madhira SL, et al. Mesenchymal cells from limbal stroma of human eye. Mol Vis. 2008;14:431–42.
47. Fuchs E, Tumbar T, Guasch G. Socializing with the neighbors: stem cells and their niche. Cell. 2004;116:769–78.
48. Moore KA, Lemischka IR. Stem cells and their niches. Science. 2006;311:1880–5.
49. Goldberg MF, Bron AJ. Limbal palisades of Vogt. Trans Am Ophthalmol Soc. 1982;80:155-17.
50. Dua HS, Shanmuganathan VA, Powell-Richards AO, Tighe PJ, Joseph A. Limbal epithelial crypts: a novel anatomical structure and a putative limbal stem cell niche. Br J Ophthalmol. 2005;89(5):529–32.
51. Shortt AJ, Secker GA, Munro PM, Khaw PT, Tuft SJ, Daniels JT. Characterization of the limbal epithelial stem cell niche: novel imaging techniques permit in vivo observation and targeted biopsy of limbal epithelial stem cells. Stem Cells. 2007;25(6):1402–9.
52. Gipson IK, Spurr-Michaud S, Tisdale A, Keough M. Reassembly of the anchoring structures of the corneal epithelium during wound repair in the rabbit. Invest Ophthalmol Vis Sci. 1989;30:425–34.
53. Gipson IK. The epithelial basement membrane zone of the limbus. Eye. 1989;3:132–40.
54. Ljubimov AV, Burgeson RE, Butkowski RJ, Michael AF, Sun TT, Kenney MC. Human corneal basement memberane heterogeneity: topographical differences in the expression of type 4 collagen and laminin isoforms. Lab Invest. 1995;72:461–73.
55. Kolega J, Manabe M, Sun TT. Basement membrane hetrogeneity and variation in corneal epithelial differentiation. Differentiation. 1989;42:54–63.
56. Li W, Hayashida Y, Chen YT, Tseng SC. Niche regulation of corneal epitheliam stem cells at the limbus. Cell Res. 2007;17(1):26–36.
57. Mann I. A study of epithelial regeneration in the living eye. Br J Ophthalmol. 1944;28:26–40.
58. Blazejewska EA, Schlötzer-Schrehardt U, Zenkel M, Bachmann B, Chankiewitz E, Jacobi C, et al. Corneal limbal microenvironment can induce transdifferentiation of hair follicle stem cells into corneal epithelial-like cells. Stem Cells. 2009;27:642–52.
59. Mariappan I, Kacham S, Purushotham J, Maddileti S, Siamwala J, Sangwan VS. Spatialdistribution of niche and stem cells in ex vivo human limbal cultures. Stem Cells Transl Med. 2014;3(11):1331–41.
60. Thoft RA, Wiley LA, Sundarraj N. The multipotential cells of the limbus. Eye. 1989;3:109–13.
61. Dua HS. Stem cells of the ocular surface: scientific principles and clinical applications. Br J Ophthalmol. 1995;79:968–9.

62. Li Y, Inoue T, Takamatsu F, Kobayashi T, Shiraishi A, Maeda N, et al. Differences between niche cells and limbal stromal cells in maintenance of corneal limbal stem cells. Invest Ophthalmol Vis Sci. 2014;55:1453–62.
63. Chen SY, Hayashida Y, Chen MY, Xie HT, Tseng SC. A new isolation method of human limbal progenitor cells by maintaining close association with their niche cells. Tissue Eng Part C Methods. 2011;17:537–48.
64. Li GG, Zhu YT, Xie HT, Chen SY, Tseng SCG. Mesenchymal stem cells derived from human limbal niche cells. Invest Ophthalmol Vis Sci. 2012;53(9):5686-5679.
65. Espana EM, Kawakita T, Romano A, Di Pascuale M, Smiddy R, Liu CY, et al. Stromal niche controls the plasticity of limbal and corneal epithelial differentiation in a rabbit model of recombined tissue. Invest Ophthalmol Vis Sci. 2003;44:5130–5.
66. Bloomfield SE, Jakobiec FA, Theodore FH. Contact lens induced keratopathy: a severe complication extending the spectrum of keratoconjunctivitis in contact lens wearers. Ophthalmology. 1984;91:290–4.
67. Martin R. Corneal conjunctivalisation in long-standing contact lens wearers. Clin Exp Optom. 2007;90:26–30.
68. Jeng BH, Halfpenny CP, Meisler DM, Stock EL. Management of focal limbal stem cell deficiency associated with soft contact lens wear. Cornea. 2011;30:18–23.
69. D'Aversa G, Luchs JL, Fox MJ, Rosenbaum PS, Udell IJ. Advancing wave-like epitheliopathy. Clinical features and treatment. Ophthalmology. 1997;104:962–9.
70. Medical Advisory Secretariat. Limbal stem cell transplantation: an evidence-based analysis. Ontario Health Technol Assess Ser. 2008;8(7).
71. Avunduk AM, Tekelioglu Y. Therapeutic use of limbal stem cells. Curr Stem Cell Res Ther. 2006;1(2):231–8.
72. Amirjamshidi H, Milani BY, Sagha HM, Movahedan A, Shafiq MA, Lavker RM et al. Limbal fibroblast conditioned media: a non-invasive treatment for limbal stem cell deficiency. Mol Vis. 2011;8(17):658–66.

Liver Stem Cell Niche

Tohru Itoh

1 Introduction

The stem cell niches serve as the microenvironment that regulates tissue resident stem cell activity in a particular tissue/organ. Thus, the characteristics of a stem cell niche can be described on the premise that the nature of the cognate stem cell population should be clearly defined. This is actually a somewhat difficult task in the field of the liver biology, as no certifiable resident stem cell population has been established yet in this organ. In tissues/organs whose functional parenchymal cells have relatively short half-lives on the order of days-to-weeks and thus should be replenished continuously, such as the hematopoietic system, intestinal epithelia, and skin, the role of tissue stem cell populations to support their homeostatic maintenance can be assumed quite naturally and has indeed been well established. In contrast, hepatocytes, the parenchymal cells in the liver, are known to have long half-lives and their normal turnover occurs slowly over a period of more than several months [1, 2], which makes the functional relevance, as well as experimental identification, of any stem cell population in this organ complicated.

2 Liver Stem/Progenitor Cells in Development and Homeostasis

The liver is a central organ for metabolism and detoxification and receives a portal blood flowing from the gastrointestinal tract and other visceral organs, thereby serving as the primary barrier to treat food-derived chemicals and toxins. Most of the

T. Itoh (✉)
Laboratory of Cell Growth and Differentiation, Institute of Molecular and Cellular Biosciences, The University of Tokyo, 1-1-1 Yayoi, Bunkyo-ku, Tokyo 113-0032, Japan
e-mail: itohru@iam.u-tokyo.ac.jp

K. Turksen (ed.), *Tissue-Specific Stem Cell Niche*,
Stem Cell Biology and Regenerative Medicine,
DOI 10.1007/978-3-319-21705-5_5

liver's functions are executed by the parenchymal hepatocytes, an epithelial-type cell population that constitutes around 80 % of the organ mass. Thus, the "stem cell" that becomes an object in the field of liver biology typically points to the one dominating the hepatocyte lineage. There is another epithelial-type cell population in the liver, namely, biliary epithelial cells (BECs), also referred to as cholangiocytes. The bile duct constituted by BECs serves as the conduit system to collect the bile produced by hepatocytes and drain it out of the liver to the duodenum. Thus, hepatocytes and BECs lie on the contiguous epithelial linings in the liver tissue.

During the course of fetal liver development, these two epithelial type cells originate from the common precursor population, called hepatoblasts. Based on this bi-lineage differentiation potential as well as vigorous proliferative activity, hepatoblasts are often remarked as fetal liver stem cells [3]. However, there is no clear evidence for their long-term self-renewal and continuous presence until the adult stage, and it is more likely that they rather correspond to a progenitor cell population that exists only transiently in the developing liver. Nevertheless, the bi-lineage differentiation potential for hepatocytes and BECs generally stands as the hallmark of liver stem/progenitor cells [4].

As mentioned earlier, hepatocytes have a relatively long half-life and there is no clear sign for the existence or requirement of any resident stem cell population to support their daily replenishment. Though still under debate, it is likely to be a more favorably accepted view that homeostatic maintenance of the hepatic parenchymal tissue under normal condition is achieved solely by the proliferation of mature hepatocytes. Genetic lineage tracing experiments in mice have been employed in recent years by several different groups to address this issue, and most, if not all, of the results obtained thus far collectively and strongly supports the notion that hepatocytes are derived only from the pre-labeled hepatocytes [5–9]. This does not necessarily negate the possibility, however, that there are some immature subpopulations of hepatocytes that are specially appointed and/or reserved to produce more mature hepatocytes.

3 Liver Stem/Progenitor Cells in Regeneration

The liver is inherently susceptible to a wide variety of damage imposed by toxins and chemicals brought about via the portal venous blood. In order to counter this hazardous situation and keep its function adequately, the liver has a strong capacity to regenerate upon various kinds of injury. While considered as fully differentiated and post-mitotic cells, the parenchymal hepatocytes actually possess high proliferative activity on call, at least in vivo. Once a part of liver parenchymal tissue is lost or damaged, the remaining hepatocytes promptly enter the cell cycle to replenish and regenerate the organ mass and function [10, 11]. This type of regenerative process is referred to as the compensatory proliferation and is best represented experimentally by the partial hepatectomy (PHx) models in rodents, where surgical removal of selected liver lobes results in hypertrophic growth of the remaining ones. In this case, more specifically, hepatocytes themselves first exhibit hypertrophy and, in case this is not sufficient to fully compensate the lost liver mass, then undergo mitosis to further increase the cell number [12, 13]. Importantly, this type of regeneration does not require contribution of any stem/progenitor cell population either within or outside the liver. Again, genetic lineage tracing experiments in mice thus far have collectively support the idea that pre-existing hepatocytes are the predominant source of new hepatocytes upon PHx [5–7, 9].

The situations differ considerably, however, in case the liver is more severely and/or chronically damaged. The hepatocyte-mediated regeneration process cannot sufficiently match the need for new hepatocytes when massive loss of the parenchymal tissue occurs acutely or the proliferative capacity of hepatocytes itself is interfered. It has long been postulated that putative stem/progenitor cell population, which is not evident under the normal condition, emerges in the liver and contribute to the parenchymal and biliary regeneration [14, 15] (Fig. 1). Typically, such cells

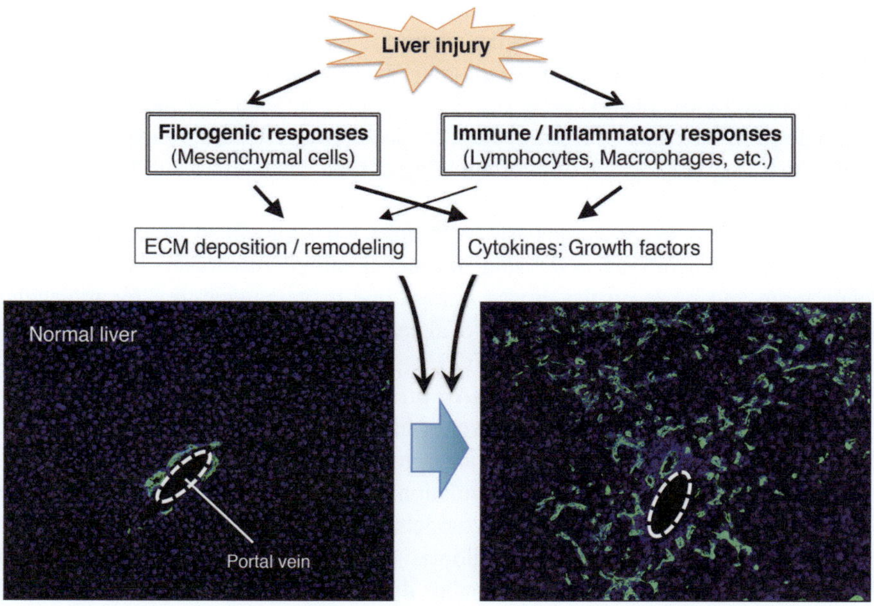

Fig. 1 In the normal liver, bile ducts consisting of biliary epithelial cells (BECs) localize adjacent to the portal vein (*bottom left panel* immunostaining with the BEC marker CK19, *green*). Upon liver injury, enigmatic cells expressing BEC markers emerge ectopically in the peri-portal region and then expand outward to the parenchymal region (*bottom right panel*). Such kind of cells has long been considered, though not formally proven, to include putative stem/progenitor cell populations with bi-lineage differentiation potential to hepatocytes and BECs and thereby contributing to liver regeneration. In this chapter, the term liver progenitor cells (LPCs) is employed to describe them based solely on their characteristic immunophenotype and histological localization. Liver injury provokes inflammatory and fibrogenic responses, which coordinately dictate induction, expansion and differentiation of LPCs through deposition and remodeling of ECM components as well as providing various cytokines and growth factors as critical niche signals

can be defined by relatively high nuclear-to-cytoplasm ratio, a hallmark of stem/progenitor cells, and intermediate immunophenotype with expression of both hepatocytic (e.g., Albumin) and biliary (e.g., CK19) markers. Expression of fetal hepatoblast markers (e.g., Afp, Dlk1), as well as stem/progenitor-associated markers (e.g., Nestin, CD44, Lgr5) is also suggested in several cases, though the characteristics may vary considerably depending on the animal model, injury phenotype and method used for analysis [16]. In accord with their somewhat ambiguous nature, they are called in several different ways, such as intermediate hepatobiliary cells, ductular hepatocytes, atypical ductular cells, liver (or hepatic) progenitor cells, liver (or hepatic) stem cells, and so forth. Practically, these stem/progenitor-like cells are experimentally characterized or clinically diagnosed based on their positive staining for BEC markers (e.g., CK19, CK7, EpCAM) and emergence and expansion ectopically in the parenchymal region in injured livers, often forming a cluster with duct-like and/or cord-like structures. This phenomenon is also referred to as ductular reaction.

Historically, rat models for carcinogen-induced liver injury protocols stands as the golden standard model and have been extensively used to induce and characterize these stem/progenitor-like cells, where the term "oval cells" is coined based on their characteristic cellular morphology [17]. Cumulative evidence based on the detailed histological observations strongly supports the idea that the rat oval cells originate from the canal of Hering, the interface between hepatocytes and BECs, as well as that they are capable of differentiating to both hepatocytes and BECs in vivo [18–20].

While our knowledge on the basic liver biology has long been fueled for decades by many studies using rats as the standard model animal, researchers have begun to focus more and more on the mouse system due to its amenability for molecular genetics approach. The most reliable "oval cell induction" model established in rats is the 2-acetylaminofluorene (2-AAF)/PHx protocol, in which hepatocyte proliferation is completely blocked by 2-AAF prior to PHx-induced regenerative stimulus [19]. The same protocol, however, was found not applicable in mice [21]. Nevertheless, several chronic liver injury protocols can indeed induce phenotypically similar stem/progenitor-like populations in this animal as well. Among the most extensively used mouse models are the administration of a 3,5-diethoxycarbonyl-1,4-dihydro-collidine (DDC)-containing diet [22] or a choline-deficient ethionine-supplemented diet (CDE) [23]. The DDC-induced injury is considered to be targeted primarily to the biliary compartment and serves as a model for sclerosing cholangitis and biliary fibrosis [24], while the CDE protocol induces fatty liver and is sometimes used as a model for non-alcoholic steatohepatitis. As the disease etiologies as well as the phenotypes of the induced cells are not necessarily the same between in rats and in mice, indiscriminate use of the same terminology "oval cell" should be avoided in mouse injury models. In this chapter, the term "LPC (liver progenitor cell)" is chosen to be used hereafter to broadly describe all the various injury-activated, putative liver stem/progenitor cell populations regardless of species or injury models, including the rat oval cells. Notably, the definition of LPCs here is based solely on their characteristics of BEC marker expression and ectopic emergence in the parenchymal region, but not on their bi-lineage differentiation potential to both hepatocytes and BECs or functional contribution to liver regeneration. Many studies have investigated the proliferation and differentiation potentials of LPCs in vitro in order to identity putative stem cell population in the liver [4]. When subjected to appropriate culture conditions with some cytokines, growth factors, and/or extracellular matrix components added, BEC marker-positive cell populations isolated form the liver under injury condition, and even those from uninjured condition (i.e., genuine BECs), usually contain highly proliferative cells that are capable of forming colonies and being continuously propagated, a potential hallmark of the tissue stem cell activity. Moreover, those clonogenic cells exhibit a differentiation potential to be induced to hepatocytes or BECs, depending on the culture conditions. Thus, not only LPCs but also the biliary epithelial tissue in the liver can be regarded as the compartment where the tissue stem cells reside, as long as they are defined based on their activity in vitro.

More recently, the development of the Cre-loxP system-mediated genetic marking and lineage tracing systems in mice prompted many liver biologists to track the fate of LPCs, in order to demonstrate and confirm their genuine bi-potential stem cell activity in vivo. As LPCs express various molecular markers that are common to BECs as mentioned earlier, driver mouse strains engineered to express Cre (or its tamoxifen-responsive variants) under the control of the BEC marker gene promoters have been employed, such as CK19-CreER, HNF1β-CreER, Sox9-CreERT2 and Osteopontin-CreERT2. Using these drivers in combination with Cre-inducible reporter strains, contribution of LPCs (or BECs) to hepatocytes can be traced to examine one part of their anticipated bi-lineage differentiation potential, while that to BECs cannot be assessed due to the expression of the same markers. Lineage-labeled hepatocytes were observed in some cases, particularly in the CDE diet protocol followed by a toxin-free recovery period, indicative of in vivo differentiation of LPCs to hepatocytes [5, 7]. However, relative contribution of the labeled hepatocytes to the entire liver parenchymal population was very limited. Moreover, no significant contribution of LPCs to newly formed hepatocytes was detected in other injury models or even in the case of CDE model in other studies [5–9, 25]. In accord with these results, experimental settings using complimentary labeling approaches, where hepatocytes instead of LPCs/BECs are initially labeled, demonstrated that pre-existing hepatocytes were the dominant source for newly formed hepatocytes upon liver injuries. Although the use of different experimental settings by different studies (e.g., Cre-driver strains, reporter strains, age and sex of mice, and liver injury models and duration) may have posed difficulties on making fully coherent interpretation of those results, they collectively and comprehensively suggest that contribution of LPCs to liver parenchymal regeneration as hepatocyte precursors can be practically negligible, at least in mouse models, thereby strongly refuting the role of LPCs as genuine liver stem/progenitor cells. One possible explanation is that the liver injury protocols employed in these studies are not sufficiently severe for LPCs to manifest their true differentiation capability, considering the fact that the 2-AAF treatment in the rat oval cell induction model can induce complete blockade of hepatocyte proliferation [21]. Notably, in the cases of more radical injury models with massive hepatocyte loss that are developed in zebrafish, predominant contribution of the BEC/LPC lineage-labeled cells has been reported [26, 27]. Future studies developing and employing some appropriate injury setting in mice should clarify this point. In addition, it could become possible in the near future to evaluate the exact differentiation capability of the rat oval cells by using the in vivo lineage tracing approach, in view of the rapidly emerging genome engineering technologies today.

Although it is still not formally demonstrated, or rather being negated by mouse studies, whether LPCs play an exact role as a source producing new hepatocytes under liver injury condition, several lines of evidence collectively support the notion that they do play a beneficial role in the regenerative process. It has been shown that the degree of the LPC expansion and ductular reaction directly correlates with severity of liver disease in human patients [28], implying its pathophysiological relevance. Mutant mice lacking one of the essential signaling components for the LPC niche signals (see below) have defects in inducing LPCs and concomitantly show exacerbated liver damage and decreased survival [29, 30]. More recently, genetic cell ablation experiments was conducted in mice using the Foxl1-Cre transgenic line, which expresses the Cre recombinase specifically in LPCs upon liver injury but not in BECs under normal condition, in combination with a Cre-inducible diphtheria toxin receptor expressing strain (ROSA-iDTR) [31]. Administration of diphtheria toxin to the Foxl1-Cre;ROSA-iDTR compound mice after CDE-induced liver injury and LPC induction successfully eliminated considerable parts of LPCs as well as their descendant hepatocytes, and resulted in prolonged injury and impaired liver function. These studies together suggest that the LPC induction is not just an incidental tissue malformation that occurs secondary to injuries but rather an active and adaptive homeostatic response to counter injury and stimulate regenerative response. Elucidating the cryptic functions of LPCs in liver regeneration, other than the cellular source of hepatocytes, is of significant importance.

4 Tissue Dynamics for Liver Progenitor Cell Expansion upon Injury

The facts that LPCs share many common markers with BECs as well as that they can be observed in tissue sections occasionally as clusters of cells with lumen-like structures have implied their structural and histological relationship with bile ducts. Precise histological examination on the rat oval cells in the 2-AAF/PHx model suggests that they do form ductular structures, whose lumen is directly connected to the biliary tree as well as to bile canaliculi formed by hepatocytes [32]. In order to test whether the clusters of LPCs are connected with each other or to the pre-existing bile duct, retrograde ink injection to the biliary system from the extrahepatic bile duct was performed in mouse models of liver injury with LPC induction [33]. The results indicated that, in all the models tested, LPCs always form contiguous tubular structure connected to the pre-existing bile duct and together constitute a united conduit system, instead of existing as isolated or disjointed units. By further examining the ink-casted tubular sturucture at three-dimensional (3D) level, the study has revealed that the phenomenon known as the LPC induction/expansion or ductular reaction, which has traditionally been observed in histopathological examinations of two-dimensional liver tissue sections, essentially represents the expansion and remodeling of the biliary tree branches from the viewpoint of 3D histodynamics. Expansion of the biliary tree occurred in an ordered fashion from the periportal zone of the hepatic lobule into the parenchymal region along the time course of liver injury.

Interestingly, when comparisons were made between various mouse injury models such as DDC, CDE, carbon tetrachloride (CCl_4) and thioacetamide (TAA), structural changes of the biliary tree were found quite diverse and well corresponded to the parenchymal injury patterns. That is, in centrilobular injury models such as TAA or CCl_4, where hepatocyte damage is restricted to those surrounding the central vein, the biliary tree extended directly with apparent tropism toward the distant injured area. In contrast, damage in the periportal area induced biliary branches to arborize and expand locally and wrap around the portal vein. The dynamic and diverse reconstruction of the biliary architecture corresponding to different types of liver injury may represent an adaptive response of the tissue [33, 34]. Although the functional roles and underlying mechanisms for this still need to be elucidated, this type of tissue plasticity should constitute an essential component to substantiate robust regenerative activity of the liver,

Notably, the expansion of the biliary branches does not necessarily mean that they all arise from the pre-existing BECs. Theoretically, they can also arise from neighboring cells by means such as cell type conversion or trans-differentiation from hepatocytes or any other types of cells. In fact, recent studies have shown that hepatocytes can convert to BECs under certain types of liver injury conditions, thereby contributing to a part of the ductular reaction [34–37].

5 Signals Regulating the Liver Progenitor Cell Expansion

Although the exact contribution of LPCs to liver regeneration as the genuine stem/progenitor cell population, differentiating to and producing hepatocytes, still remains obscure, they certainly account for a functionally relevant component and play a substantial role in promoting the regenerative process. While characterization of LPCs is still on the way, understanding of the underlying mechanisms for their induction and regulation has made significant progress. Obviously, emergence and expansion of LPCs upon liver injury is not an autonomous process within these cells but involves various other types of cells, which interact either directly or indirectly with them. Apart from the regenerative process induced upon PHx where there is essentially no "injury" to the remnant liver tissue, acute and chronic liver injury conditions with hepatocyte damage or loss usually accompany provocation of inflammatory and fibrogenic responses. Accordingly, various kinds of immunomodulatory cytokines, extracellular matrices (ECMs), as well as other mesenchymal cell-derived signals are produced and propagated, which cooperatively shape the specific milieu regulating the LPC activity.

Among various types of inflammatory/immune cells, the roles of lymphocytes [38, 39] as well as macrophages [40, 41] in the LPC regulation have been documented. In addition, involvement of several inflammatory cytokines, such as tumor necrosis factor (TNF)-alpha, lymphotoxin-beta, interferon-alpha, interferon-gamma, and interleukin-6 for induction and regulation of LPCs has been reported [42–44], although the exact modes of their actions remain not fully clarified. One of the cytokine whose relevant role in the LPC regulation is well established is TNF-related weak inducer of apoptosis (TWEAK). Forced expression of this cytokine in the mouse liver by using a transgenic model or adenoviral gene transfer led to induction of LPCs [45]. Conversely, in the mice lacking the TWEAK receptor Fn14, as well as in those treated with a neutralizing anti-TWEAK monoclonal antibody, the LPC response upon DDC or CDE regimen was significantly suppressed [45, 46]. This suppression, however, was only in a partial or transient manner, implicating a role of some other signals for LPC activation. Interestingly, it has been known that transplantation of bone marrow-derived cells can pose some beneficial effects on liver injury including resolution of fibrosis and improvement of liver function, possibly through stimulating tissue progenitor cell activation and subsequent regeneration [47, 48]. A recent study in mice revealed that the macrophage fraction in bone marrow cells were responsible for LPC activation in the engrafted liver by producing TWEAK [49]. The study further demonstrated that administration of recombinant TWEAK was sufficient to induce LPC activation and ductular reaction, thereby implicating its potential therapeutic use.

Interleukin-22 (IL-22) is a member of the interleukin-10 family cytokines produced by several kinds of immune cells such as helper T cells, NK cells, and NKT cells. Interestingly, expression of IL-22R1, a component of the functional heterodimeric receptor complex for IL-22, is restricted to epithelial-type cells, such as keratinocytes, bronchial epithelial cells, intestinal epithelial cell, and hepatocytes, suggesting a role of this cytokine in epithelial homeostasis in response to immune regulation. Indeed, previous studies have shown that IL-22 acts directly on and serves as a survival factor for hepatocytes upon hepatocellular injury conditions such as concanavalin A-induced hepatitis in mice [50–52]. More recently, the role of this cytokine in LPC regulation has also been suggested [53]. Expression of IL-22 in the liver was increased in patients with hepatitis B virus infection and correlated with the grade of inflammation and proliferation of LPCs. Although up-regulation of the endogenous IL-22 gene expression has not been clearly observed in animal models, transgenic mice with hepatic overexpression of IL-22 or mice infected with an IL-22-expressing adenovirus had increased proliferation of LPCs upon DDC diet-induced injury. Thus, IL-22 can be a candidate molecule linking injury-associated inflammatory response to the LPC expansion. Further studies are needed to clarify whether IL-22 broadly stimulates proliferation and/or survival of hepatic epithelial-lineage cells (i.e., hepatocytes, BECs, and LPCs) in a similar way, or it imposes any specific function on LPCs.

With regard to fibrogenic response in injured liver, the central player engaged in collagen production is activated myofibroblast cells, typically characterized by the expression of alpha-smooth muscle actin (αSMA), which are derived from quiescent mesenchymal cells in response to pro-fibrogenic signals such as TGF-β. Among several kinds of mesenchymal cell populations in the liver, hepatic stellate cells (HSCs) are the most extensively characterized and their role in fibrogenesis and fibrosis is well documented. A recent genetic fate tracing study in mice have demonstrated that HSCs are the predominant contributors to liver fibrosis irrespective of its etiology including models of toxic (e.g., CCl_4 and TAA), cholestatic (DDC), and fatty liver disease (related to the CDE model) [54], while other studies have suggested contributions of multiple different mesenchymal cell populations to collagen production according to injury models [55, 56]. Mesenchymal cells such as HSCs have long been postulated to physically interact with LPCs and provide an array of signals and growth factors on them [20]. In addition to immune cells and mesenchymal cells, ECM components have been suggested to play an important role in regulating LPCs [57, 58]. In particular, LPCs are usually surrounded by a laminin-rich matrix and are in intimate contact with macrophages and myofibroblasts, thereby forming a stereotypical "LPC niche" structure in many types of chronic liver injury in rodents and humans [59]. In vitro studies have shown that laminin supports maintenance of LPCs in an undifferentiated phenotype and their biliary specification. Remodeling of ECM components, including degradation of the collagen matrix and formation of the laminin-rich niche structure, is required to mount a proper activation of the LPC response [60].

It has been shown that a population of mesenchymal cells expressing thymus cell antigen 1 (Thy1; also known as CD90), which is distinct from stellate cells and may be related to portal fibroblasts, reside in close proximity to oval cells in rat liver [61]. In models of oval cell-inducing liver injury, the $Thy1^+$ cells are activated to express αSMA, extensively proliferate, and and express growth factors and cytokines, including TWEAK, IL-1β, IL-6, IL-15, and hepatocyte growth factor (HGF), that can potentially affect oval cells. A recent study in mice has revealed that the Thy1-expressing mesenchymal cell population indeed plays a critical role as a "niche" to stimulate LPC activation [30]. Upon DDC-induced liver injury as well as in several other liver disease models, $Thy1^+$ cells were found to expand in the periportal region along with and in close proximity to LPCs. Searching for candidate signaling molecules involved in their interaction identified a member of the fibroblast growth factor family ligand, FGF7, as a signal emanating from the $Thy1^+$ cells. Notably, this FGF family ligand in general is produced by mesenchymal-type cells and in turn acts on epithelial-type cells [62]. Accordingly, its cognate receptor FGFR2b, an epithelial-specific variant form of FGFR2, was confirmed to be expressed on LPCs. Genetic loss-of-function and gain-of-function experiments using knockout and transgenic mice, respectively, revealed that FGF7 is both necessary and sufficient to induce the LPC response in the adult mouse liver. Intriguingly, overexpression of this growth factor in the course of DDC-induced injury significantly ameliorated hepatocyte injury and cholestatic disorders, suggesting its role in both hepatocyte and biliary regeneration via activation of LPCs.

HGF is another key growth factor involved in regulation of the LPC response. While its essential role as a primary mitogen for hepatocytes during the process of compensatory hypertrophy had long been known, it remained obscure whether this growth factor also played a relevant role in regulating LPCs. A study using conditional knockout mice for the HGF receptor c-Met demonstrated that the LPC response was significantly suppressed in these mice upon DDC-induced liver injury, resulting in more extensive liver damage and increased mortality [29]. Although it has been shown that in vivo administration of HGF can augment the extent of the LPC response pre-induced by liver injury, it is not clear whether it is also capable of stimulating de novo induction of the response as TWEAK or FGF7 does. In a study using LPC-derived clones in vitro, the HGF/c-Met signal has been shown to induce differentiation of those stem/progenitor-like cells toward hepatocytes via signaling pathways involving AKT and STAT3 [63]. Interestingly, this effect of HGF is antagonized by epidermal growth factor receptor (EGFR) signaling via induction of Notch1. Signaling by the EGFR—Notch1 axis instead promotes BEC differentiation in vitro, and the loss of EGFR leads to impaired differentiation of LPCs toward biliary lineage upon DDC-induced liver injury in vivo. Notably, Notch signal is well known to play fundamental roles in inducing differentiation of fetal hepatoblasts to the BEC lineages, which reasonably supports the idea that the same signal is also involved in biliary differentiation of LPCs.

Wnt/β-catenin signal, a well-known pathway playing critical roles in regulating stem/progenitor cells in many tissues and organs, has also been reported by several groups to be involved in LPC regulation in the liver. Under various liver injury conditions with LPC response in rat, mouse, and human, expression of several Wnt family genes has been observed [64–67]. Concomitant activation of the downstream β-catenin pathway is induced in LPCs. There are 19 members in the Wnt ligand family, and it is not consistent among those papers which of the ligands are expressed and may play a relevant role.

Intriguingly, Wnt/β-catenin signaling in conjunction with the Notch signaling pathway plays a critical role in regulating differentiation of LPCs. A previous study by Spee et al. sought to investigate the characteristics of the LPC niche in parenchymal and biliary human liver disease [68]. Specifically, they compared gene expression profiles of the $CK7^+$ LPCs collected by laser-capture microdissection from patient livers with acute necrotising hepatitis (AH), cirrhosis after hepatitis C infection, and primary biliary cirrhosis (PBC). The study revealed a distinct expression pattern of several surface markers and progenitor-associated genes among them. With regard to the LPC regulatory signals, the Wnt pathway components were commonly expressed in all disease groups. In contrast, expression of the Notch signaling pathway components as well as activation of this pathway was predominantly observed in PBC, well consistent with the suggested role of Notch signal in promoting biliary differentiation of LPCs. A following study in mice has unveiled the detailed mode of interaction between these two signaling pathways [69]. In chronically injured liver, periportal myofibroblasts expressing αSMA, which might correspond to the $Thy1^{+}$ cells mentioned above [30], provide the Notch ligand Jagged1. This acts on the Notch receptor expressed on LPCs to activate the downstream signaling pathway, leading to differentiation of LPCs to BECs. This Notch-dependent signal for biliary differentiation from the αSMA^{+} myofibloblast niche is dominant when the liver is suffered from biliary injury as is the case with the DDC model. When hepatocyte death was induced in other injury models such as the CDE protocol, cellular debris derived from injured hepatocytes are engulfed by macrophages, which leads to activation of the macrophage and stimulate production of Wnt3a. This canonical Wnt molecule acts on LPCs to induce β-catenin signaling and expression of a Notch signal inhibitor, Numb. This eventually results in suppression of the default Notch signaling for biliary differentiation in LPCs and in turn stimulates their differentiation to hepatocytes. Thus, the balance between two distinct niche structures with different signals is critical to shape the outcome of activated LPCs to induce proper regenerative response according to the nature of liver injury.

6 Concluding Remarks

Recent progress in understanding the cellular and molecular frameworks for LPC regulation have started to reveal the "LPC niche" structures, composed of specific cell populations supporting LPC activities as well as the responsible signals mediating their interaction. Further elucidating the entire network of the signals as well as the ways to manipulate their activities should pave the way for fully understanding the mechanisms of liver regeneration and hold promise for future development of efficient therapeutic strategies to treat patients with liver disease.

References

1. Macdonald RA. "Lifespan" of liver cells. Autoradio-graphic study using tritiated thymidine in normal, cirrhotic, and partially hepatectomized rats. Arch Intern Med. 1961;107:335–43.
2. Magami Y, Azuma T, Inokuchi H, Kokuno S, Moriyasu F, Kawai K, et al. Cell proliferation and renewal of normal hepatocytes and bile duct cells in adult mouse liver. Liver. 2002;22 (5):419–25.
3. Tanaka M, Itoh T, Tanimizu N, Miyajima A. Liver stem/progenitor cells: their characteristics and regulatory mechanisms. J Biochem. 2011;149(3):231–9.
4. Miyajima A, Tanaka M, Itoh T. Stem/progenitor cells in liver development, homeostasis, regeneration, and reprogramming. Cell Stem Cell. 2014;14(5):561–74.
5. Espanol-Suner R, Carpentier R, Van Hul N, Legry V, Achouri Y, Cordi S, et al. Liver progenitor cells yield functional hepatocytes in response to chronic liver injury in mice. Gastroenterology. 2012;143(6):1564–75.

6. Malato Y, Naqvi S, Schurmann N, Ng R, Wang B, Zape J, et al. Fate tracing of mature hepatocytes in mouse liver homeostasis and regeneration. J Clin Investig. 2011;121(12):4850–60.
7. Rodrigo-Torres D, Affo S, Coll M, Morales-Ibanez O, Millan C, Blaya D, et al. The biliary epithelium gives rise to liver progenitor cells. Hepatology (Baltimore, MD) 2014;60(4):1367–77.
8. Tarlow BD, Finegold MJ, Grompe M. Clonal tracing of Sox9+ liver progenitors in mouse oval cell injury. Hepatology (Baltimore, MD). 2014;60(1):278–89.
9. Yanger K, Knigin D, Zong Y, Maggs L, Gu G, Akiyama H, et al. Adult hepatocytes are generated by self-duplication rather than stem cell differentiation. Cell Stem Cell. 2014;15 (3):340–9.
10. Michalopoulos GK. Liver regeneration. J Cell Physiol. 2007;213(2):286–300.
11. Michalopoulos GK, DeFrances MC. Liver regeneration. Science. 1997;276(5309):60–6.
12. Itoh T, Miyajima A. Liver regeneration by stem/progenitor cells. Hepatology (Baltimore, MD). 2014;59(4):1617–26.
13. Miyaoka Y, Ebato K, Kato H, Arakawa S, Shimizu S, Miyajima A. Hypertrophy and unconventional cell division of hepatocytes underlie liver regeneration. Curr Biol. 2012;22 (13):1166–75.
14. Fausto N. Liver regeneration and repair: hepatocytes, progenitor cells, and stem cells. Hepatology (Baltimore, MD). 2004;39(6):1477–87.
15. Roskams TA, Libbrecht L, Desmet VJ. Progenitor cells in diseased human liver. Semin Liver Dis. 2003;23(4):385–96.
16. Jelnes P, Santoni-Rugiu E, Rasmussen M, Friis SL, Nielsen JH, Tygstrup N, et al. Remarkable heterogeneity displayed by oval cells in rat and mouse models of stem cell-mediated liver regeneration. Hepatology (Baltimore, MD). 2007;45(6):1462–70.
17. Farber E. Similarities in the sequence of early histological changes induced in the liver of the rat by ethionine, 2-acetylamino-fluorene, and 3'-methyl-4-dimethylaminoazobenzene. Cancer Res. 1956;16(2):142–8.
18. Evarts RP, Hu Z, Omori N, Omori M, Marsden ER, Thorgeirsson SS. Precursor-product relationship between oval cells and hepatocytes: comparison between tritiated thymidine and bromodeoxyuridine as tracers. Carcinogenesis. 1996;17(10):2143–51.
19. Evarts RP, Nagy P, Marsden E, Thorgeirsson SS. A precursor-product relationship exists between oval cells and hepatocytes in rat liver. Carcinogenesis. 1987;8(11):1737–40.
20. Paku S, Schnur J, Nagy P, Thorgeirsson SS. Origin and structural evolution of the early proliferating oval cells in rat liver. Am J Pathol. 2001;158(4):1313–23.
21. Michalopoulos GK. The liver is a peculiar organ when it comes to stem cells. Am J Pathol. 2014;184(5):1263–7.
22. Preisegger KH, Factor VM, Fuchsbichler A, Stumptner C, Denk H, Thorgeirsson SS. Atypical ductular proliferation and its inhibition by transforming growth factor beta1 in the 3,5-diethoxycarbonyl-1,4-dihydrocollidine mouse model for chronic alcoholic liver disease. Lab Invest. 1999;79(2):103–9.
23. Akhurst B, Croager EJ, Farley-Roche CA, Ong JK, Dumble ML, Knight B, et al. A modified choline-deficient, ethionine-supplemented diet protocol effectively induces oval cells in mouse liver. Hepatology (Baltimore, MD). 2001;34(3):519–22.
24. Fickert P, Stoger U, Fuchsbichler A, Moustafa T, Marschall HU, Weiglein AH, et al. A new xenobiotic-induced mouse model of sclerosing cholangitis and biliary fibrosis. Am J Pathol. 2007;171(2):525–36.
25. Schaub JR, Malato Y, Gormond C, Willenbring H. Evidence against a stem cell origin of new hepatocytes in a common mouse model of chronic liver injury. Cell Rep. 2014;8(4):933–9.
26. Choi TY, Ninov N, Stainier DY, Shin D. Extensive conversion of hepatic biliary epithelial cells to hepatocytes after near total loss of hepatocytes in zebrafish. Gastroenterology. 2014;146(3):776–88.
27. He J, Lu H, Zou Q, Luo L. Regeneration of liver after extreme hepatocyte loss occurs mainly via biliary transdifferentiation in zebrafish. Gastroenterology. 2014;146(3):789–800 e8.

28. Lowes KN, Brennan BA, Yeoh GC, Olynyk JK. Oval cell numbers in human chronic liver diseases are directly related to disease severity. Am J Pathol. 1999;154(2):537–41.
29. Ishikawa T, Factor VM, Marquardt JU, Raggi C, Seo D, Kitade M, et al. Hepatocyte growth factor/c-Met signaling is required for stem-cell-mediated liver regeneration in mice. Hepatology (Baltimore, MD). 2012;55(4):1215–26.
30. Takase HM, Itoh T, Ino S, Wang T, Koji T, Akira S, et al. FGF7 is a functional niche signal required for stimulation of adult liver progenitor cells that support liver regeneration. Genes Dev. 2013;27(2):169–81.
31. Shin S, Upadhyay N, Greenbaum LE, Kaestner KH. Ablation of Foxl1-Cre-labeled hepatic progenitor cells and their descendants impairs recovery of mice from liver injury. Gastroenterology. 2015;148(1):192-202 e3.
32. Paku S, Nagy P, Kopper L, Thorgeirsson SS. 2-acetylaminofluorene dose-dependent differentiation of rat oval cells into hepatocytes: confocal and electron microscopic studies. Hepatology (Baltimore, MD). 2004;39(5):1353–61.
33. Kaneko K, Kamimoto K, Miyajima A, Itoh T. Adaptive remodeling of the biliary architecture underlies liver homeostasis. Hepatology (Baltimore, MD). 2015;61(6):2056–66.
34. Nagahama Y, Sone M, Chen X, Okada Y, Yamamoto M, Xin B, et al. Contributions of hepatocytes and bile ductular cells in ductular reactions and remodeling of the biliary system after chronic liver injury. Am J Pathol. 2014;184(11):3001–12.
35. Sekiya S, Suzuki A. Hepatocytes, rather than cholangiocytes, can be the major source of primitive ductules in the chronically injured mouse liver. Am J Pathol. 2014;184(5):1468–78.
36. Tarlow BD, Pelz C, Naugler WE, Wakefield L, Wilson EM, Finegold MJ, et al. Bipotential adult liver progenitors are derived from chronically injured mature hepatocytes. Cell Stem Cell. 2014;15(5):605–18.
37. Yanger K, Zong Y, Maggs LR, Shapira SN, Maddipati R, Aiello NM, et al. Robust cellular reprogramming occurs spontaneously during liver regeneration. Genes Dev. 2013;27(7):719–24.
38. Knight B, Akhurst B, Matthews VB, Ruddell RG, Ramm GA, Abraham LJ, et al. Attenuated liver progenitor (oval) cell and fibrogenic responses to the choline deficient, ethionine supplemented diet in the BALB/c inbred strain of mice. J Hepatol. 2007;46(1):134–41.
39. Strick-Marchand H, Masse GX, Weiss MC, Di Santo JP. Lymphocytes support oval cell-dependent liver regeneration. J Immunol. 2008;181(4):2764–71.
40. Van Hul N, Lanthier N. Espanol Suner R, Abarca Quinones J, van Rooijen N, Leclercq I. Kupffer cells influence parenchymal invasion and phenotypic orientation, but not the proliferation, of liver progenitor cells in a murine model of liver injury. Am J Pathol. 2011;179 (4):1839–50.
41. Viebahn CS, Benseler V, Holz LE, Elsegood CL, Vo M, Bertolino P, et al. Invading macrophages play a major role in the liver progenitor cell response to chronic liver injury. J Hepatol. 2010;53(3):500–7.
42. Akhurst B, Matthews V, Husk K, Smyth MJ, Abraham LJ, Yeoh GC. Differential lymphotoxin-beta and interferon gamma signaling during mouse liver regeneration induced by chronic and acute injury. Hepatology (Baltimore, MD). 2005;41(2):327–35.
43. Knight B, Yeoh GC, Husk KL, Ly T, Abraham LJ, Yu C, et al. Impaired preneoplastic changes and liver tumor formation in tumor necrosis factor receptor type 1 knockout mice. J Exp Med. 2000;192(12):1809–18.
44. Yeoh GC, Ernst M, Rose-John S, Akhurst B, Payne C, Long S, et al. Opposing roles of gp130-mediated STAT-3 and ERK-1/ 2 signaling in liver progenitor cell migration and proliferation. Hepatology (Baltimore, MD). 2007;45(2):486–94.
45. Jakubowski A, Ambrose C, Parr M, Lincecum JM, Wang MZ, Zheng TS, et al. TWEAK induces liver progenitor cell proliferation. J Clin Investig. 2005;115(9):2330–40.
46. Tirnitz-Parker JE, Viebahn CS, Jakubowski A, Klopcic BR, Olynyk JK, Yeoh GC, et al. Tumor necrosis factor-like weak inducer of apoptosis is a mitogen for liver progenitor cells. Hepatology (Baltimore, MD). 2010;52(1):291–302.

47. Houlihan DD, Newsome PN. Critical review of clinical trials of bone marrow stem cells in liver disease. Gastroenterology. 2008;135(2):438–50.
48. Thomas JA, Pope C, Wojtacha D, Robson AJ, Gordon-Walker TT, Hartland S, et al. Macrophage therapy for murine liver fibrosis recruits host effector cells improving fibrosis, regeneration, and function. Hepatology (Baltimore, MD). 2011;53(6):2003–15.
49. Bird TG, Lu WY, Boulter L, Gordon-Keylock S, Ridgway RA, Williams MJ, et al. Bone marrow injection stimulates hepatic ductular reactions in the absence of injury via macrophage-mediated TWEAK signaling. Proc Natl Acad Sci USA. 2013;110(16):6542–7.
50. Radaeva S, Sun R, Pan HN, Hong F, Gao B. Interleukin 22 (IL-22) plays a protective role in T cell-mediated murine hepatitis: IL-22 is a survival factor for hepatocytes via STAT3 activation. Hepatology (Baltimore, MD). 2004;39(5):1332–42.
51. Zenewicz LA, Yancopoulos GD, Valenzuela DM, Murphy AJ, Karow M, Flavell RA. Interleukin-22 but not interleukin-17 provides protection to hepatocytes during acute liver inflammation. Immunity. 2007;27(4):647–59.
52. Zhang Y, Cobleigh MA, Lian JQ, Huang CX, Booth CJ, Bai XF, et al. A proinflammatory role for interleukin-22 in the immune response to hepatitis B virus. Gastroenterology. 2011;141 (5):1897–906.
53. Feng D, Kong X, Weng H, Park O, Wang H, Dooley S, et al. Interleukin-22 promotes proliferation of liver stem/progenitor cells in mice and patients with chronic hepatitis B virus infection. Gastroenterology. 2012;143(1):188–98 e7.
54. Mederacke I, Hsu CC, Troeger JS, Huebener P, Mu X, Dapito DH, et al. Fate tracing reveals hepatic stellate cells as dominant contributors to liver fibrosis independent of its aetiology. Nat Commun. 2013;4:2823.
55. Iwaisako K, Jiang C, Zhang M, Cong M, Moore-Morris TJ, Park TJ, et al. Origin of myofibroblasts in the fibrotic liver in mice. Proc Natl Acad Sci USA. 2014;111(32):E3297–305.
56. Magness ST, Bataller R, Yang L, Brenner DA. A dual reporter gene transgenic mouse demonstrates heterogeneity in hepatic fibrogenic cell populations. Hepatology (Baltimore, MD). 2004;40(5):1151–9.
57. Van Hul NK, Abarca-Quinones J, Sempoux C, Horsmans Y, Leclercq IA. Relation between liver progenitor cell expansion and extracellular matrix deposition in a CDE-induced murine model of chronic liver injury. Hepatology (Baltimore, MD). 2009;49(5):1625–35.
58. Vestentoft PS, Jelnes P, Andersen JB, Tran TA, Jorgensen T, Rasmussen M, et al. Molecular constituents of the extracellular matrix in rat liver mounting a hepatic progenitor cell response for tissue repair. Fibrogenesis Tissue Repair. 2013;6(1):21.
59. Lorenzini S, Bird TG, Boulter L, Bellamy C, Samuel K, Aucott R, et al. Characterisation of a stereotypical cellular and extracellular adult liver progenitor cell niche in rodents and diseased human liver. Gut. 2010;59(5):645–54.
60. Kallis YN, Robson AJ, Fallowfield JA, Thomas HC, Alison MR, Wright NA, et al. Remodelling of extracellular matrix is a requirement for the hepatic progenitor cell response. Gut. 2011;60(4):525–33.
61. Yovchev MI, Zhang J, Neufeld DS, Grozdanov PN, Dabeva MD. Thymus cell antigen-1-expressing cells in the oval cell compartment. Hepatology (Baltimore, MD). 2009;50(2):601–11.
62. Steiling H, Werner S. Fibroblast growth factors: key players in epithelial morphogenesis, repair and cytoprotection. Curr Opin Biotechnol. 2003;14(5):533–7.
63. Kitade M, Factor VM, Andersen JB, Tomokuni A, Kaji K, Akita H, et al. Specific fate decisions in adult hepatic progenitor cells driven by MET and EGFR signaling. Genes Dev. 2013;27(15):1706–17.
64. Apte U, Thompson MD, Cui S, Liu B, Cieply B, Monga SP. Wnt/beta-catenin signaling mediates oval cell response in rodents. Hepatology (Baltimore, MD). 2008;47(1):288–95.
65. Hu M, Kurobe M, Jeong YJ, Fuerer C, Ghole S, Nusse R, et al. Wnt/beta-catenin signaling in murine hepatic transit amplifying progenitor cells. Gastroenterology. 2007;133(5):1579–91.

66. Itoh T, Kamiya Y, Okabe M, Tanaka M, Miyajima A. Inducible expression of Wnt genes during adult hepatic stem/progenitor cell response. FEBS Lett. 2009;583(4):777–81.
67. Yang W, Yan HX, Chen L, Liu Q, He YQ, Yu LX, et al. Wnt/beta-catenin signaling contributes to activation of normal and tumorigenic liver progenitor cells. Cancer Res. 2008;68 (11):4287–95.
68. Spee B, Carpino G, Schotanus BA, Katoonizadeh A, Vander Borght S, Gaudio E, et al. Characterisation of the liver progenitor cell niche in liver diseases: potential involvement of Wnt and Notch signalling. Gut. 2010;59(2):247–57.
69. Boulter L, Govaere O, Bird TG, Radulescu S, Ramachandran P, Pellicoro A, et al. Macrophage-derived Wnt opposes Notch signaling to specify hepatic progenitor cell fate in chronic liver disease. Nat Med. 2012;18(4):572–9.

The Elusive Pancreatic Stem Cell

Caitlin M. Braitsch and Ondine Cleaver

1 Introduction

Diabetes is a metabolic disease characterized by dysregulation of blood sugar homeostasis, which is becoming increasingly prevalent in the United States and worldwide [1, 2]. Diabetes presents in two different forms: Type 1 diabetes (T1D), which results from autoimmune destruction of pancreatic beta cells, and Type 2 diabetes (T2D), a condition in which cells fail to respond to insulin signaling (http://www.who.int/diabetes/). Currently there exists no cure for either type of diabetes, and standard treatments involve insulin injections and management of associated side effects. Diabetes leads to loss of quality of life for individuals and strains the healthcare system financially. There is an urgent need for a cure for diabetes. In the case of T1D, the focus has turned to therapies based on replacement or regeneration of beta cells. However, obtaining sufficient transplantable and functional beta cells for the diabetic population has not yet been possible. One promising approach that has attracted attention is the generation of beta cells from human embryonic stem cells (hESCs). hESC-derived beta cells could theoretically replace those lost to autoimmune destruction. However, generating and implanting fully functional beta cells into patients has not yet been achieved. A second theoretical approach would be to stimulate expansion of putative pancreatic stem (or progenitor) cells, thereby generating new beta cells within the diabetic pancreas. Alternatively, residual differentiated beta cells could be stimulated to proliferate, thus expanding the endogenous beta cell population. Both regenerative approaches could outpace beta cell destruction and renew insulin production in diabetic patients. The viability of regenerative approaches, however, is contingent upon a number of questions: Are there stem cells in the developing or the adult pancreas that could be stimulated to

C.M. Braitsch · O. Cleaver (✉)
Department of Molecular Biology, University of Texas Southwestern Medical Center, 5323 Harry Hines Blvdd, Dallas, TX 75390, USA
e-mail: ondine.cleaver@utsouthwestern.edu

K. Turksen (ed.), *Tissue-Specific Stem Cell Niche*,
Stem Cell Biology and Regenerative Medicine,
DOI 10.1007/978-3-319-21705-5_6

produce more beta cells? Can differentiated beta cells be coaxed into replicating in a therapeutically relevant manner?

This review will cover the key events underlying normal pancreas development, as well as the possibilities regarding resident or facultative stem cells. Recent studies focused on the role of pancreatic multipotent progenitor cells (MPCs) during embryonic development will be discussed. Efforts to generate insulin-producing beta cellsto treat patients with diabetes will also be compared. We will discuss recent investigations into the controversial topic of pancreatic progenitor cells and regenerative approaches to treat diabetes, as well as recent leaps in the use of iPS-to-beta cell technology. To date, the existence of a pancreatic stem cell, either embryonic or adult, has remained frustratingly elusive. Therefore, we will discuss what is known, what questions remain open, and explore the alternative approaches that are currently stimulating the field of pancreas biology.

2 Origins of Beta Cells During Embryonic Development

The mature pancreas comprises many cell types, including: exocrine, ductal, endocrine, and stromal/mesenchymal (including vascular, neural, and stromal) lineages [3]. Endocrine cells include alpha, beta, delta, pancreatic polypeptide (PP), and epsilon cells, which aggregate to form the Islets of Langerhans. These cell types secrete glucagon, insulin, somatostatin, PP, and ghrelin hormones, respectively. Exocrine, ductal, and endocrine cell lineages are derived from a common progenitor endodermal epithelium during embryonic development, while pancreatic vascular and mesenchymal lineages originate from adjacent mesoderm [3]. The dynamics by which the pancreatic progenitor epithelium transforms and then expands to form the mature pancreas remain poorly understood. However, it is clear that the progenitor epithelium ultimately gives rise to an epithelial ‘tree’, of primarily exocrine function, which comprises approximately 95 % of pancreatic mass. The endocrine islets embedded within the densely packed epithelial network constitute about 1–2 % of the total cell number. Beta cell proliferation slows after birth, with a very low turnover rate in adulthood [4]. Seminal work by Stanger and others showed that total pancreas mass depends on the number of progenitor cells in the early pancreatic bud, and this number is determined early in embryonic development [5]. The discrete morphogenetic events and signals required to generate a mature pancreas have been discussed extensively in other reviews [6–8] and will be summarized here.

2.1 The Early Pancreatic Bud

In the early embryo, the pancreatic epithelium arises from the foregut endoderm [3, 9]. Endodermal fate mapping experiments using chick embryos have demonstrated that the dorsal pancreas is derived from the midline endoderm, while the ventral pancreas stems from two lateral areas of endoderm [10, 11]. Emergence of the dorsal pancreas precedes that of the ventral pancreas in the chick embryo. The opposite phenomenon occurs in mouse embryos, in which the ventral pancreas emerges at embryonic day 8.5 (or E8.5), prior to the dorsal pancreas [12]. Ventral pancreatic progenitor cells are first observed just posterior to the liver diverticulum, in the ventro-lateral endoderm. Following closure of the anterior gut tube, the dorsal and ventral pancreatic 'placodes' emerge via epithelial columnarization by E8.75 [13], at which time pancreatic transcription factors such as Pdx1, Hlxb9, and Nkx6.1 begin to be expressed [14, 15]. During initial pancreatic bud formation, many signals such as retinoic acid, FGF, and BMP are required for pancreas specification [6, 16, 17]. In addition, the pancreatic bud receives signals from adjacent tissues, including the notochord, dorsal aorta, and lateral plate mesoderm [8]. The notochord, which lies dorsal to the pancreatic epithelium during much of its early development, secretes factors like Activin and FGF2 required to repress neuroectodermal signals like Sonic hedgehog, thereby permitting pancreatic gene expression [18, 19]. Soon thereafter, midline fusion of the dorsal aortae separates the dorsal pancreas from the notochord [20]. The dorsal aortic endothelium provides crucial signals to direct isletogenesis and insulin expression [20, 21]. At E9.5 splanchnic lateral plate mesoderm displaces the dorsal aorta and surrounds the pancreatic epithelial bud [3]. Splanchnic mesenchyme, which envelops the budding pancreas, is essential to cell fate specification, epithelial morphogenesis, and cell proliferation throughout embryonic development [22–24]. Together, these studies underscore the complex and changing nature of the pre-pancreatic niche, in which tissues surrounding the pre-pancreatic epithelium grow and change coordinately with the developing pancreas, continuously providing spatiotemporal signals required for specification of pancreatic progenitor cells.

2.2 Dynamics of the Pancreatic Epithelium

Under the influence of signals from surrounding tissues, the dorsal pancreatic bud undergoes morphogenetic changes during development. The pancreatic epithelium thickens and evaginates between E9.0-E10.5 and undergoes dramatic epithelial reorganization. These transformations include transient epithelial stratification and loss of apicobasal polarity around E10.0-E11.5 (Fig. 1a) [3, 13]. During this time, the pancreatic bud comes to resemble a 'bag' of epithelial cells, all relatively homogenous in appearance. The only identifiable distinction amongst bud cells is that the outer peripheral (or 'cap' cells) are columnar and express laminin along their outer/basal surface, while the inner 'body' cells are more cuboidal and apolar [13]. As the pancreatic epithelium stratifies, MPCs are specified and allocated to distinct budding 'tip' structures as the bud expands (after about E11.0) [13, 25]. How and where MPCs emerge within the stratified pancreatic bud, however, remains unknown.

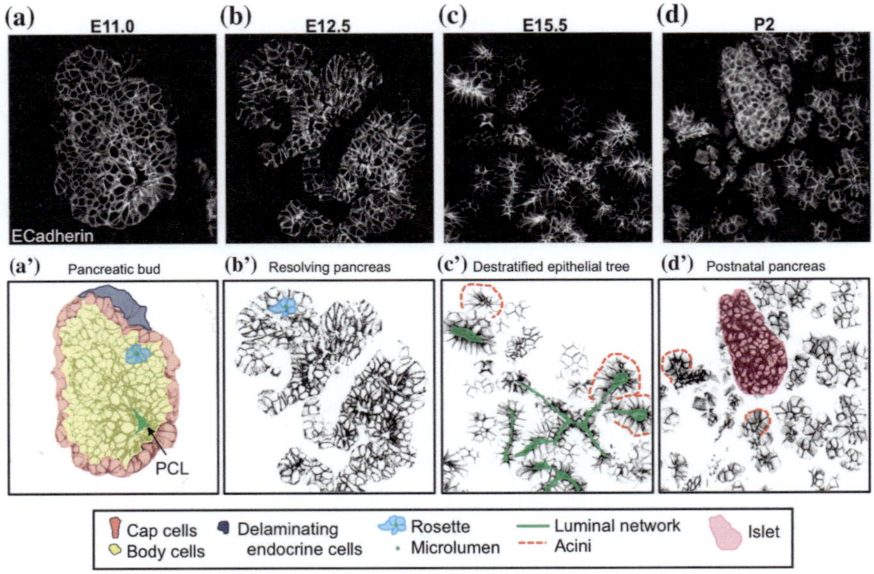

Fig. 1 The pancreatic epithelium undergoes dramatic morphogenetic changes during embryonic development. **a–d** E-Cadherin (white) is expressed throughout the pancreatic epithelium, as detected by immunofluorescent analysis. **a′–d′** Schematic representations of the pancreas. **a′** At embryonic day 11 (E11.0), the pancreatic bud consists of an inner mass of stratified, apolar cells (body cells, *yellow*), surrounded by an outer layer (cap cells, *red*). A small subset of body cells encloses the primary central lumen, a cavity continuous with the gut tube (*green*). Primary transition endocrine cells (*gray*) begin a process of cluster budding (extrusion of groups of cells). **b′** At E12.5, the epithelium begins to resolve from multiple to single layers. During this process, pancreatic epithelial cells regain polarity and apically constrict, forming rosette-like structures (*blue*) surrounding a microlumen (*green*). **c′** By E15.5, the pancreas has formed a destratified, branched epithelium, which forms a luminal network (*green*) with connecting acini (*red dotted lines*). **d′** In the postnatal day 2 (P2) pancreas, endocrine cells are actively delaminating, progressively forming islets of Langerhans (*pink*) adjacent to the pancreatic epithelium

Almost as soon as the epithelium stratifies around E10.5, it begins to destratify back to a monolayer. These rearrangements progressively form growing epithelial branches between E12.5-15.5. This occurs via a process of asynchronous apical constriction of groups of epithelial cells within the stratified epithelium, which leads to the formation of rosette-like structures (Fig. 1b). These events culminate in the opening of distinct microlumens at the centers of the rosettes, which in turn connect to form a continuous luminal network [13, 23, 26]. Epithelial cells regain apical polarity upon formation of rosettes, and this process is required for proper distribution of MPCs [23, 27]. The small GTPase Rho family protein Cdc42 and the Rho GTPase activating protein Stard13 both regulate cell polarity proteins and cytoskeletal dynamics. Pancreas-specific deletion of Cdc42 or Stard13 leads to aberrant microlumen formation with abnormal distribution of MPCs. These studies demonstrate how failure of epithelial architecture leads to disruption of cell shape, rosette formation and tip domain morphogenesis. These failures in turn impair progenitor proliferation and cell fate. Therefore, an important question remains: where and when do MPCs emerge within the early pancreatic bud and what are the dynamics of their expansion and ultimate differentiation into pancreatic lineages?

2.3 Pancreatic Branching

Central to the events shaping the developing pancreas is the transformation of the stratified bud into a branched tubular network, or 'branching morphogenesis'. Following initial stratification and bud formation, pancreas development unfolds rapidly via complicated morphogenetic changes that are only now beginning to be understood [13, 27]. During this time, the epithelium undergoes remodeling and 'destratification', resulting in an epithelial tree of tubular branches by E15.5 (Fig. 1c). For these branches to take shape, the epithelium must transform its initial multi-layered organization into continuous, mono-layered tubes [7]. In their final form, pancreatic branches consist of tubular branches lined by cuboidal ductal cells, terminating in intercalated ducts surrounded by plump acinar columnar cells, with centroacinar cells at the juncture of the two cell types. To date, the precise geometry of the cellular rearrangements and proliferation involved in the transformation and branching of the pancreas, as well as the molecular underpinnings of these events, have not been elucidated.

Morphogenesis of pancreatic branches depends on proper regulation of epithelial dynamic movements and cell shape, as shown in the Cdc42 and Stard13 ablation models [23, 27]. Loss of Cdc42 in pancreatic epithelial cells results in disruption of proper cell polarity, which in turn impacts destratification, luminal network formation, and branch formation. In addition, cell lineage allocation is also disrupted, yielding increased acinar and decreased endocrine cell differentiation. Similarly, loss of Stard13 tip domain organization leads to hampered proliferation and abrogated branch formation. These results demonstrate how pancreatic architecture is intimately dependent on proper epithelial dynamics.

2.4 Primary and Secondary Transitions

During mammalian pancreas developmentas described above, endocrine cell specification and differentiation can be divided into two major phases. During the 'primary transition', an initial population of endocrine cells blebs off the early stratified bud as small clusters starting around E10.0 [28, 29]. These atypical cells are referred to as primary transition endocrine cells. The first phase of pancreatogenesis occurs as the initial stratified bud forms, lineages sort out, and morphogenetic programs are launched. The first insulin-expressing cells also begin to appear during this time; however, they often co-express glucagon and do not express all markers of mature β-cells. In addition, they are not physiologically equivalent to adult β-cells [30]. As primary transition endocrine cells emerge from the stratified epithelium, pancreatic branching initiates around E10.5-E11.5. Via poorly understood events, the pancreatic epithelium remodels into bipotent 'trunk' and multipotent MPC containing 'tip' domains. These events presage and initiate the period known as the 'secondary transition'. This second phase of pancreatic development is characterized by a burst of widespread proliferation and endocrine cell delamination from the central region of the pancreatic bud (E12.5–E16.5). During this time of rapid pancreatic growth, endocrine cells emerge individually, rather than in clusters, and then form islets in a non-clonal manner by coalescence. Tip progenitor fate becomes restricted to the acinar lineage following the secondary transition, while central epithelial cells become allocated to either duct or endocrine fate.

2.5 *Developmental Trajectory of Pancreatic Lineages*

Our understanding of cellular mechanisms and molecular regulation guiding beta cell fate during pancreatic epithelial morphogenesis in the developing embryo has been guided by loss of function studies, as well as lineage tracing experiments. Using reporter mouse lines for pancreatic genes, studies have elucidated the developmental trajectory of cell populations within the pancreas (Fig. 2). Of particular interest is identifying those cells that give rise to beta cells in the early bud, in order to uncover their supportive or instructive niche. Here, we examine key pancreatic genes and the cell populations that express them at different time points, by surveying lineage tracing experiments carried out using various pancreatic reporter mouse lines. A theme that emerges from these studies is that there exists broad overlap of most key pancreatic genes in the early stratified pancreatic bud, while subsets of genes become restricted to particular lineages as development proceeds.

2.5.1 Pdx1

Pancreatic duodenal homeobox 1 (Pdx1) is among the earliest transcription factors expressed in the early pancreatic bud epithelium, starting from approximately E8.75 through mid-gestation, and is essential for pancreas development in both mouse and human embryos [31–34]. Mice lacking Pdx1 are born apancreatic, displaying only a severely abrogated dorsal 'ductule' but no ventral bud [35]. Although Pdx1 expression initiates throughout the bud, it becomes restricted to beta cells postnatally. Conditional deletion of Pdx1 after initial epithelial expression (at E11.5 and E12.5) demonstrated its requirement in acinar cells, as Pdx1 null acinar cells develop poorly and exhibit a profound reduction of the acinar transcription factor

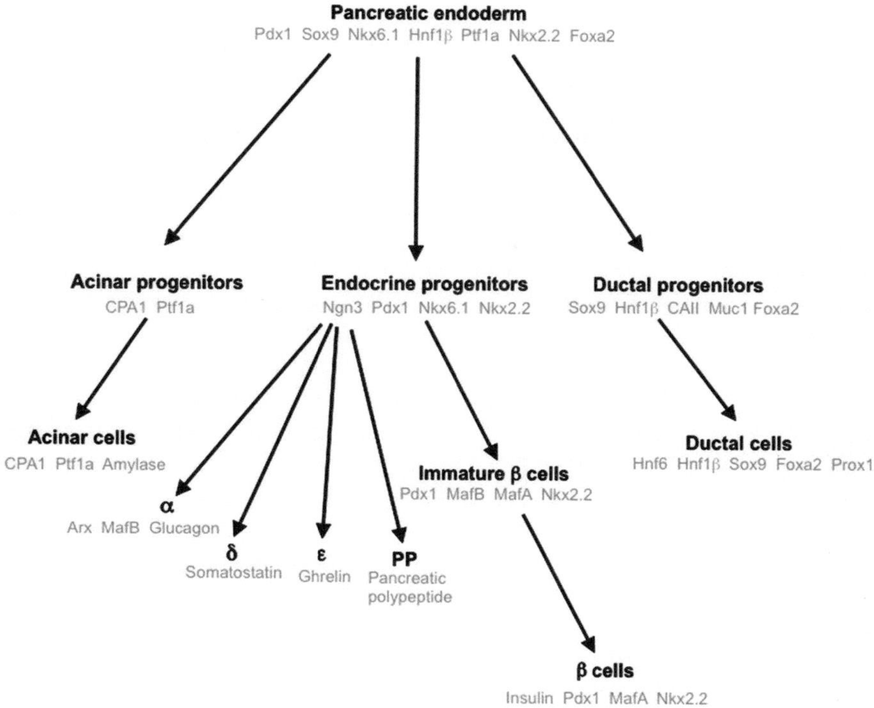

Fig. 2 An overview of pancreatic lineage specification during embryonic development. The pancreatic endoderm produces acinar, endocrine, and ductal progenitors, which then differentiate further into functional acinar and ductal cells, as well as endocrine cells. Beta cells first acquire an immature phenotype, followed by beta cell maturation. Selected expression profiles shown are non-exhaustive (transcription factors, plus other genetic markers). Please see these reviews for further detail [6, 126]

Ptf1a/p48 expression [36]. Conversely, overexpression of Pdx1 can cause liver progenitors to shift towards a pancreatic fate, demonstrating its critical importance to pancreatic lineages [37, 38]. Human PDX1 haploinsufficiency is associated with a form of maturity onset diabetes of the young (MODY4) [39], which indicates the importance of Pdx1 in regulating genes in mature β cells, including *Ins2*, *Glut2*, *Gck*, and *IAPP*. Important insights into pancreatic lineages have come from seminal lineage tracings of Pdx1 expressing cells. Gu and colleagues indelibly labeled the Pdx1 expressing progenitor epithelium using both Pdx1Cre and Pdx1CreER^{T2} systems [40]. With early labeling, the Pdx1^{+} population gave rise to all three primary pancreatic lineages, including duct, acinar, and endocrine cells, while inducibly labeling after birth identified endocrine but not ductal cells. These experiments demonstrate that the early progenitor epithelium widely expresses Pdx1, which becomes gradually restricted to the endocrine lineage.

2.5.2 Ngn3

Like Pdx1, Neurogenin3 (Ngn3) is also expressed throughout the pancreas progenitor epithelium during initial bud formation [41]. However, in contrast to the early Pdx1 lineage, Ngn3 quickly restricts a subset of the epithelium and Ngn3-expressing cells only give rise to endocrine cells. Ngn3 lineage restriction was demonstrated using an Ngn3CreER^{T2} reporter mouse [40]. Pulse tamoxifen induction either early (E8.5) or late (E12.5 or postnatally) showed that Ngn3^{+} progenitors gave rise solely to endocrine cellsat all stages examined. Indeed, mature endocrine cells of all types (α, β, δ, PP, and ε cells) arise from a subset of the pancreatic epithelium that transiently expresses Ngn3 [40]. Ngn3 is critical to the specification of all endocrine cells as loss of Ngn3 function results in total loss of endocrine cells [42]. In addition, Magenheim, Dor, and colleagues showed that fine control of Ngn3 expression was critical in balancing ductal versus endocrine fate, since experimental deletion of Ngn3 in epithelial cells shifted their fate from endocrine to ductal lineage [43]. Epithelial cells with low *Ngn3* expression failed to acquire endocrine cell fate, indicating that *Ngn3* dosage dictates net endocrine cell allocation [44]. Grapin-Botton and colleagues also demonstrated that epithelial cells experienced 'sequential competence', leading Ngn3-expressing progenitors to give rise first to α cells, then β and δ cells, and finally PP cells [45]. Together these studies show that Ngn3 regulates endocrine cell specification, with the progenitor epithelium being exquisitely sensitive to Ngn3 levels in the developing pancreas.

2.5.3 Hes1

Hes1, a basic helix-loop-helix (bHLH) transcription factor and downstream target gene of the Notch signaling pathway, is expressed in early pancreatic MPCs [46, 47]. Early Hes1 expression is seen throughout the pancreatic epithelium, and becomes increasingly restricted to the exocrine lineage. After birth Hes1 expression persists only in a few differentiated terminal ducts and centroacinar cells [47]. Lineage tracing experiments by Murtaugh and colleagues, using a Hes1CreER^{T2} knock-in allele, showed that Hes1-expressing cells labeled at E9.5 were multipotent, but by E15.5 the Hes1 lineage lost its multipotency, giving rise only to acinar and ductal cell types [47]. Ectopic Notch signaling early in development prevented progenitor cell differentiation into the exocrine lineage; instead, Notch-expressing cells remained in a progenitor-like state. When ectopic Notch was induced later during embryonic development at E13.5, ductal cells differentiated at the expense of acinar cell fate. By E15.5, ectopic Notch did not affect exocrine differentiation. These data indicate that Notch signaling has dual roles. At first Notch represses differentiation, but at later time points it promotes exocrine cell fate. An intriguing site of high Hes1 expression includes the terminal duct and centroacinar cells, the latter representing a cell type hypothesized to act as a resident stem cell that may give rise to endocrine cells [48]. Consequently, lineage tracing experiments using Hes1CreER^{T2} labeling were carried out to examine this possibility in the resting or injured adult pancreas [49]. In quiescent pancreas tissue, numerous labeled cells were identified in the islets; however, these were primarily endothelial cells, with rarely labeled glucagon$^+$ alpha cells, but no beta cells. Following pancreatic ductal ligation (PDL), which causes local beta cell proliferation and potential neogenesis, lineage tracing revealed that Hes1$^+$ duct and centroacinar cells did not act as resident stem cells in the pancreas. By contrast, Hes1-expressing descendants of crypt cells, the stem cells of the intestinal epithelium, were broadly labeled. Together, these findings demonstrate that the early Hes lineage contributes to all pancreatic cell types, while expression is later limited to a subset of the exocrine compartment.

2.5.4 Sox9

Sox9 (sex determining region Y protein 9) is another transcription factor expressed in the early pancreatic epithelium during initial budding and epithelial stratification. Sox factors have generally been implicated in stem cell biology [50]. For example, in the nervous system, Sox factors are markers of neuronal stem cells [51]. Lineage tracing experiments in the pancreas have shown that Sox9$^+$ cells are multipotent throughout much of embryonic development, but not adulthood [52]. Sander and colleagues demonstrated that Sox9 is initially expressed throughout the pancreatic epithelium, but gradually becomes restricted to ducts by late gestation. Using an inducible Sox9CreER^{T2} BAC transgenic mouse line, these authors showed that the Sox9 lineage still gives rise to some acinar and endocrine cells, even at E14.5 and later when Sox9 expression is primarily restricted to the ducts. These data may indicate that the Sox9$^+$ population serves as a progenitor source for all pancreatic lineages until birth. Postnatally, when Sox9 is solely expressed in pancreatic ductal epithelium, Sox9$^+$ ductal cells can form rare non-beta endocrine cells, demonstrating a significant reduction in potential. When challenged with PDL, pancreatic Sox9-expressing cells can give rise to Ngn3$^+$ cells within the ducts, but these never progress to mature endocrine cells. Another study by Furuyama and others also found no contribution of Sox9$^+$ cells to the endocrine compartment following partial pancreatectomy (PPx), cerulean induced pancreatitis, streptozotocin (STZ) treatment, or PDL [53]. Therefore, Sox9$^+$ cells serve as multipotent progenitors during pancreatic development at least until birth, but the potency of Sox9-expressing cells in adulthood following injury remains unclear.

2.5.5 Hnf1β

Similar to transcription factors described thus far, Hnf1β or TCF2), a POU homeobox transcription factor, is expressed throughout the early pancreatic bud epithelium [54], but becomes progressively restricted to ducts by adulthood. Hnf1β is associated with human disease MODY5 and a number of pancreas atrophies. Using an Hnf1βCreER^{T2} lineage tracing approach, Ferrer and colleagues showed that the Hnf1β$^+$ population gave rise to all three pancreatic lineages between E11.5

and E13.5, with relatively equivalent contributions to ductal, acinar, alpha, and beta cells. Following the secondary transition, between E13.5 and E16.5, Hnf1β1$^+$ cells only produced ductal and endocrine cells [55, 56]. After birth, the Hnf1β—expressing cell population became further restricted in fate, as they only contributed to ductal cells. Tamoxifen induction of reporter mice either at E18.0 or after birth labeled Cytokeratin19$^+$ (ductal) cells only, with no overlap of endocrine or acinar lineages, even after PDL. In addition, Hnf1β$^+$ cells did not contribute to beta cells following beta cell ablation and regeneration approaches (alloxan-EGF/gastrin treatment). Comparison of Hnf1βCreER^{T2} findings with Sox9CreER^{T2} results indicated that Hnf1β$^+$ cells may lose their multipotency earlier or that recombination does not occur as readily in that system, potentially due to lower Hnf1β expression.

2.5.6 Ptf1a

Ptf1a (or P48) encodes a bHLH transcription factor that constitutes the DNA binding subunit of the trimeric PTF1 complex, which is critical to pancreas development. Ptf1a is essential for development of the ventral pancreas and for outgrowth of the dorsal bud, as well as for maintenance of exocrine gene programs, including elastase 1 and amylase expression. Ptf1a has been proposed to be involved in shifting MPC to proacinar fate at branch tips, based on the switching of Pft1a coregulatory binding partners RBP-J and RBP-JL [57]. Using a Ptf1aCre reporter line, the Wright group demonstrated that Ptf1a-expressing cells in the early pancreas contribute to endocrine cells, ducts, and acini [58]. Consistent with this expression pattern, Ptf1a is required for all pancreatic cell types, as ablation of Ptf1a results in pancreas agenesis and loss of pancreatic gene expression [58, 59]. Pulse labeling of Ptf1a expressing cells further refined our understanding of progenitor potential in the pancreatic bud. Lineage tracing using an inducible Ptf1aCreER^{T2} system showed that in the embryonic pancreas, Ptf1a-expressing cells contributed widely to all pancreatic cell lineages following tamoxifen injection at E11.5. However, by E13.5, labeling of the Ptf1a lineage led to a dramatic decline in contribution to ductal and endocrine lineages. After E15.5, there was no further contribution to the endocrine compartment by the Ptf1a$^+$ population, with only ductal and acinar cells labeled [60]. No endocrine cells were labeled with Ptf1aCreER^{T2} after induction, suggesting that Pft1a$^+$ cells only contribute to acini postnally. In contrast, pancreas injury with PDL resulted in significant transdifferentiation of Pft1a$^+$ acinar cells into ductal cells with a subset becoming competent to express Ngn3. Some of these Pft1a$^+$/Ngn3$^+$ transdifferentiated cells properly formed insulin-expressing endocrine cells. Combined PDL and STZ treatment enhanced this acinar-to-duct-to-endocrine conversion, demonstrating what the authors of the study termed "restoration of an early-organogenesis-like MPC competence in adult acinar cells".

2.5.7 Cpa1

In 2007, an important study identified the growing branch 'tips' of the pancreatic epithelium as a critical locale that generates all epithelial-derived lineages of the embryonic pancreas [25]. These 'tip' cells were characterized by specific expression of carboxypeptidase A1 (Cpa1), a zinc metalloprotease (Fig. 3b). Zhou, Melton, and colleagues identified $Cpa1^{+}$ tips as segregating within the epithelium during development. $Cpa1^{+}$ tip cells co-express many key transcription factors, including Ptf1a and c-Myc, and constitute a subset of $Pdx1^{+}$ cells. To test the significance of this observation, an inducible $Cpa1CreER^{T2}$ reporter line was used to transiently label this tip population at different time points during organogenesis. Importantly, $Cpa1^{+}$ cells were found to be multipotent before E14, in that $Cpa1^{+}$ cells gave rise in a step-wise fashion to ductal, acinar, and endocrine cells as well as to new $Cpa1^{+}$ cells, suggesting they are capable of self-renewal. The results showed that tips contribute to all three main pancreatic lineages during early bud formation (when labeled from E11.0 through E12.5). However, after labeling at E14-15, $Cpa1^{+}$ cells only contributed to acinar cells, demonstrating that tip cell multipotency is lost after this time. These findings were the first to identify an architecturally defined region that gave rise to distinct progenitors with defined fates.

2.5.8 $CAII^{-}$

Most studies presented above suggest that pancreatic progenitors solely reside in the embryonic pancreas. However, a few studies suggest that cells within the adult pancreas retain the ability to contribute to multiple lineages. Specifically, a source for new beta cells has long been suggested to reside in the pancreatic ducts, as endocrine cells are known to originate from protodifferentiated bipotential duct-like progenitors in central regions of the pancreas. In addition, endocrine cells are often observed either within, or nearby ducts. One such study was carried out by Inada, Bonner-Weir and colleagues [61]. In this study, the authors addressed the potential of duct cells using a human carbonic anhydrase II (CAII) promoter fragment [36]. In mice, CAII expression is restricted to the ductal lineage by E18.5. Lineage tracing experiments using an inducible CAIICre or $CAIICreER^{T2}$ line found that $CAII^{+}$ cells acted as a pool of facultative progenitors that can give rise to islet endocrine cells and acini, in addition to ductal cells, both postnatally under resting conditions or following PDL. However, Cre expression was not restricted to the ductal compartment in these mice, which may suggest alternative cells of origin for newly formed beta cells [62].

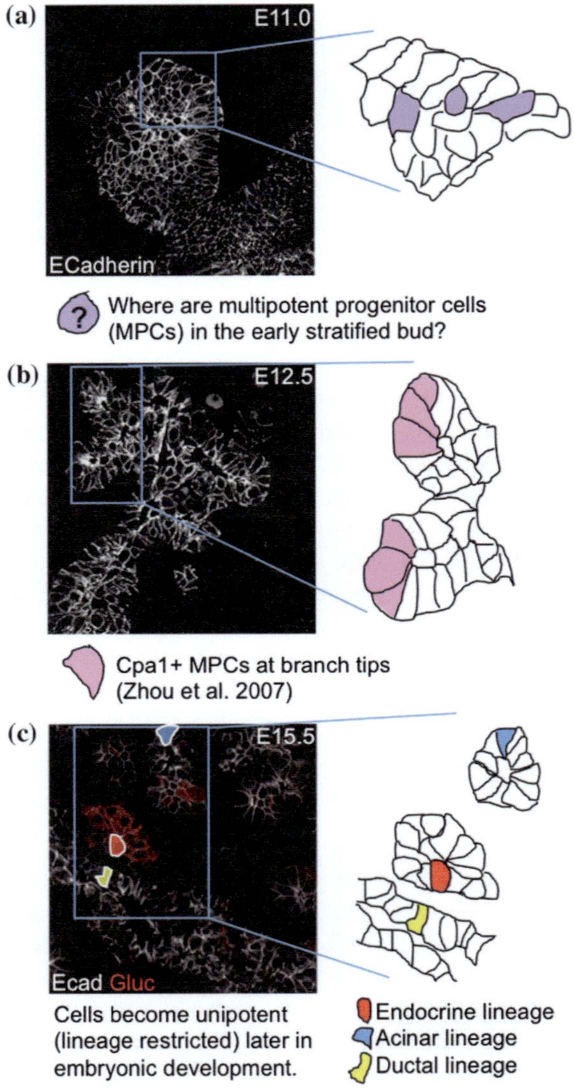

Fig. 3 Potential of pancreatic epithelial cells during embryonic development. Epithelium is visualized by E-Cadherin immunofluorescence (*white*). **a** In the early pancreatic bud (from E10.0 through ~ E12.5), multipotent progenitor cells (MPCs) are present, but their location(s) within the bud remain unknown. **b** As the pancreatic epithelium resolves into branches (here shown at E12.5), epithelial cells located at the branch tips have been identified as MPCs, are multipotential, and are characterized by expression of Cpa1 (25). **c** By E15.5, pancreatic epithelial cells largely become unipotent, with fates restricted to endocrine, ductal, or acinar lineages under normal conditions. Alpha cell precursors are delineated by glucagon immunofluorescence (*red*)

2.6 Pancreatic Progenitors—Where Are They?

Lineage tracing studies such as those described above, as well as our evolving understanding of the morphogenetic events underlying pancreas development, have both shaped our understanding of pancreatic 'stem' or 'progenitor' cells. Generally speaking, pancreatic bud cells appear to have the potential to contribute to all lineages early, but this potential is restricted following the secondary transition. We can point to the early stratified bud and state with relative certainty that the cells therein give rise to both exocrine (ductal/acinar) and endocrine compartments. However, when and where these important and potent cells can be found during pancreas formation or in its resting state is only beginning to be elucidated. It remains a challenge to point exactly to individual cells with specific lineage commitments in the early bud (Fig. 3). We are only beginning to understand how cells sort out within the relatively amorphous epithelium. In addition, we know little about whether specific ‘niches’ exist to foster specific progenitor cell fate decisions. As pancreatic branching begins, we know that ‘tip’ cells labeled using inducible Cpa1 reporter lines give rise to all pancreatic lineages prior to the secondary transition, but only yield acinar cells afterwards. Hence, the early bud and branch ‘tips’ in the early embryo are the only cells identified as containing ‘stem’ or ‘progenitor’ cells. It is important to point out, however, that $Cpa1^{+}$ cells are also identifiable within more central regions of the pancreatic epithelium, raising the question as to whether progenitors may be more widespread in the early bud. After the secondary transition, while the central pancreatic epithelium contains cells that give rise to both duct and endocrine cells, we cannot point with assurance to a specific location containing pancreatic progenitor cells. The field of pancreas development and regeneration can look to other tissues with easily identifiable stem cells, such as those in intestinal crypts that give rise to intestinal villi epithelium, and hope that 1 day we may be able to identify pancreatic progenitors with such anatomical precision and assurance. Whether underlying developmental pathways will be able to be emulated to generate new pancreatic cell types in patients suffering from debilitating diseases such as diabetes or pancreatic cancer remains at this point mere conjecture.

2.7 *Insights from Embryonic Development of the Pancreas*

Together, developmental and lineage tracing studies suggest that pancreatic progenitors exist solely in the early embryonic bud. Multipotency is lost progressively during late pancreas development, and after birth there exists little cell fate plasticity. The majority of lineage tracing systems support the argument that postnatal pancreatic lineages are unipotent with limited ability to expand. Facultative stem or progenitor cells are defined as unipotent cells capable of acquiring multipotency under conditions other than homeostasis, e.g., following injury [63]. Importantly, a resident population of facultative endocrine progenitor cells has not been unequivocally identified. Instead, hope can be placed on the potential of populations of existing differentiated cells, such as beta cells or acinar cells. Beta cells can expand by induced reentry into the cell cycle, while acinar cells can undergo transdifferentiation and contribute to new beta cell generation. These possibilities will be further discussed below.

3 Adult Beta Cells: Sources and Questions

3.1 Adult Beta Cells from Endogenous Sources

In contrast to beta cell neogenesis in the embryo, the origin of new beta cells in adults has been an area of intense debate. While embryonic generation of new beta cells is rapid and robust, it declines markedly after weaning in mice [64]. New beta cells have been shown to arise during adulthood, albeit much more infrequently [65]. In addition, the possible sources of new adult beta cells remain controversial. There are clear examples of the generation of new beta cells in adulthood. For example, when the body experiences increasing metabolic physiological demands during pregnancy and obesity, an increase in the number of beta cells is triggered [64]. The question arises as to how these new beta cells are stimulated to emerge and from where. Do they arise from hidden endogenous progenitors we have yet to discover? Or, do differentiated beta cells merely proliferate to expand their numbers (replication)? It is also conceivable that non-beta cells may contribute to the increased number of beta cells seen in adult islets when they are needed. Does this type of transdifferentiation occur normally during maintenance of the pancreas or under injury or disease conditions? These are the questions being asked across the fields of pancreas biology, development and regeneration (Fig. 4).

Answers to these questions may be found in the study of various models of pancreas expansion. One example is pancreas growth during pregnancy in mice, where islet mass is increased and is a process recently shown to be dependent on serotonin levels [66]. To date, the source of these new beta cells during pregnancy is unclear [67]. However, the Kaestner group showed that islet expansion and beta cell hypertrophy occurred during pregnancy, and these processes induce both proliferative and survival pathways in response to increased metabolic demands [68]. Humans also increase beta cell mass during pregnancy [69], although the adaptive increase in beta cell mass is less pronounced in humans compared to pregnant mice. Further investigation of beta cell hypertrophy is needed, as these mechanisms could yield insights into potential diabetic therapies in the future.

Another system used to examine endocrine expansion is nondiabetic mouse models of obesity, where increased insulin resistance leads to a compensatory increase in beta cell mass [65]. Nondiabetic obese humans with insulin resistance also have expanded beta cell mass with altered islet morphology [70, 71]. The possible origins of these new beta cells include beta cell replication or beta cell differentiation from other cell types. Nonetheless, the observation that new beta cells can appear in adulthood leads to the tantalizing possibility that the mechanisms spurring beta cell generation are in place and could be harnessed therapeutically, enhanced or even recapitulated. In order to achieve this goal, however, we must determine when new beta cells appear, and by what mechanism.

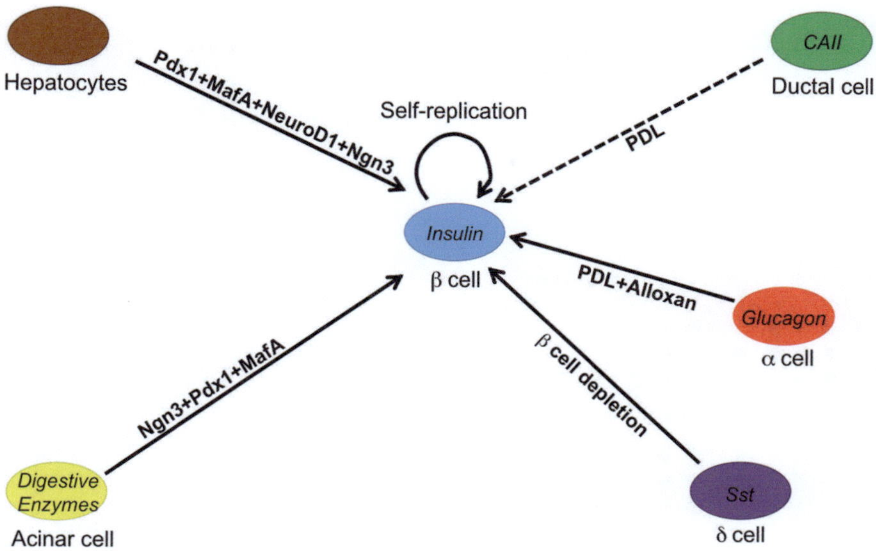

Fig. 4 Evidence supports mechanisms of cellular interconversion and self-replication, as depicted in this model of beta cell production in the mature pancreas. Acinar cells (*yellow*), characterized by secretion of digestive enzymes, can transdifferentiate into insulin$^+$ beta cells (*blue*) following cellular reprogramming with Ngn3, Pdx1, and MafA transcription factors [81, 83]. Ductal cells expressing carbonic anhydrase II (CAII) (*green*) may transdifferentiate into endocrine beta cells, as well as acinar cells, following pancreatic ductal ligation (PDL) [61, 98], although conflicting data has been obtained using other reporters such as Sox9 or Hnf1β [56, 100]. Reprogramming of hepatocytes (*brown*) occurs following transduction with pancreatic transcription factors Pdx1, MafA, NeuroD1, and Ngn3 [91, 93]. Other endocrine cells (Glucagon$^+$ alpha cells (*red*) [85, 147] and Somatostatin$^+$ delta cells (*purple*) [148]) can be stimulated to transdifferentiate into beta cells. Substantial evidence suggests that beta cell neogenesis predominantly occurs via self-replication [72–74]

3.2 *Beta Cell Replication*

While many in the field have looked to various extra-pancreatic tissues as a source for beta cells, others have asked whether new beta cells might simply come from old beta cells, implying that simple self-replication may be the predominant form of beta cell expansion in adults. An elegant study by Dor and others showed that total beta cell mass increases 6.5-fold from postnatal stages (three months) to adult (12 months) [72], indicating that most beta cells in the adult pancreas are formed postnatally. These beta cells predominantly enlarge islets constructed in the embryo. In order to determine whether new beta cells in adults are derived from progenitor cells or differentiated beta cells, pulse-chase lineage tracing studies were performed (Fig. 5a). Specifically, labeling was carried out using a tamoxifen-inducible Rat Insulin Promoter Cre Estrogen Receptor (RIPCreER^{T2}) mouse model with a reporter, in order to label insulin-expressing cells. In normal adult pancreata, as well as the PPx model of pancreas regeneration, all islets contained similar proportions of

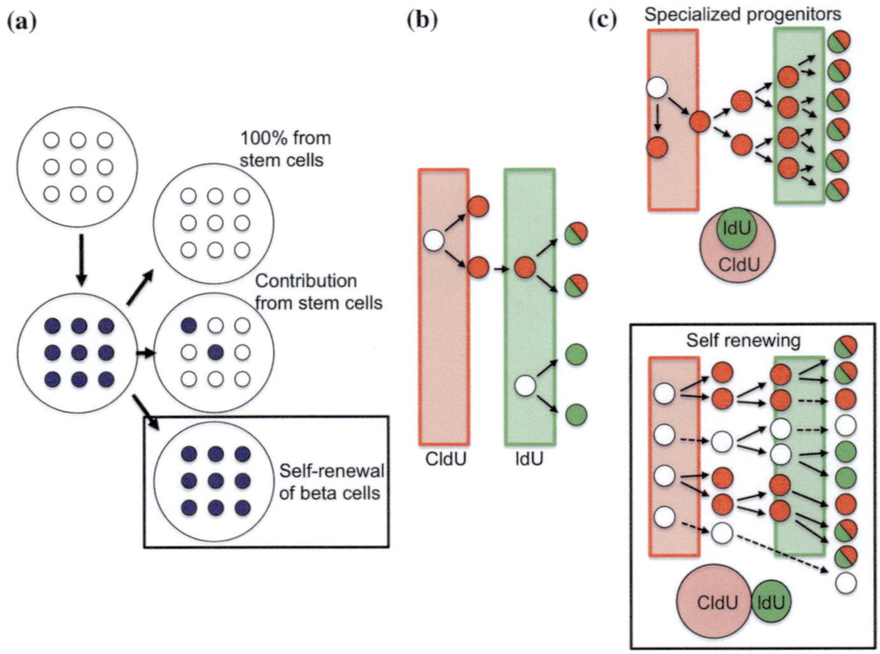

Fig. 5 Pulse-chase lineage tracing experiments demonstrated the origin of new beta cells in adult mice. **a** Beta cells are indelibly labeled (pulse), followed by a chase period. If a new islet is derived entirely from stem cells, then the islet will contain no labeled cells ("100 % from stem cells"). If the stem cells maintain the beta cell population during homeostasis, then the proportion of labeled beta cells within the islet will gradually decrease ("Contribution from stem cells"). If beta cells are maintained via self-replication, then the proportion of labeled cells will remain constant ("self-renewal of beta cells"). *Blue circles* represent labeled beta cells within islets. Adapted from [72]. **b** A DNA analog-based lineage tracing system was used to label the first and second rounds of cell division with CldU and IdU, respectively. If sequential rounds of cell division occur, then cells will be co-labeled. **c** Models of adult beta cell division suggest that specialized progenitors will be co-labeled via sequential cell division, whereas beta cell self-renewal will result in random labeling with CldU and/or IdU, or neither. **b**, **c** Adapted from [73]

labeled beta cells over time. These findings suggested that islet neogenesis does not occur during normal tissue maintenance (homeostasis) or during regeneration. Rather, new beta cells arose by way of self-duplication of extant insulin-expressing cells in the adult pancreas. In support of this idea, lineage tracing experiments using DNA analog incorporation showed that adult murine beta cells self-renew during homeostasis and regeneration (Fig. 5b, c) [73]. Long-term bromodeoxyuridine (BrdU) incorporation studies indicate that adult mouse beta cells proliferate at a very slow rate: on average 1 out of every 1400 beta cells turns over per day [74]. In a clinical trial studying short term BrdU incorporation in human patients, turnover of human beta cells occurred in the first three decades of life, with no significant cell turnover observed in patients older than 30; these data were supported by radiocarbon dating of human beta cells, in which the C^{14} content correlated with beta cell "birth dates" of less than 30 years [75]. Therefore, beta cell self-replication may be the main mechanism by which new beta cells are generated in the adult rodent pancreas, with a lack of beta cell turnover in adult humans (over 30 years of age).

There have been recent efforts to find methods to stimulate beta cell replication as a therapeutic basis for treatment of diabetes. A report in 2013 raised the possibility that angiopoietin like protein 8 (Angptl8), also called betatrophin or lipasin, might represent a viable approach, in that treatment of diabetic mice with Angptl8 was found to promote beta cell replication and improve beta cell function (glucose tolerance) in mice [76]. However, Gusarova and others recently showed that in their hands Angptl8 overexpression in adult mice did not affect glucose tolerance or beta cell mass [77]. Upon further investigation, the Melton group found that Angptl8 increased beta cell proliferation, but to a much lower extent than originally reported, and response to Angptl8 was highly variable from mouse to mouse [78]. Notably, glucose infusion alone stimulated beta cell proliferation and increased beta cell area [77]. Together these findings underscore some of the inherent limitations in using diabetic mouse models to find agents that increase beta cell number and thereby treat diabetes, with the aim of developing therapies to human diabetes. Indeed, identifying agonists that induce beta cell replication in humans has historically proven difficult. Many agonists which have been found to increase beta cell mass in rodents ultimately fail to abrogate diabetes in human patients [79]. Therefore, replacement therapies, including transplantation of newly generated, encapsulated beta cells or islets, may still be the most feasible approach for successfully treating and even curing diabetes.

3.3 Beta Cell Transdifferentiation

Beyond eyeing replication of endogenous progenitor or differentiated beta cells as sources of new beta cells, there is growing interest in the possibility that beta cells may be generated via transdifferentiation from other functional cell lineages, such as exocrine or ductal cells (Fig. 4). Transdifferentiation is defined as a process entailing a shift of cell fate, or the transformation of differentiated cells (not stem cells) into a different cell type [80]. So far, groups investigating this idea have arrived at different conclusions. While some studies have shown that new beta cells arise in the adult pancreas predominantly via proliferation of differentiated cells [72], others have shown that new beta cells can arise via transdifferentiation from different lineages, such as acinar [81–83] or ductal cells [61], under certain conditions. Lineage tracing studies using transgenic mouse models have been useful for identifying source(s) of new adult beta cells; however, these results have provided conflicting data regarding whether pancreatic and other cell lineages have the capacity to transdifferentiate [84–87]. Whether new beta cells arise from neighboring cell types, from expansion of existing beta cells, or from unknown endogenous pancreatic stem cells, remains a major point of contention. Below, we will discuss the range of findings regarding beta cell generation, either during homeostasis or injury response, which have stimulated questions in the field.

3.4 Acinar Origin

One promising source of beta cells is the most abundant cell type of the pancreas, the exocrine acinar cells. Reprogramming of acinar cells toward the endocrine lineage was accomplished experimentally using injected adenoviral constructs, forcing the co-expression of crucial transcription factors Ngn3, Pdx1, and MafA [83]. These experiments resulted in a subset of reprogrammed cells that expressed insulin and exhibited other characteristics of beta cells, albeit in extra-islet locations scattered within the adult pancreas. These artificial beta-like cells were derived from the acinar lineage, as determined by lineage tracing in tamoxifen-inducible Cpa1CreER^{T2};R26R transgenic mice. In the STZ-induced diabetic mouse model, co-transfection of Ngn3, Pdx1, and MafA augmented glucose tolerance and improved fasting blood glucose levels [83]. These results indicate bona fide insulin production and secretion. Induced beta cells did not incorporate BrdU, thereby ruling out exocrine dedifferentiation to a common progenitor that might expand and ultimately re-differentiate into beta cells. While the authors could not rule out transient dedifferentiation, this study demonstrated the possibility that acinar cells were poised and competent to shift their fate from exocrine to endocrine.

A different approach to stimulating transdifferentiation towards the beta cell fate involves treatment of exocrine tissue with exogenous glucose or growth factors. Like manipulation of pancreatic transcription factors, this method could also lead to acinar transformation into beta cells. Studies in rats showed that beta cell mass can expand by beta cell neogenesis in response to chronic hyperglycemia, and that new beta cells in this model arise from acinar cell transdifferentiation [88]. Similarly, treatment of diabetic mice with epidermal growth factor (EGF) and ciliary neurotrophic factor (CNTF) leads to reprogramming of acinar cells to express Ngn3 and then insulin [89]. Ngn3 expression is normally restricted to endocrine progenitor cells that differentiate into beta cells during embryonic developmental stages; Ngn3 expression is downregulated postnatally and thought to be largely absent from the adult pancreas [41]. Therefore, induction of Ngn3 in acinar cells suggested dedifferentiation into a progenitor-like cell, indicating possible acinar-to-beta cell transdifferentiation. Importantly, EGF and CNTF-treated mice are capable of ameliorating hyperglycemia. These experiments further demonstrated that acinar cells can be pushed to change fate and take on key characteristics of endogenous beta cells. The questions of whether acinar cells normally do this under pathological conditions, and whether this has therapeutic potential, remain unanswered.

3.5 Hepatobiliary Origin

Unlike the pancreas, the adult mammalian liver retains a striking ability to regenerate. The liver can fully regenerate following resection of up to 75 % in adult mice and humans [2]. Given this inherent regenerative capacity, reprogramming hepatocytes into pancreatic beta cells could provide a feasible alternative source of new beta cells for treating diabetes. Several labs have utilized Pdx1, or a combination of transcription factors (Pdx1, MafA, NeuroD1, and/or Ngn3), to reprogram hepatocytes into insulin-producing beta/endocrine cells [90–93]. Ectopic expression of Pdx1 by hepatocytes leads to glucose sensing and insulin secretion from hepatocytes in vivo [90, 94]. Transient ectopic Pdx1 leads to autoinduction of Pdx1 expression in hepatocytes, which persists for up to eight months, at which time STZ-induced diabetic mice are capable of stabilizing blood sugar [90]. Ectopic expression of Pdx1 leads to hepatocyte dedifferentiation, as indicated by changes in cellular gene expression, followed by re-differentiation into a pancreatic phenotype [95]. Interestingly, unlike pancreatic beta cells, Pdx1-induced hepatocytes are protected from autoimmune attack, perhaps due to the protective effects of the enzymatic activity in hepatocytes [96]. Protection from beta cell-directed autoimmune attack makes hepatobiliary reprogramming a realistic clinical approach to treating diabetes; however, inducing Pdx1 expression currently requires viral vector infection, which is not a viable approach for human therapies.

3.6 Ductal Origin: Beta Cell Regeneration?

Another potential source of new beta cells is the adult pancreatic ductal epithelium. The idea that ductal cells can regenerate the pancreas and provide new beta cells has been interrogated using pancreatic injury models, namely PDL and PPx. Seminal studies addressing this question have used the PDL as a classic model of regeneration in rodents. PDL is physiologically and clinically significant, as it resembles the most common pancreatic injury in humans, the blockage of the main pancreatic duct [64]. Blockage of the duct in humans and mice eventually leads to acute pancreatitis, with beta cell dysfunction and hyperglycemia [97]. Several studies have shown that new endocrine cells arise from ducts in the adult pancreas, following PDL as well as other types of injury [60, 61, 98, 99]. The study by Bonner-Weir's group mentioned above, which utilized the human CAII promoter to label and lineage trace duct cells in the pancreas, directly addressed this possibility. At postnatal stages, CAII expression is largely restricted to the ducts, with only a few CAII-derived acini, alpha, and beta cells [61]. Following PDL, however, the number of CAII-derived cells detected in islets and acini was significantly increased, suggesting that the ductal epithelium contributes to new beta cells following PDL.

Along those lines, Xu and colleagues found that PDL could induce Ngn3 expression in a subset of ductal cells [99], which was suggestive of dedifferentiation into a progenitor-like cell and supported the possibility that ductal-to-beta cell transdifferentiation occurred. However, a contrasting report showed that *Ngn3* mRNA is expressed in the uninjured adult mouse pancreas, bringing into question the significance of Ngn3 assessments in the Xu study [100]. This difference in results was attributed to technical differences in quantitative PCR experiments. Therefore, a key question remains: does Ngn3 expression (or activation) mark beta cell neogenesis? Transient Ngn3 expression has historically been believed to indicate committed, unipotential beta cell progenitors. However the finding that low levels of *Ngn3* expression are detectable in adult beta cells raises the possibility of potentially different roles for Ngn3 in different contexts.

Ductal-to-beta cell transdifferentiation following PDL also is controversial in that several groups have been unable to replicate these data [47, 52, 56, 100, 101]. Rankin and others found increased ductal cell proliferation 7 days after PDL, but no associated increase in beta cell mass or area [100]. Differences in results were attributed to differences in quantification methods; i.e., beta cell mass and area were quantified throughout the entire pancreas, rather than sampling only a small part of the pancreas. In addition, pancreata were examined 7 days post-ligation, whereas other groups quantified beta cell mass and observed beta cell hyperplasia 3 days post-ligation. Therefore, determining the extent of beta cell neogenesis following PDL may depend on the timing of histological analysis. In support of this idea, Pan and others reported that acinar-to-endocrine conversion post-PDL was only observed after two months of recovery [60].

Several groups have shown that cells expressing developmental markers of MPCs do not contribute to beta cell neogenesis following PDL. For instance, $Hes1^+$ cells, initially representing early pancreatic progenitors but eventually restricted to terminal ducts and centroacinar cells in the adult pancreas, were not found to contribute to the beta cell population following injury [47]. Similar observations were made when lineage tracing using the transcription factor Sox9, which marks the entire progenitor epithelium, but then becomes restricted to ductal cells [52]. Following PDL, $Sox9^+$ cells did not give rise to beta cells in the regenerating pancreas, and Ngn3 expression was only detected in Sox9-labeled ductal cells. Another early marker of the progenitor epithelium in the early pancreas, Hnf1β, becomes restricted to ductal cells in adults, and $Hnf1\beta^+$ cells do not contribute to islet β cells, even following PDL [56]. Together, these studies underscore how determining whether ductal cells transdifferentiate into beta cells following PDL is contingent upon the ductal-specific promoters driving Cre/$CreER^{T2}$ used for analysis, as well as the methods and time points analyzed post-ligation.

3.7 *Lobe Regeneration*

While generation of specific pancreatic lineages has been addressed in various models, gross coordinated growth of pancreatic tissue is also a therapeutic possibility. Growth of entire new lobes of the pancreas has been reported. Indeed, the PPx model of pancreatic injury, in which a defined percentage of the pancreas is resected, has been associated both with lobe regeneration and beta cell expansion in adult rodents. Following up to 90 % resection of the adult rodent pancreas, both endocrine and exocrine cells increase rates of cellular proliferation, thereby facilitating regeneration of all pancreatic lineages. Bonner-Weir and others demonstrated that post-PPx, ductal cells give rise to large areas of the pancreas apparent as new lobes, by way of ductal cell proliferation and differentiation [102]. These newly formed pancreatic lobes contain the appropriate proportion of islets, acinar, and ductal cells, which indicates pancreatic regeneration following PPx in adult rodents. These studies suggested that some type of pancreatic progenitors, possibly a version of MPCs, is present in pancreatic ducts of adult rats. However, other lineage tracing studies have reached contradictory conclusions. While beta cell proliferation (and therefore regeneration) does increase following PPx, the lineage of new islets was found to be derived from extant beta cells post-PPx [72]. Importantly, unlike rodent models of diabetes, humans do not appear to undergo beta cell regeneration following PPx [103]. Examination of adult human pancreata following 50 % resection found that PPx does not stimulate beta cell regeneration in adult humans [103]. Therefore, perspective must be retained when elucidating regenerative mechanisms from rodent experiments, as they may display mechanistic differences with the human pancreas and require caution in developing treatments for human patients with diabetes.

Evidence of beta cell regeneration following pancreatic injury stimulated the field to find ways to exploit this regenerative mechanism. However, the beta cell regenerative response to pancreatic injury is possibly more robust in rodent models of diabetes, as compared to human patients, reminding us that studies of beta cell regeneration/neogenesis in mice and rats must be carefully interpreted when applied to clinical therapies [103]. Therefore, keeping in mind the ultimate goal of curing T1D, recent efforts have shifted focus from regeneration to replacement.

4 Laboratory Generation of New Beta Cells

Current therapies for treating T1D include insulin therapy for regulating blood glucose levels, as well as whole pancreas and islet transplantation [104], but each of these options places a considerable burden on the patient. Monitoring and regulating blood glucose levels via insulin injection requires constant vigilance. While there have been considerable therapeutic advances in insulin therapy, including insulin pumps, continuous blood glucose monitors, and closed loop systems, current treatment protocols cannot completely protect patients from hyper- and hypoglycemia, along with their inherent complications [105]. Whole pancreas and islet transplantation procedures restore endogenous glucose-stimulated insulin secretion (GSIS) to T1D patients, sometimes leading to insulin independence [106]. However, patients are generally considered for whole pancreas transplants only when receiving a concurrent kidney transplant, which substantially limits the eligible cohort. In 2000, Shapiro and colleagues developed the 'Edmonton Protocol' [107], which has since been adapted and used as a basis for islet transplantation. This alternative transplant option is still considered to be an experimental procedure, with less than 700 islet transplants performed worldwide between 1999 and 2010 [108]. Three years post-islet transplant, only 44 % of recipients maintained insulin independence. With both whole pancreas and islet transplants, there is potential for serious complications from the requisite immunosuppressive therapy. Even if transplantation became fully optimized and the process routine, there would still remain the insurmountable issue of persistent shortage of donor organs. Therefore, alternatives to whole pancreas and islet transplantation are still being sought.

Ultimately, for T1D patients to achieve insulin independence via beta cell replacement, beta cells need to be acquired or generated on a large scale to provide a tractable source for transplantation. Investigators have tackled this problem using human cell populations that can be directed toward pancreatic endoderm and ultimately, insulin-producing beta cells. However, transplanting cells obtained from donors or generated in the laboratory introduces unavoidable issues of immune rejection. Specifically, transplanted cells will present foreign major histone compatibility (MHC) proteins to recipient T cells leading to an immune response, cytotoxicity, and eventual graft failure, even with immunosuppressive drug therapy [109]. Hence, one conceptually attractive approach involves coaxing a patient's own cells into becoming beta cells, thereby circumventing this problem altogether.

4.1 Bone Marrow Origin

Some groups looked to human mesenchymal stem cells (MSCs) in the hopes of guiding them toward a pancreatic endocrine cell fate [110, 111]. The practical advantage of using these cells is that they would not require immunosuppression therapy [92]. MSCs comprise the stromal compartment of the bone marrow (BM), which would be a useful and accessible cellular source for laboratory generation of new beta cells [92]. Several labs attempted to differentiate MSCs into insulin-producing cells. While some reports have shown that MSCs could be directed toward an endocrine cell fate [110, 112], others contend that MSCs did not themselves differentiate into beta cells. Rather, it was proposed that MSCs may promote endogenous beta cell regeneration. MSC-derived cells have been shown to ameliorate hyperglycemia in STZ-induced diabetic mice [113, 114]. Therefore, BM-derived MSC results have been controversial, as experiments have been difficult to reproduce [111, 115, 116]. Consequently, despite the drawbacks of immunosuppressive therapy and potential for teratoma formation, investigators increasingly turned to ESCs as a viable resource for generating insulin-producing cells.

4.2 Laboratory Generated Stem Cells

Utilizing the vast potential of hESCs has proven to be more successful [117]. Directed differentiation protocols guiding hESCs toward a pancreatic endocrine cell fate have flourished in the last decade [118], with repeated improvements and optimization of methods yielding higher percentages of desired cell types. Multiple pharmaceutical companies and academic laboratories have tackled the challenge of guiding ESCs and iPSCs toward a glucose-responsive, insulin-secreting mature beta cell fate. This difficult prospect has faced (and still faces) numerous obstacles, but years of research have yielded elegant, multi-step in vitro protocols for deriving pancreatic endocrine cells that successfully attenuate hyperglycemia following transplantation into immunodeficient, diabetic mice [119–124]. The recent history of deriving pancreatic endocrine cells from iPSCs is beyond the scope of this review. (See elegant reviews by Schiesser and Wells, as well as Bouwens et al., for further detail [118, 125]). This section will discuss current methods and prospects for directing ESCs toward beta cell fate, and the potential of both ESC and iPSC approaches, with the goal of generating patient-specific beta cells for replacement therapy for T1D.

4.3 hESCs

Pancreatic endocrine cells have been successfully derived from mouse ESCs (mESCs) as well as hESCs, both of which result in cells capable of ameliorating hyperglycemia in diabetic mouse models [126, 127]. Directing ESCs toward a beta cell fate requires defined progression through a series of specific stages, which recapitulate pancreatic development in vitro [128]. Generally, ESCs are directed to differentiate into mesendoderm, followed by definitive endoderm (DE), primitive gut (PG), posterior foregut, pancreatic progenitor (PP), and finally endocrine precursor cells (Fig. 6). Each stage requires treatment with specific growth factors and inhibitors, approximating the signaling pathway milieu that guides beta cell development in vivo, and each stage is characterized by expression of specific transcription factors [129]. It should be noted that research groups have repeatedly optimized their directed differentiation protocols, which may comprise anywhere from four to seven stages of differentiation in order to obtain endocrine cells [123, 130, 131]. Subsequently, when the endocrine precursor cells are implanted into immunodeficient mice, the cells will continue to mature into beta cells within

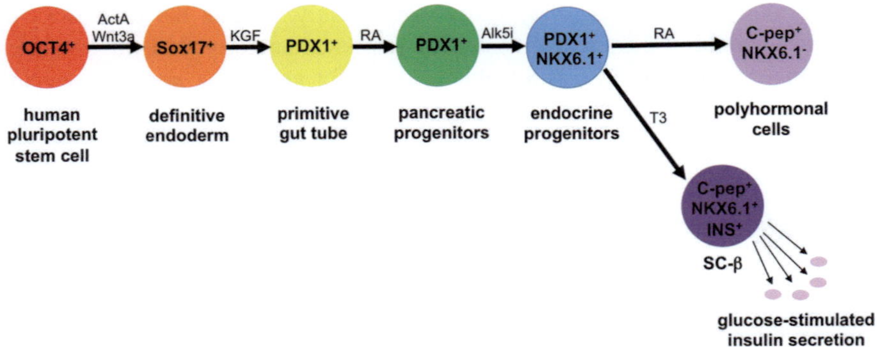

Fig. 6 Directed differentiation protocols guide human pluripotent stem cells (hPSCs) in vitro toward endocrine and beta cell fate. Specific media components trigger defined signaling pathways that guide hPSC differentiation toward the beta cell fate. This sequential culturing process requires 23–35 days in culture. Please note that this model depicts an oversimplified directed differentiation protocol, and the actual protocols that have been published by numerous laboratories are more complex. Please see Schiesser and Wells [118] for details. $OCT4^+$ hPSCs are directed toward definitive endoderm by treatment with Activin A (ActA) and Wnt3a, among others. $SOX17^+$ definitive endoderm, treated with Keratinocyte Growth Factor (KGF), then acquires $PDX1^+$ primitive gut tube cell fate. Pancreatic progenitors are promoted by treatment with Retinoic Acid (RA). Treating pancreatic progenitors with the TGFβR inhibitor Alk5i then promotes NKX6.1 expression in endocrine progenitors. Cells are further differentiated into endocrine cells, and then polyhormonal cells positive for insulin, insulin+glucagon, or insulin+somatostatin by treatment with RA [119, 129, 137, 149–151]. Differentiation media components were improved by Pagliuca, Melton, and others [130]. In this optimized protocol, endocrine cells express PDX1, NKX6.1, and C-peptide, and treatment with thyroid hormone T3 and other soluble factors further differentiates the cells into functional beta cells (SC-β) capable of secreting insulin in response to glucose challenge in vivo

3–4 months [132]. However, the events and underlying mechanisms driving this in vivo maturation process remain poorly understood. Initially, when beta cell maturation was attempted in vitro, the pancreatic cells have proven to be abnormal or poorly differentiated; e.g., differentiated endocrine cells were reportedly polyhormonal, or were epigenetically abnormal in terms of transcriptome and chromatin structure [119, 133]. These characteristics led to the assumption that maturation did not occur normally. Therefore, crucial, unknown signals stimulate the final important aspects of beta cell maturation from hESC-derived cells in vivo, suggesting that the three-dimensional microenvironment and morphogenesis is key for beta cell maturation and survival.

Because ESC-derived beta cells mature properly in vivo, but not in vitro, attempts have been made to differentiate these ESC-derived cells in the recipient, this case in mice. Pharmaceutical companies like Viacyte were already developing macroencapsulation devices to house and protect the pancreatic progenitor cells that, once implanted, can fully differentiate into endocrine cells including beta cells [134]. Importantly, encapsulation of implanted cells also prevents autoimmune destruction of transplanted cells. However, the porous nature of the capsule does allow diffusion of smaller molecules, as well as the requisite vascularization of the implant membrane, thereby facilitating diffusion of insulin and other secreted factors into the blood stream [122]. Phase I clinical trials of encapsulated pancreatic progenitor cells in a small cohort of human patients began in 2015, and the trials will clarify the likelihood of the encapsulation approach working in humans, as it does in mice.

Another pharmaceutical company, BetaLogics, has tweaked the in vitro ESC differentiation protocol in order to correct skewing of cell fates during culture, including the generation of polyhormonal (insulin^{+}/glucagon^{+}/somatostatin^{+}) cells [123]. A recently developed protocol from this group directs human ESCs through a seven-stage protocol that eventually yields mature beta cells. These were shown to express and secrete insulin in response to glucose stimulation, with no expression of glucagon or somatostatin. In addition, these differentiated beta cells rapidly reversed diabetes in mice following transplantation.

4.4 iPSCs

iPSCs are perhaps a more attractive resource for beta cell replacement in diabetic patients than ESCs, since iPSC-derived endocrine cells are by definition generated from the patient's own cells. This characteristic thereby bypasses issues of MHC-based immune rejection, as well as ethical issues arising from the use of human embryonic tissues [135]. Notably, however, this approach does not completely circumvent the need for immunosuppression therapy in patients with T1D, given that T1D is an autoimmune disorder where endogenous beta cells are targeted by the immune system. Therefore, the encapsulation device described above, or a similar protective technology, will still be required for any beta cell replacement therapy in the context of T1D.

Many studies, more than can feasibly be cited in this review, have derived endocrine progenitor cells, or cells similar to fetal beta cells, from hESCs and iPSCs with a wide variety of methods, and with widely different efficiencies [119, 129, 133, 136–138]. A recent protocol published by the Melton group reportedly supports the generation of bona fide glucose-sensing beta cells, which have the ability to secrete the appropriate amount of insulin multiple times in vivo in mice, even in diabetic mouse models (Fig. 6) [130]. This iPSC to beta cell methodology yields cells that have the ability to secrete the appropriate amount of insulin multiple times in vivo in mice, even in diabetic mouse models. In this study, a suspension-based culture system was used to stimulate hESCs and human iPSCs (hiPSCs) to differentiate through the stereotypical stages of DE, PG, and then PP cell fate. Importantly, unlike previous protocols that have yielded polyhormonal (PH) endocrine cells, the refined differentiation protocol produced cells capable of responding appropriately with GSIS in vitro. This protocol was optimized by testing over 150 compounds in more than 70 combinations, and yielded "SC-β" cells that expressed insulin, but not glucagon or other endocrine hormones. SC-β cells functionally resembled isolated primary cadaveric beta cells more closely than hESC-derived PH endocrine cells [130]. Transplantation of SC-β cells restored normoglycemia to an immunodeficient model of diabetes within two weeks. With these advances in hiPSC-derived beta cell function in vitro, as well as SC-β cells' ability to incorporate rapidly into the hormone-response system in vivo, this seminal study indicates that the field is approaching the major goal of treating T1D successfully with beta cell replacement therapy.

What patient tissues would be used as starting material for generation of iPSC cells? Many such protocols have proposed using differentiated cells such as keratinocytes or blood cells. However, one disadvantage of deriving iPSCs from these types of mature, differentiated cell types is that it requires the introduction of integrating retroviral vectors. These vectors carry the requisite pluripotency-associated factors (originally OCT3/4, SOX2, KLF4, and c-MYC) into somatic cells, in order to reprogram the cell [139]. Integration of reprogramming constructs into the genome increases the risk of mutagenesis and tumor development, particularly in the case of proto-oncogene c-MYC [140]. However, a recent report demonstrated the efficacy of using the Sendai virus as an RNA-based gene delivery system, rather than a DNA-based retrovirus, to reprogram cells derived from T1D patients [141]. Using an RNA-based vector like the Sendai virus would decrease the likelihood of integration into the host genome. Therefore, it is clear the safety of iPSC technology is still evolving, and will likely continue to improve in the future.

4.5 Beta Cell Maturation

As described above, protocols deriving beta cells from hPSCs are continually improving. However, one crucial step remains to be fully mastered: achieving beta cell maturation. MafA, a basic leucine zipper transcription factor, is a hallmark of mature beta cells. In vivo, MafA and MafB are expressed later in embryonic development, specifically in endocrine precursor cells [142]. MafB is required to promote *MafA* transcription, as well as *Pdx1*, in immature beta cells during development. MafA is expressed starting at E13.5 and is not required for islet formation in the embryo. In adult mice, MafA and MafB are expressed specifically in beta and alpha cells, respectively. Interestingly, $MafA^{-/-}$ adult mice are diabetic with impaired GSIS [143], indicating that MafA is required for beta cell function and homeostasis in the adult pancreas. In addition, a study of hESC-derived pancreatic progenitor cells transplanted into adult rats found that $Nkx6.1^{lo}$ cells lacking nuclear MafA have impaired insulin expression [131], which indicates that MafA expression correlates with mature beta cell functionality. Whether the SC-β cells described above express MafA is currently unknown [130].

A recent study from the Dor group reported that neonatal beta cells are unable to proliferate in response to hyperglycemia [144]. After weaning, beta cell maturation is completed, at which time compensatory beta cell proliferation can occur.

Together, these data indicate that there is a critical beta cell maturation program during embryonic development of the pancreas, and this program is completed after weaning. It should be noted that unlike in mouse, MafA and MafB expression overlap in both alpha and beta cells in the adult human pancreas [145], which provides further evidence of inherent differences between human and mouse pancreas biology that will have to be fully understood and integrated into our current knowledge base regarding islet cell biology.

4.6 Organoids

A final approach considered here may entail the generation pancreas "organoids". Organoids are three-dimensional balls of cells grown in culture that recapitulate many aspects of normal in vivo development. The promise of organoids was demonstrated with the generation of human gut organoids (hGOs) in vitro [146]. In a recent study from the Wells group, hESCs or iPSCs were incubated with the BMP antagonist noggin, followed by retinoic acid, leading to the formation of three dimensional foregut spheroids, and subsequent posterior foregut specification. Upon further culture, the foregut spheroids stratified and differentiated into a glandular epithelium that closely recapitulated stomach organogenesis in vivo. After 34 days in culture, hGOs contained a characteristic stem cell niche. These data indicate that differentiation of endodermal tissue from ESCs/iPSCs can proceed without recapitulation of every developmental cue seen in the embryo, implying that the stomach stem cell niche is capable of self-assembly. Utilizing similar techniques to generate pancreatic organoids is an intriguing possibility that warrants inquiry.

5 Conclusions

To date, evidence for an incontrovertible adult pancreatic stem cell has been lacking. Lineage tracing experiments in rodent pancreata indicate that new beta cells do arise in the adult pancreas, in response to physiological changes such as pregnancy or obesity, or following injury. However, these new beta cells are derived predominantly via self-replication from existing beta cells, or possibly by way of transdifferentiation from other cell types. Hence, there is a frustrating paucity of evidence supporting the existence of a resident pancreatic stem cell in the adult rodent pancreas, and even less so for stem cells in adult human pancreata.

To date, the search for pancreatic stem cells in the adult has been merely tantalizing, but not altogether fruitful. We have learned a great deal from pancreatic progenitors in the embryo, however it appears that these cells retain their potential as stem cells only under the context of embryonic development. As a consequence, researchers have turned toward beta cell replacement as a potential treatment for diabetes. Beta cell replacement in the form of encapsulated pancreatic ES-derived beta cells appears to be a viable therapeutic option. Phase I clinical trials of encapsulated beta cell technology will begin in 2015 (Viacyte). Encapsulating and implanting a patient's own iPSC-derived endocrine cells may be the T1D treatment of the future. Importantly, the recognition of the stepwise process required to generate a functional beta cell has been instructed by our knowledge of the normal developmental events that occur in the embryonic pancreas. However, the process of guiding iPSCs toward an endocrine cell fate requires extensive cell culture and further improvements to recapitulate pancreatic endocrine cell differentiation. Much work remains ahead to fully elucidate and recapitulate the steps involved in establishing beta cell fate; however, as a field we are making slow but steady progress towards making a clinical impact.

Acknowledgments Our apologies to those whose work we were not able to cite due to space restrictions. This work was supported by JDRF 3-PDF-2014-202-A-N to CB, and NIH R01 grant DK079862 to OC.

References

1. Zorn AM, Wells JM. Vertebrate endoderm development and organ formation. Annu Rev Cell Dev Biol. 2009;25:221–51.
2. Michalopoulos GK, DeFrances MC. Liver regeneration. Science. 1997;276(5309):60–6.
3. Jorgensen MC, Ahnfelt-Ronne J, Hald J, Madsen OD, Serup P, Hecksher-Sorensen J. An illustrated review of early pancreas development in the mouse. Endocr Rev. 2007;28(6):685–705.
4. Bonner-Weir S. Perspective: Postnatal pancreatic beta cell growth. Endocrinology. 2000;141 (6):1926–9.
5. Stanger BZ, Tanaka AJ, Melton DA. Organ size is limited by the number of embryonic progenitor cells in the pancreas but not the liver. Nature. 2007;445(7130):886–91.
6. Cano DA, Soria B, Martin F, Rojas A. Transcriptional control of mammalian pancreas organogenesis. Cell Mol Life Sci: CMLS. 2014;71(13):2383–402.
7. Shih HP, Wang A, Sander M. Pancreas organogenesis: from lineage determination to morphogenesis. Annu Rev Cell Dev Biol. 2013;29:81–105.
8. Slack JM. Developmental biology of the pancreas. Development. 1995;121(6):1569–80.
9. Wells JM, Melton DA. Vertebrate endoderm development. Annu Rev Cell Dev Biol. 1999;15:393–410.
10. Kumar M, Jordan N, Melton D, Grapin-Botton A. Signals from lateral plate mesoderm instruct endoderm toward a pancreatic fate. Dev Biol. 2003;259(1):109–22.
11. Matsushita S, Ishii Y, Scotting PJ, Kuroiwa A, Yasugi S. Pre-gut endoderm of chick embryos is regionalized by 1.5 days of development. Dev Dyn.: an official publication of the American Association of Anatomists. 2002;223(1):33–47.
12. Miki R, Yoshida T, Murata K, Oki S, Kume K, Kume S. Fate maps of ventral and dorsal pancreatic progenitor cells in early somite stage mouse embryos. Mech Dev. 2012;128(11–12):597–609.
13. Villasenor A, Chong DC, Henkemeyer M, Cleaver O. Epithelial dynamics of pancreatic branching morphogenesis. Development. 2010;137(24):4295–305.
14. Li H, Arber S, Jessell TM, Edlund H. Selective agenesis of the dorsal pancreas in mice lacking homeobox gene Hlxb9. Nat Genet. 1999;23(1):67–70.

15. Pedersen JK, Nelson SB, Jorgensen MC, Henseleit KD, Fujitani Y, Wright CV, et al. Endodermal expression of Nkx6 genes depends differentially on Pdx1. Dev Biol. 2005;288 (2):487–501.
16. Chen Y, Pan FC, Brandes N, Afelik S, Solter M, Pieler T. Retinoic acid signaling is essential for pancreas development and promotes endocrine at the expense of exocrine cell differentiation in Xenopus. Dev Biol. 2004;271(1):144–60.
17. Martin M, Gallego-Llamas J, Ribes V, Kedinger M, Niederreither K, Chambon P, et al. Dorsal pancreas agenesis in retinoic acid-deficient Raldh2 mutant mice. Dev Biol. 2005;284 (2):399–411.
18. Hebrok M, Kim SK, Melton DA. Notochord repression of endodermal Sonic hedgehog permits pancreas development. Genes Dev. 1998;12(11):1705–13.
19. Kim SK, Hebrok M, Melton DA. Notochord to endoderm signaling is required for pancreas development. Development. 1997;124(21):4243–52.
20. Lammert E, Cleaver O, Melton D. Induction of pancreatic differentiation by signals from blood vessels. Science. 2001;294(5542):564–7.
21. Lammert E, Gu G, McLaughlin M, Brown D, Brekken R, Murtaugh LC, et al. Role of VEGF-A in vascularization of pancreatic islets. Curr Biol : CB. 2003;13(12):1070–4.
22. Duvillie B, Attali M, Bounacer A, Ravassard P, Basmaciogullari A, Scharfmann R. The mesenchyme controls the timing of pancreatic beta-cell differentiation. Diabetes. 2006;55 (3):582–9.
23. Kesavan G, Sand FW, Greiner TU, Johansson JK, Kobberup S, Wu X, et al. Cdc42-mediated tubulogenesis controls cell specification. Cell. 2009;139(4):791–801.
24. Landsman L, Nijagal A, Whitchurch TJ, Vanderlaan RL, Zimmer WE, Mackenzie TC, et al. Pancreatic mesenchyme regulates epithelial organogenesis throughout development. PLoS Biol. 2011;9(9):e1001143.
25. Zhou Q, Law AC, Rajagopal J, Anderson WJ, Gray PA, Melton DA. A multipotent progenitor domain guides pancreatic organogenesis. Dev Cell. 2007;13(1):103–14.
26. Harding MJ, McGraw HF, Nechiporuk A. The roles and regulation of multicellular rosette structures during morphogenesis. Development. 2014;141(13):2549–58.
27. Petzold KM, Naumann H, Spagnoli FM. Rho signalling restriction by the RhoGAP Stard13 integrates growth and morphogenesis in the pancreas. Development. 2013;140(1):126–35.
28. Pictet RL, Clark WR, Williams RH, Rutter WJ. An ultrastructural analysis of the developing embryonic pancreas. Dev Biol. 1972;29(4):436–67.
29. Rutter WJ, Kemp JD, Bradshaw WS, Clark WR, Ronzio RA, Sanders TG. Regulation of specific protein synthesis in cytodifferentiation. J Cell Physiol. 1968;72(2):Suppl 1–18.
30. Herrera PL. Adult insulin- and glucagon-producing cells differentiate from two independent cell lineages. Development. 2000;127(11):2317–22.
31. Guz Y, Montminy MR, Stein R, Leonard J, Gamer LW, Wright CV, et al. Expression of murine STF-1, a putative insulin gene transcription factor, in beta cells of pancreas, duodenal epithelium and pancreatic exocrine and endocrine progenitors during ontogeny. Development. 1995;121(1):11–8.
32. Offield MF, Jetton TL, Labosky PA, Ray M, Stein RW, Magnuson MA, et al. PDX-1 is required for pancreatic outgrowth and differentiation of the rostral duodenum. Development. 1996;122(3):983–95.
33. Stoffers DA, Thomas MK, Habener JF. Homeodomain protein IDX-1: a master regulator of pancreas development and insulin gene expression. Trends Endocrinol Metab. 1997;8 (4):145–51.
34. Ohlsson H, Karlsson K, Edlund T. IPF1, a homeodomain-containing transactivator of the insulin gene. EMBO J. 1993;12(11):4251–9.
35. Jonsson J, Carlsson L, Edlund T, Edlund H. Insulin-promoter-factor 1 is required for pancreas development in mice. Nature. 1994;371(6498):606–9.
36. Hale MA, Kagami H, Shi L, Holland AM, Elsasser HP, Hammer RE, et al. The homeodomain protein PDX1 is required at mid-pancreatic development for the formation of the exocrine pancreas. Dev Biol. 2005;286(1):225–37.

37. Afelik S, Chen Y, Pieler T. Combined ectopic expression of Pdx1 and Ptf1a/p48 results in the stable conversion of posterior endoderm into endocrine and exocrine pancreatic tissue. Genes Dev. 2006;20(11):1441–6.
38. Horb ME, Shen CN, Tosh D, Slack JM. Experimental conversion of liver to pancreas. Curr Biol. 2003;13(2):105–15.
39. Stoffers DA, Zinkin NT, Stanojevic V, Clarke WL, Habener JF. Pancreatic agenesis attributable to a single nucleotide deletion in the human IPF1 gene coding sequence. Nat Genet. 1997;15(1):106–10.
40. Gu G, Dubauskaite J, Melton DA. Direct evidence for the pancreatic lineage: NGN3 + cells are islet progenitors and are distinct from duct progenitors. Development. 2002;129 (10):2447–57.
41. Villasenor A, Chong DC, Cleaver O. Biphasic Ngn3 expression in the developing pancreas. Dev Dyn : an official publication of the American Association of Anatomists. 2008;237 (11):3270–9.
42. Gradwohl G, Dierich A, LeMeur M, Guillemot F. neurogenin3 is required for the development of the four endocrine cell lineages of the pancreas. Proc Natl Acad Sci USA. 2000;97(4):1607–11.
43. Magenheim J, Klein AM, Stanger BZ, Ashery-Padan R, Sosa-Pineda B, Gu G, et al. Ngn3(+) endocrine progenitor cells control the fate and morphogenesis of pancreatic ductal epithelium. Dev Biol. 2011;359(1):26–36.
44. Wang S, Yan J, Anderson DA, Xu Y, Kanal MC, Cao Z, et al. Neurog3 gene dosage regulates allocation of endocrine and exocrine cell fates in the developing mouse pancreas. Dev Biol. 2010;339(1):26–37.
45. Johansson KA, Dursun U, Jordan N, Gu G, Beermann F, Gradwohl G, et al. Temporal control of neurogenin3 activity in pancreas progenitors reveals competence windows for the generation of different endocrine cell types. Dev Cell. 2007;12(3):457–65.
46. Esni F, Ghosh B, Biankin AV, Lin JW, Albert MA, Yu X, et al. Notch inhibits Ptf1 function and acinar cell differentiation in developing mouse and zebrafish pancreas. Development. 2004;131(17):4213–24.
47. Kopinke D, Brailsford M, Shea JE, Leavitt R, Scaife CL, Murtaugh LC. Lineage tracing reveals the dynamic contribution of Hes1 + cells to the developing and adult pancreas. Development. 2011;138(3):431–41.
48. Cleveland MH, Sawyer JM, Afelik S, Jensen J, Leach SD. Exocrine ontogenies: on the development of pancreatic acinar, ductal and centroacinar cells. Semin Cell Dev Biol. 2012;23(6):711–9.
49. Rovira M, Scott SG, Liss AS, Jensen J, Thayer SP, Leach SD. Isolation and characterization of centroacinar/terminal ductal progenitor cells in adult mouse pancreas. Proc Natl Acad Sci USA. 2010;107(1):75–80.
50. Sarkar A, Hochedlinger K. The sox family of transcription factors: versatile regulators of stem and progenitor cell fate. Cell Stem Cell. 2013;12(1):15–30.
51. Suh H, Consiglio A, Ray J, Sawai T, D'Amour KA, Gage FH. In vivo fate analysis reveals the multipotent and self-renewal capacities of Sox2 + neural stem cells in the adult hippocampus. Cell Stem Cell. 2007;1(5):515–28.
52. Kopp JL, Dubois CL, Schaffer AE, Hao E, Shih HP, Seymour PA, et al. Sox9 + ductal cells are multipotent progenitors throughout development but do not produce new endocrine cells in the normal or injured adult pancreas. Development. 2011;138(4):653–65.
53. Furuyama K, Kawaguchi Y, Akiyama H, Horiguchi M, Kodama S, Kuhara T, et al. Continuous cell supply from a Sox9-expressing progenitor zone in adult liver, exocrine pancreas and intestine. Nat Genet. 2011;43(1):34–41.
54. Haumaitre C, Barbacci E, Jenny M, Ott MO, Gradwohl G, Cereghini S. Lack of TCF2/vHNF1 in mice leads to pancreas agenesis. Proc Natl Acad Sci USA. 2005;102 (5):1490–5.

55. Maestro MA, Boj SF, Luco RF, Pierreux CE, Cabedo J, Servitja JM, et al. Hnf6 and Tcf2 (MODY5) are linked in a gene network operating in a precursor cell domain of the embryonic pancreas. Hum Mol Genet. 2003;12(24):3307–14.
56. Solar M, Cardalda C, Houbracken I, Martin M, Maestro MA, De Medts N, et al. Pancreatic exocrine duct cells give rise to insulin-producing beta cells during embryogenesis but not after birth. Dev Cell. 2009;17(6):849–60.
57. Masui T, Long Q, Beres TM, Magnuson MA, MacDonald RJ. Early pancreatic development requires the vertebrate suppressor of hairless (RBPJ) in the PTF1 bHLH complex. Genes Dev. 2007;21(20):2629–43.
58. Kawaguchi Y, Cooper B, Gannon M, Ray M, MacDonald RJ, Wright CV. The role of the transcriptional regulator Ptf1a in converting intestinal to pancreatic progenitors. Nat Genet. 2002;32(1):128–34.
59. Krapp A, Knofler M, Frutiger S, Hughes GJ, Hagenbuchle O, Wellauer PK. The p48 DNA-binding subunit of transcription factor PTF1 is a new exocrine pancreas-specific basic helix-loop-helix protein. EMBO J. 1996;15(16):4317–29.
60. Pan FC, Bankaitis ED, Boyer D, Xu X, Van de Casteele M, Magnuson MA, et al. Spatiotemporal patterns of multipotentiality in Ptf1a-expressing cells during pancreas organogenesis and injury-induced facultative restoration. Development. 2013;140(4):751–64.
61. Inada A, Nienaber C, Katsuta H, Fujitani Y, Levine J, Morita R, et al. Carbonic anhydrase II-positive pancreatic cells are progenitors for both endocrine and exocrine pancreas after birth. Proc Natl Acad Sci USA. 2008;105(50):19915–9.
62. Ziv O, Glaser B, Dor Y. The plastic pancreas. Dev Cell. 2013;26(1):3–7.
63. Yanger K, Stanger BZ. Facultative stem cells in liver and pancreas: fact and fancy. Dev Dyn: an official publication of the American Association of Anatomists. 2011;240(3):521–9.
64. Bonner-Weir S, Li WC, Ouziel-Yahalom L, Guo L, Weir GC, Sharma A. Beta-cell growth and regeneration: replication is only part of the story. Diabetes. 2010;59(10):2340–8.
65. Bouwens L, Rooman I. Regulation of pancreatic beta-cell mass. Physiol Rev. 2005;85(4):1255–70.
66. Kim H, Toyofuku Y, Lynn FC, Chak E, Uchida T, Mizukami H, et al. Serotonin regulates pancreatic beta cell mass during pregnancy. Nat Med. 2010;16(7):804–8.
67. Toselli C, Hyslop CM, Hughes M, Natale DR, Santamaria P, Huang CT. Contribution of a non-beta-cell source to beta-cell mass during pregnancy. PLoS ONE. 2014;9(6):e100398.
68. Rieck S, White P, Schug J, Fox AJ, Smirnova O, Gao N, et al. The transcriptional response of the islet to pregnancy in mice. Mol Endocrinol. 2009;23(10):1702–12.
69. Butler AE, Cao-Minh L, Galasso R, Rizza RA, Corradin A, Cobelli C, et al. Adaptive changes in pancreatic beta cell fractional area and beta cell turnover in human pregnancy. Diabetologia. 2010;53(10):2167–76.
70. Ferrannini E, Nannipieri M, Williams K, Gonzales C, Haffner SM, Stern MP. Mode of onset of type 2 diabetes from normal or impaired glucose tolerance. Diabetes. 2004;53(1):160–5.
71. Mezza T, Muscogiuri G, Sorice GP, Clemente G, Hu J, Pontecorvi A, et al. Insulin resistance alters islet morphology in nondiabetic humans. Diabetes. 2014;63(3):994–1007.
72. Dor Y, Brown J, Martinez OI, Melton DA. Adult pancreatic beta-cells are formed by self-duplication rather than stem-cell differentiation. Nature. 2004;429(6987):41–6.
73. Teta M, Rankin MM, Long SY, Stein GM, Kushner JA. Growth and regeneration of adult beta cells does not involve specialized progenitors. Dev Cell. 2007;12(5):817–26.
74. Teta M, Long SY, Wartschow LM, Rankin MM, Kushner JA. Very slow turnover of beta-cells in aged adult mice. Diabetes. 2005;54(9):2557–67.
75. Perl S, Kushner JA, Buchholz BA, Meeker AK, Stein GM, Hsieh M, et al. Significant human beta-cell turnover is limited to the first three decades of life as determined by in vivo thymidine analog incorporation and radiocarbon dating. J Clin Endocrinol Metab. 2010;95(10):E234–9.
76. Yi P, Park JS, Melton DA. Betatrophin: a hormone that controls pancreatic beta cell proliferation. Cell. 2013;153(4):747–58.

77. Gusarova V, Alexa Corey A, Na E, Stevis Panayiotis E, Xin Y, Bonner-Weir S, et al. ANGPTL8/Betatrophin does not control pancreatic beta cell expansion. Cell. 159(3):691–6.
78. Yi P, Park J-S, Melton DA. Perspectives on the activities of ANGPTL8/Betatrophin. Cell. 159(3):467–8.
79. Stewart AF. Betatrophin versus bitter-trophin and the elephant in the room: time for a new normal in beta-cell regeneration research. Diabetes. 2014;63(4):1198–9.
80. Lane SW, Williams DA, Watt FM. Modulating the stem cell niche for tissue regeneration. Nat Biotechnol. 2014;32(8):795–803.
81. Akinci E, Banga A, Greder LV, Dutton JR, Slack JM. Reprogramming of pancreatic exocrine cells towards a beta (beta) cell character using Pdx1, Ngn3 and MafA. Biochem J. 2012;442 (3):539–50.
82. Al-Adsani A, Burke ZD, Eberhard D, Lawrence KL, Shen CN, Rustgi AK, et al. Dexamethasone treatment induces the reprogramming of pancreatic acinar cells to hepatocytes and ductal cells. PLoS ONE. 2010;5(10):e13650.
83. Zhou Q, Brown J, Kanarek A, Rajagopal J, Melton DA. In vivo reprogramming of adult pancreatic exocrine cells to beta-cells. Nature. 2008;455(7213):627–32.
84. Lu J, Jaafer R, Bonnavion R, Bertolino P, Zhang CX. Transdifferentiation of pancreatic alpha-cells into insulin-secreting cells: From experimental models to underlying mechanisms. World J Diabetes. 2014;5(6):847–53.
85. Thorel F, Nepote V, Avril I, Kohno K, Desgraz R, Chera S, et al. Conversion of adult pancreatic alpha-cells to beta-cells after extreme beta-cell loss. Nature. 2010;464 (7292):1149–54.
86. Yechoor V, Liu V, Espiritu C, Paul A, Oka K, Kojima H, et al. Neurogenin3 is sufficient for transdetermination of hepatic progenitor cells into neo-islets in vivo but not transdifferentiation of hepatocytes. Dev Cell. 2009;16(3):358–73.
87. Yi F, Liu GH. Izpisua Belmonte JC. Rejuvenating liver and pancreas through cell transdifferentiation. Cell Res. 2012;22(4):616–9.
88. Lipsett M, Finegood DT. beta-cell neogenesis during prolonged hyperglycemia in rats. Diabetes. 2002;51(6):1834–41.
89. Baeyens L, Lemper M, Leuckx G, De Groef S, Bonfanti P, Stange G, et al. Transient cytokine treatment induces acinar cell reprogramming and regenerates functional beta cell mass in diabetic mice. Nat Biotechnol. 2014;32(1):76–83.
90. Ber I, Shternhall K, Perl S, Ohanuna Z, Goldberg I, Barshack I, et al. Functional, persistent, and extended liver to pancreas transdifferentiation. J Biol Chem. 2003;278(34):31950–7.
91. Kaneto H, Nakatani Y, Miyatsuka T, Matsuoka TA, Matsuhisa M, Hori M, et al. PDX-1/VP16 fusion protein, together with NeuroD or Ngn3, markedly induces insulin gene transcription and ameliorates glucose tolerance. Diabetes. 2005;54(4):1009–22.
92. Meivar-Levy I, Aviv V, Ferber S. Adult cell reprogramming: Using nonpancreatic cell sources to generate surrogate beta cells for treatment of diabetes. In: Efrat S, editor. Stem cell therapy for diabetes. Stem cell biology and regenerative medicine. Totowa: Humana Press; 2010. p. 183–202.
93. Wang AY, Ehrhardt A, Xu H, Kay MA. Adenovirus transduction is required for the correction of diabetes using Pdx-1 or Neurogenin-3 in the liver. Mol Therapy : the journal of the American Society of Gene Therapy. 2007;15(2):255–63.
94. Ferber S, Halkin A, Cohen H, Ber I, Einav Y, Goldberg I, et al. Pancreatic and duodenal homeobox gene 1 induces expression of insulin genes in liver and ameliorates streptozotocin-induced hyperglycemia. Nat Med. 2000;6(5):568–72.
95. Meivar-Levy I, Sapir T, Gefen-Halevi S, Aviv V, Barshack I, Onaca N, et al. Pancreatic and duodenal homeobox gene 1 induces hepatic dedifferentiation by suppressing the expression of CCAAT/enhancer-binding protein beta. Hepatology. 2007;46(3):898–905.
96. Shternhall-Ron K, Quintana FJ, Perl S, Meivar-Levy I, Barshack I, Cohen IR, et al. Ectopic PDX-1 expression in liver ameliorates type 1 diabetes. J Autoimmun. 2007;28(2–3):134–42.

97. Nealon WH, Thompson JC. Progressive loss of pancreatic function in chronic pancreatitis is delayed by main pancreatic duct decompression. A longitudinal prospective analysis of the modified puestow procedure. Ann Surg. 1993;217(5):458–66; discussion 66–8.
98. Wang RN, Kloppel G, Bouwens L. Duct- to islet-cell differentiation and islet growth in the pancreas of duct-ligated adult rats. Diabetologia. 1995;38(12):1405–11.
99. Xu X, D'Hoker J, Stange G, Bonne S, De Leu N, Xiao X, et al. Beta cells can be generated from endogenous progenitors in injured adult mouse pancreas. Cell. 2008;132(2):197–207.
100. Rankin MM, Wilbur CJ, Rak K, Shields EJ, Granger A, Kushner JA. beta-Cells are not generated in pancreatic duct ligation-induced injury in adult mice. Diabetes. 2013;62 (5):1634–45.
101. Xiao X, Chen Z, Shiota C, Prasadan K, Guo P, El-Gohary Y, et al. No evidence for beta cell neogenesis in murine adult pancreas. J Clin Investig. 2013;123(5):2207–17.
102. Bonner-Weir S, Baxter LA, Schuppin GT, Smith FE. A second pathway for regeneration of adult exocrine and endocrine pancreas. A possible recapitulation of embryonic development. Diabetes. 1993;42(12):1715–20.
103. Menge BA, Tannapfel A, Belyaev O, Drescher R, Muller C, Uhl W, et al. Partial pancreatectomy in adult humans does not provoke beta-cell regeneration. Diabetes. 2008;57 (1):142–9.
104. Aathira R, Jain V. Advances in management of type 1 diabetes mellitus. World journal of diabetes. 2014;5(5):689–96.
105. Gillespie KM. Type 1 diabetes: pathogenesis and prevention. CMAJ : Can Med Assoc J = journal de l'Association medicale canadienne. 2006;175(2):165–70.
106. Ryan EA, Paty BW, Senior PA, Bigam D, Alfadhli E, Kneteman NM, et al. Five-year follow-up after clinical islet transplantation. Diabetes. 2005;54(7):2060–9.
107. Shapiro AM, Lakey JR, Ryan EA, Korbutt GS, Toth E, Warnock GL, et al. Islet transplantation in seven patients with type 1 diabetes mellitus using a glucocorticoid-free immunosuppressive regimen. New England J Med. 2000;343(4):230–8.
108. Barton FB, Rickels MR, Alejandro R, Hering BJ, Wease S, Naziruddin B, et al. Improvement in outcomes of clinical islet transplantation: 1999-2010. Diabetes Care. 2012;35(7):1436–45.
109. Drukker M. Immunological considerations for cell therapy using human embryonic stem cell derivatives. StemBook. Cambridge (MA) 2008.
110. Ianus A, Holz GG, Theise ND, Hussain MA. In vivo derivation of glucose-competent pancreatic endocrine cells from bone marrow without evidence of cell fusion. J Clin Investig. 2003;111(6):843–50.
111. Lechner A, Yang YG, Blacken RA, Wang L, Nolan AL, Habener JF. No evidence for significant transdifferentiation of bone marrow into pancreatic beta-cells in vivo. Diabetes. 2004;53(3):616–23.
112. Tang DQ, Cao LZ, Burkhardt BR, Xia CQ, Litherland SA, Atkinson MA, et al. In vivo and in vitro characterization of insulin-producing cells obtained from murine bone marrow. Diabetes. 2004;53(7):1721–32.
113. Karnieli O, Izhar-Prato Y, Bulvik S, Efrat S. Generation of insulin-producing cells from human bone marrow mesenchymal stem cells by genetic manipulation. Stem Cells. 2007;25 (11):2837–44.
114. Li Y, Zhang R, Qiao H, Zhang H, Wang Y, Yuan H, et al. Generation of insulin-producing cells from PDX-1 gene-modified human mesenchymal stem cells. J Cell Physiol. 2007;211 (1):36–44.
115. Choi JB, Uchino H, Azuma K, Iwashita N, Tanaka Y, Mochizuki H, et al. Little evidence of transdifferentiation of bone marrow-derived cells into pancreatic beta cells. Diabetologia. 2003;46(10):1366–74.
116. Kojima H, Fujimiya M, Matsumura K, Nakahara T, Hara M, Chan L. Extrapancreatic insulin-producing cells in multiple organs in diabetes. Proc Natl Acad Sci USA. 2004;101 (8):2458–63.

117. Thomson JA, Itskovitz-Eldor J, Shapiro SS, Waknitz MA, Swiergiel JJ, Marshall VS, et al. Embryonic stem cell lines derived from human blastocysts. Science. 1998;282(5391):1145–7.
118. Schiesser JV, Wells JM. Generation of beta cells from human pluripotent stem cells: are we there yet? Ann NY Acad Sci. 2014;1311:124–37.
119. D'Amour KA, Bang AG, Eliazer S, Kelly OG, Agulnick AD, Smart NG, et al. Production of pancreatic hormone-expressing endocrine cells from human embryonic stem cells. Nat Biotechnol. 2006;24(11):1392–401.
120. Kelly OG, Chan MY, Martinson LA, Kadoya K, Ostertag TM, Ross KG, et al. Cell-surface markers for the isolation of pancreatic cell types derived from human embryonic stem cells. Nat Biotechnol. 2011;29(8):750–6.
121. Kroon E, Martinson LA, Kadoya K, Bang AG, Kelly OG, Eliazer S, et al. Pancreatic endoderm derived from human embryonic stem cells generates glucose-responsive insulin-secreting cells in vivo. Nat Biotechnol. 2008;26(4):443–52.
122. Motte E, Szepessy E, Suenens K, Stange G, Bomans M, Jacobs-Tulleneers-Thevissen D, et al. Composition and function of macroencapsulated human embryonic stem cell-derived implants: comparison with clinical human islet cell grafts. Am J Physiol Endocrinol Metab. 2014;307(9):E838–46.
123. Rezania A, Bruin JE, Arora P, Rubin A, Batushansky I, Asadi A, et al. Reversal of diabetes with insulin-producing cells derived in vitro from human pluripotent stem cells. Nat Biotechnol. 2014;32(11):1121–33.
124. Szot GL, Yadav M, Lang J, Kroon E, Kerr J, Kadoya K, et al. Tolerance induction and reversal of diabetes in mice transplanted with human embryonic stem cell-derived pancreatic endoderm. Cell Stem Cell. 2015;16(2):148–57.
125. Bouwens L, Houbracken I, Mfopou JK. The use of stem cells for pancreatic regeneration in diabetes mellitus. Nat Rev Endocrinol. 2013;9(10):598–606.
126. Jensen J. Gene regulatory factors in pancreatic development. Dev Dyn: an official publication of the American Association of Anatomists. 2004;229(1):176–200.
127. Soria B, Roche E, Berna G, Leon-Quinto T, Reig JA, Martin F. Insulin-secreting cells derived from embryonic stem cells normalize glycemia in streptozotocin-induced diabetic mice. Diabetes. 2000;49(2):157–62.
128. Abdelalim EM, Emara MM. Advances and challenges in the differentiation of pluripotent stem cells into pancreatic beta cells. World J Stem Cell. 2015;7(1):174–81.
129. Nostro MC, Sarangi F, Ogawa S, Holtzinger A, Corneo B, Li X, et al. Stage-specific signaling through TGFbeta family members and WNT regulates patterning and pancreatic specification of human pluripotent stem cells. Development. 2011;138(5):861–71.
130. Pagliuca FW, Millman JR, Gurtler M, Segel M, Van Dervort A, Ryu JH, et al. Generation of functional human pancreatic beta cells in vitro. Cell. 2014;159(2):428–39.
131. Rezania A, Bruin JE, Xu J, Narayan K, Fox JK, O'Neil JJ, et al. Enrichment of human embryonic stem cell-derived NKX6.1-expressing pancreatic progenitor cells accelerates the maturation of insulin-secreting cells in vivo. Stem Cells. 2013;31(11):2432–42.
132. Rezania A, Bruin JE, Riedel MJ, Mojibian M, Asadi A, Xu J, et al. Maturation of human embryonic stem cell-derived pancreatic progenitors into functional islets capable of treating pre-existing diabetes in mice. Diabetes. 2012;61(8):2016–29.
133. Xie R, Everett LJ, Lim HW, Patel NA, Schug J, Kroon E, et al. Dynamic chromatin remodeling mediated by polycomb proteins orchestrates pancreatic differentiation of human embryonic stem cells. Cell Stem Cell. 2013;12(2):224–37.
134. Schulz TC, Young HY, Agulnick AD, Babin MJ, Baetge EE, Bang AG, et al. A scalable system for production of functional pancreatic progenitors from human embryonic stem cells. PLoS ONE. 2012;7(5):e37004.
135. Stepniewski J, Kachamakova-Trojanowska N, Ogrocki D, Szopa M, Matlok M, Beilharz M, et al. Induced pluripotent stem cells as a model for diabetes investigation. Sci Rep. 2015;5:8597.

136. Cheng X, Ying L, Lu L, Galvao AM, Mills JA, Lin HC, et al. Self-renewing endodermal progenitor lines generated from human pluripotent stem cells. Cell Stem Cell. 2012;10 (4):371–84.
137. Hrvatin S, O'Donnell CW, Deng F, Millman JR, Pagliuca FW, DiIorio P, et al. Differentiated human stem cells resemble fetal, not adult, beta cells. Proc Natl Acad Sci USA. 2014;111 (8):3038–43.
138. Narayanan K, Lim VY, Shen J, Tan ZW, Rajendran D, Luo SC, et al. Extracellular matrix-mediated differentiation of human embryonic stem cells: differentiation to insulin-secreting beta cells. Tissue Eng Part A. 2014;20(1–2):424–33.
139. Takahashi K, Tanabe K, Ohnuki M, Narita M, Ichisaka T, Tomoda K, et al. Induction of pluripotent stem cells from adult human fibroblasts by defined factors. Cell. 2007;131 (5):861–72.
140. Modlich U, Navarro S, Zychlinski D, Maetzig T, Knoess S, Brugman MH, et al. Insertional transformation of hematopoietic cells by self-inactivating lentiviral and gammaretroviral vectors. Mol Ther: the journal of the American Society of Gene Therapy. 2009;17(11):1919–28.
141. Kudva YC, Ohmine S, Greder LV, Dutton JR, Armstrong A, De Lamo JG, et al. Transgene-free disease-specific induced pluripotent stem cells from patients with type 1 and type 2 diabetes. Stem Cell Trans Med. 2012;1(6):451–61.
142. Artner I, Hang Y, Mazur M, Yamamoto T, Guo M, Lindner J, et al. MafA and MafB regulate genes critical to beta-cells in a unique temporal manner. Diabetes. 2010;59(10):2530–9.
143. Zhang C, Moriguchi T, Kajihara M, Esaki R, Harada A, Shimohata H, et al. MafA is a key regulator of glucose-stimulated insulin secretion. Mol Cell Biol. 2005;25(12):4969–76.
144. Stolovich-Rain M, Enk J, Vikesa J, Nielsen FC, Saada A, Glaser B, et al. Weaning triggers a maturation step of pancreatic beta cells. Dev Cell. 2015;32(5):535–45.
145. Bonnavion R, Jaafar R, Kerr-Conte J, Assade F, van Stralen E, Leteurtre E, et al. Both PAX4 and MAFA are expressed in a substantial proportion of normal human pancreatic alpha cells and deregulated in patients with type 2 diabetes. PLoS ONE. 2013;8(8):e72194.
146. McCracken KW, Cata EM, Crawford CM, Sinagoga KL, Schumacher M, Rockich BE, et al. Modelling human development and disease in pluripotent stem-cell-derived gastric organoids. Nature. 2014;516(7531):400–4.
147. Chung CH, Hao E, Piran R, Keinan E, Levine F. Pancreatic beta-cell neogenesis by direct conversion from mature alpha-cells. Stem Cells. 2010;28(9):1630–8.
148. Chera S, Baronnier D, Ghila L, Cigliola V, Jensen JN, Gu G, et al. Diabetes recovery by age-dependent conversion of pancreatic delta-cells into insulin producers. Nature. 2014.
149. Aguayo-Mazzucato C, Zavacki AM, Marinelarena A, Hollister-Lock J, El Khattabi I, Marsili A, et al. Thyroid hormone promotes postnatal rat pancreatic beta-cell development and glucose-responsive insulin secretion through MAFA. Diabetes. 2013;62(5):1569–80.
150. Rezania A, Riedel MJ, Wideman RD, Karanu F, Ao Z, Warnock GL, et al. Production of functional glucagon-secreting alpha-cells from human embryonic stem cells. Diabetes. 2011;60(1):239–47.
151. Thowfeequ S, Ralphs KL, Yu WY, Slack JM, Tosh D. Betacellulin inhibits amylase and glucagon production and promotes beta cell differentiation in mouse embryonic pancreas. Diabetologia. 2007;50(8):1688–97.

The Intestinal Stem Cell Niche

Celina Yuemin Chee, David M. Virshup and Babita Madan

1 Intestinal Anatomy and Physiology

The mammalian small intestine and colon comprise the parts of the gastrointestinal system that perform the crucial function of nutrient absorption from food products digested in the stomach and also form an effective barrier against xenobiotics and microbes present in the gut lumen. In accord with the important function of the intestine for survival and the mechanical and chemical stressors that the intestinal epithelium is exposed to, the intestine is also one of the most rapidly proliferating tissues in the body. The entire small intestinal epithelium is capable of turning over in 3–5 days [1–3], compared to other tissues such as the skin with a turnover time of 40–56 days [4], or heart with a turnover time in years [5, 6].

The luminal surface of the small intestine is made up of a layer of simple columnar epithelium organized into multiple finger-like projections called villi, which serve to increase its absorptive surface area. Between these villi lie the crypts of Lieberkühn, invaginations of the epithelial surface containing multipotent adult stem cells that maintain the proliferation and homeostasis of the small intestine. As these stem cells divide, they produce committed progenitors, known as transit amplifying cells, that continue to rapidly divide and move upward from the crypts to the villi where they differentiate into three major cell types: (i) enterocytes, the main absorptive cell type that comprise 80 % of the intestinal epithelium, (ii) enteroendocrine cells that produce hormones controlling intestinal function and metabolic homeostasis, and (iii) mucous-producing goblet cells that aid in the transport of material through the gut lumen [7]. The only cell type that does not undergo this upward migration is the Paneth cell, which remains at the base of the crypts within the intestinal stem cell niche, producing anti-microbial substances

C.Y. Chee · D.M. Virshup (✉) · B. Madan
Program in Cancer and Stem Cell Biology, Duke-NUS Graduate Medical School,
8 College Road, Singapore 169857, Singapore
e-mail: David.virshup@duke-nus.edu.sg

K. Turksen (ed.), *Tissue-Specific Stem Cell Niche*,
Stem Cell Biology and Regenerative Medicine,
DOI 10.1007/978-3-319-21705-5_7

such as cryptdins, lysozyme and a multitude of signaling molecules (Fig. 1a). Unlike the other differentiated cell types, the Paneth cells do not cycle rapidly and can remain in the small intestinal crypts for 3–6 weeks. When the cells at the tip of the villus die, they are shed into the intestinal lumen and removed from the body [8]. The colon is similar to the small intestine except that it does not possess villi but has a flat epithelial surface and serves primarily to absorb water from the contents of the lumen. Furthermore, Paneth cells are only found in the ascending colon and it has thus been proposed that c-kit positive secretory cells may serve the same function in rest of the colon [9]. Immediately underlying the basement membrane of the epithelial layer is the lamina propria, which is a layer of connective tissue containing fibroblasts, myofibroblasts, nerves, blood vessels and lymphatic vessels. Besides providing structural support to the intestine, it also

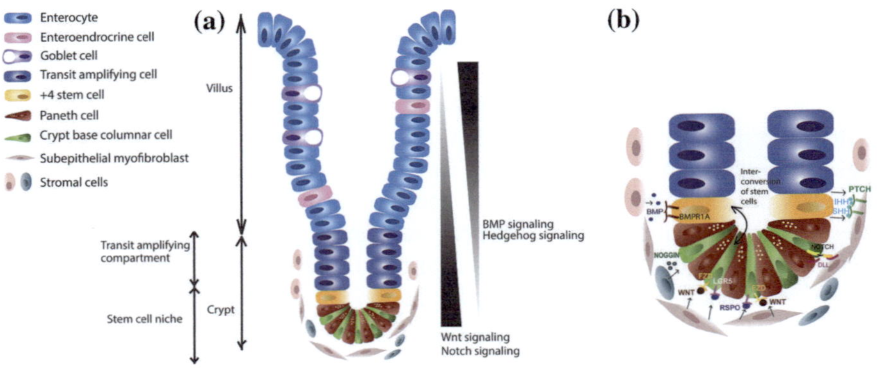

Fig. 1 **a** Two different intestinal stem cell populations exist at the base of the crypts of Lieberkühn, the crypt base columnar cells and +4 position stem cells. These stem cells are surrounded by the intestinal stem cell niche, which is made up of Paneth cells interspersed between the crypt base columnar cells, as well as the surrounding stromal cells such as subepithelial myofibroblasts. This intestinal stem cell niche provides the correct signaling milieu in vivo for the controlled proliferation and differentiation of the stem cells. Wnt signaling is required for stem cell proliferation, while Notch signaling inhibits differentiation to the secretory lineage and are thus highest at the base of the crypt. Wnt signaling is also inhibited by BMP signaling which exists in an increasing gradient along the crypt-villus axis. Hedgehog signaling has diverse effects including increased BMP signaling, inhibited Wnt signaling and restricting niche-specific subepithelial myofibroblasts to the base of the crypt. **b** The stromal component of the stem cell niche plays a crucial role in most of the signaling pathways regulating the niche. Subepithelial myofibroblasts are sources of Wnt and R-spondins (Wnt agonists), as well as BMP antagonists such as gremlim1/2 and chordin-like 1. Their corresponding Wnt receptors, FZD and LGR5 can be found on the adjacent epithelial crypt base cells while BMP receptors are found at and above the +4 position of the epithelium. Other stromal cells also contribute by secreting Noggin (BMP antagonist) at the crypt base and BMP in an increasing gradient from the crypt to the villus. Similarly, Hedgehog is secreted by epithelial cells while its receptor, Patched, is found in the mesenchyme. This signaling is important to the localization of the subepithelial myofibroblasts and formation of the crypts. Notch-1 and Notch-2 receptors are found on both +4 stem cells as well as the crypt base columnar stem cells. Despite the fact that the two stem cell populations have different characteristics and expression profiles, significant plasticity exists within the small intestine, allowing for the inter-conversion of these populations

supplies blood to the epithelium and transports away absorbed nutrients from the intestinal lumen. We describe the self-renewal of the murine small intestine and the molecular pathways that regulate the differentiation of the stem cells into various epithelial cell types.

2 Intestinal Stem Cells

The renewal of each individual crypt is driven by a very small number of tissue stem cells. The first evidence for clonality of the human intestinal and colonic crypts and the stem cell derivation of all the epithelial lineages came from two studies. The first utilized in situ hybridization of a Y chromosome-specific probe on a XO/XY individual with familial adenomatous polyposis to demonstrate that each individual intestinal crypt possesses either XO or XY cells but not both. However, the villus epithelium comprises of a mixture of XO and XY cells [10], indicating that the villi derive from the stem cells of more than one crypt. The second study looked at female subjects heterozygous for a mutation on the X-linked gene, glucose-6-phosphate dehydrogenase (G6PD). Due to X-inactivation, these individuals are functionally mosaic and thus have a mixture of cells with low or high G6PD activity in the intestinal epithelium. Both the colonic and intestinal crypts however have only one phenotype, with no evidence of mixing [11]. More recently, a study utilizing *R26R-Confetti;Lgr5-EGFP-CreERT2* mice shows that competition between stem cells at the crypt base for access to the stem cell niche maintains the stem cell population. Fate mapping studies using these mice demonstrate that crypts drift towards clonality within a period of 1–6 months. Stem cells further from the boundary of the niche experience a survival advantage and hence are more likely to colonize the crypt, consistent with the above observations [12, 13].

The key characteristics of stem cells are their capability to proliferate indefinitely to produce more stem cells, and their ability to differentiate into different cell types. Another characteristic of stem cells is quiescence that protects them from external chemical or physical stressors, as well as modifications of the genome from replicative errors or aging [14].

Two models have been proposed to explain the location and identity of small intestinal stem cells within the crypt, (i) the +4 position model and (ii) the stem cell zone model. The +4 model is based on the existence of a unique population of highly radiosensitive cells located 4 cells from the base of the crypt. This radiosensitivity has been theorized to be a safety mechanism to prevent stem cells from transmitting any DNA damage to their progeny [15]. These cells may also be highly sensitive to tamoxifen, which complicates interpretation of lineage tracing studies, as discussed later [16]. The stem cell zone model on the other hand argues that stem cells exist at the base of the crypt between positions 1–4 [17]. In vivo studies using lineage tracing and genetically engineered mice, as well as ex vivo studies using organoid assays (described below) have facilitated the identification of various markers for both of these populations.

For a rapidly proliferating tissue like the intestine, the adult stem cells have been theorized to be in two compartments, with one set remaining in a quiescent state as "back-up" while the other proliferates to maintain tissue homeostasis. Upon injury or degeneration of the active population, the quiescent population adapts its phenotype to replenish or support the proliferating cells [18, 19]. Evidence suggests that the +4 population of cells represents this quiescent stem cell population, while Lgr5, a cell marker described in greater detail below, marks the rapidly proliferating compartment.

2.1 *In Vivo Assays for Identifying Intestinal Stem Cells*

One of the most useful methods of identifying stem cell populations is to find a unique set of markers that can reliably distinguish them from other cell types. These markers facilitate the visualization, targeting and manipulation of stem cells and also allow sorting of these stem cells into pure populations for ex vivo analysis. One such analytical technique is lineage tracing whereby a label expressed in a single cell is transmitted to all of its progeny, allowing that single cell's fate and behavior to be studied without affecting the normal function of the cell or its neighbors [20]. Alternatively, label-retaining experiments use a label that is incorporated into the nuclear material or cell membrane, and is lost with successive cell divisions, allowing the cycling rate of specific cell types to be quantified.

Multiple strains of genetically engineered mice have been generated to test the effects of specific genes on intestinal homeostasis and function in vivo. These studies have largely benefited from the use of Cre recombinase, which allows for precise gene regulation by site-specific recombination between *loxP* recognition sites [21]. Moreover, use of tissue specific promoters or inducible Cre also permits spatio-temporal regulation of gene expression [22, 23].

2.2 *Ex Vivo Assays for Identifying Intestinal Stem Cells*

Recent improvements in in vitro culturing techniques of the intestinal epithelium have also contributed to our understanding of the intestinal stem cells and its niche. Short-term primary culture of the intestine has been carried out for many years [24, 25]. However, while these in vitro cultures were useful for imaging and experimentation, they could not truly recapitulate the in vivo physiology, behavior or proliferative potential of the intestinal stem cells. Two techniques have been developed to overcome this limitation. In the first, intestinal epithelial cells are cultured without any mesenchymal support, while the second utilizes an air-liquid interface and includes both epithelial and stroma elements.

The first method utilizes matrigel, a mixture of extracellular proteins includings laminins and collagens to provide both 3-dimensional support and signaling factors for cell growth. Addition of R-Spondin1 (a Wnt agonist), epidermal growth factor and Noggin (a bone morphogenic protein (BMP) antagonist) into the culture medium mimics the normal intestinal crypt environment and signaling milieu. Using this method, isolated intestinal crypts without mesenchyme and even single cells can be successfully cultured ex vivo for more than 8 months and can be serially replated without loss of replating efficiency [26]. These "organoids" are made up of a single layer of villus-like epithelium, forming multiple crypt and villus-like structures around a central lumen into which apoptotic cells are shed, with their morphology closely resembling normal intestinal physiology. All terminally differentiated cell types of the intestine can be found throughout the organoid. Follow up studies on this technique show that increased organoid forming efficiency, with increased self-renewal and reduced differentiation of the single cells, can be achieved by adding a GSK3 inhibitor, CHIR99021, and high concentrations (1–2 mM) of valproic acid [27].

An air-liquid interface is introduced in the second technique to improve oxygenation of the cells, while a collagen gel and mesenchymal cells are included in the culture instead of individual signaling molecules to simulate the in vivo 3-dimensional intestinal stem cell niche. Cultures from murine neonatal intestine are able to grow up to 350 days in culture as cystic structures with representation of all the differentiated cell types of the intestine. Crypt-like structures and villus-like projections are also observed. There is however a decrease in proliferative capacity with time and with increasing age of the intestine, either due to developing mesenchymal defects with age or intrinsic defects in the intestinal stem cells. The induction of Wnt signaling by an exogenous R-spondin fusion protein improves proliferation and also increases the presence of putative stem cells [28].

Both these techniques capture the importance of the intestinal stem cell niche in the provision of structural support and an appropriate microenvironment in vivo for normal function of intestine stem cells. Numerous applications of this technology are currently being explored for therapeutic benefit such as creating patient-specific cystic fibrosis disease models from biopsy samples or growing replacement tissue for patients with diseased or dysfunctional intestines [29, 30].

2.3 *Cell Specific Markers of the +4 Position*

The initial discovery of the +4 position crypt cell notes its distinguishing characteristics, such as label-retention, slower cell cycling time, extreme radiosensitivity, involvement in post-injury regeneration and role as an origin for crypt cell migration [31, 32]. Further study of its label-retaining phenotype reveals the many safeguards these cells possess specifically to protect their genomic integrity, such as selective segregation of the DNA template and rigorous apoptotic pathways. By selectively segregating their DNA template, the newly synthesized strand (and any introduced mutation) is passed on to the proliferating and differentiating progeny and eventually lost while the original DNA strand remains in the stem line, known as the “immortal strand” hypothesis [33]. Meanwhile, the apoptotic pathways ensure that any errors in the template strand from environmental stressors are quickly eliminated [34]. Several markers of this intestinal stem cell population including *Bmi 1, Dcamkl-1, mTert* and *Hopx* have been reported.

Bmi1, a component of Polycomb group repressor complex, is required for the self-renewal of hematopoietic and neural stem cells. Given its broad tissue distribution, it was proposed to function in regulating self-renewal of other tissues. In situ hybridization and *Bmi1-EGFP* reporter mice confirm expression of *Bmi1* in the small intestinal crypts, specifically at the +4 position. Lineage tracing using *Bmi1-IRES-CreERT2;R26R-LacZ* mice also indicates that *Bmi1+* cells are both self-renewing, and capable of differentiation into all terminal cell types of the small intestine, fulfilling the criteria of stem cells [35, 36, 61]. The *Bmi1+* cells can also generate epithelial organoids in culture. Follow-up studies using *Bmi1* as a specific marker for these cells confirm that a few *Bmi1*-expressing cells retain label and are in fact slow cycling. Furthermore, stimulation of the Wnt pathway in these cells through β-catenin induction causes adenoma formation while ablation of the *Bmi1+* cells through cell-specific expression of diphtheria toxin leads to a loss of intestinal crypts, capturing the importance of these stem cells for crypt proliferation [35, 36]. However, while *Bmi1* expression is prominent in the duodenal crypts, it is poorly expressed in the ileal crypts [36], suggesting the existence of other stem cell populations that do not express *Bmi1*.

Doublecortin and Ca^{2+}/calmodulin-dependent kinase-like 1 (Dcamkl-1), a microtubule-associated protein kinase, is also highly expressed around the +4 position (49 % of DCAMKL-1+ cells are in the +4 position and only 4 % of crypts contain DCAMKL-1+ crypt base columnar cells). Co-staining of DCAMKL-1 with proliferating cell nuclear antigen (PCNA) verifies its expression in quiescent PCNA negative, label-retaining cells [37]. While the expression of DCAMKL-1 is lost in the proliferative crypts 84 h after a dose of ionizing radiation greater than 8 Gy, its presence can be detected 7–10 days later, indicating that these cells and their appropriate niche can be regenerated within the intestine. Furthermore, pulse labeling with BrdU during this regeneration period shows that while BrdU labeled cells are present in the upper crypt and villi at day 7, only rare cells in the lower crypts are still labeled on day 10. On day 10, cells at the +4 position co-express BrdU and DCAMKL-1, but do not express PCNA, indicating a role of the DCAMKL-1+ population in regeneration after injury before their return to quiescence [37]. DCAMKL-1 positive cells isolated from a mouse small intestine are capable of generating spheroids that when dissociated and injected into nude mice form nodular structures that stain positive for glandular (cytokeratin 14), secretory (Math1) and epithelial progenitor/stem cell markers (Msi-1). DCAMKL-1 as a stem cell marker is still controversial, as lineage tracing studies have not been reported. Co-expression of DCAMKL-1 with tuft cell differentiation markers [38] and its expression in a subset of enteroendocrine cells indicates that its expression may not be limited to intestinal stem cells [39].

Other cell specific markers proposed to identify +4 stem cells include *Hopx* and *Tert*. *Hopx* is an atypical homeobox gene and analysis of *Hopx-LacZ* knock-in mice shows expression of *Hopx* in the crypts along the entire length of the intestine with the strongest expression at +4 position [40]. The location of *Hopx*+ cells in the intestinal crypt is distinct from the Lgr5+ cells. After irradiation, *Hopx*+ cells show label-retention following pulse labeling with BrdU and 14 days of regeneration. Moreover, lineage tracing studies using *Hopx-IRES-ERCre;ROSA-LacZ* mice show that *Hopx*+ cells repopulate the entire length of the crypt and villus and are capable of producing all the differentiated cell types of the intestine. Progeny of the *Hopx*+ cells persist for at least 13 months in the crypts despite the entire intestinal epithelial turnover rate being 5 days. *Hopx-IRES-ERCre/ROSA-mTmG* mice also confirm these results [40]. Most importantly, this study demonstrates a bidirectional relationship between the active and quiescent stem cells in their niches as discussed later.

The *Tert* gene encodes telomerase reverse transcriptase, a protein required to maintain telomere length and thus protect against cellular senescence in stem cells. Slow cycling cells expressing *mTert*-GFP localize to the +4 position and have strong overlap with the *Bmi1*+ cells but are distinct from the *Lgr5*+ cells [41]. Similar to Bmi1 and Hopx, lineage tracing studies using *mTert-CreER;ROSA26 LacZ* mice show that *mTert* expressing cells contribute to the regeneration of the intestinal epithelium and production of all four differentiated cell types after injury. However, in contrast to the original model, these cells are described to be radiation-resistant [41]. These studies suggest that *mTert* may mark an independent, quiescent and radiation resistant pool of intestinal stem cells.

An alternative method to label quiescent or very slowly dividing cells uses transient transgenic expression of a fluorescent histone (H2B-GFP). This marks a slow-cycling population of cells at the +4 position of the small intestinal crypt that do not express the proliferation markers Ki67 and phospho-histone H3. Interestingly and consistent with the findings in *mTert*-expressing cells, this population is also radiation resistant [42].

2.4 Cell Specific Markers of the Stem Cell Zone Model

Leucine-rich repeat-containing G-protein coupled receptor 5 (*Lgr5*) is the first specific genetic marker for the stem cell zone model [43]. Wnt signaling is known to be important in intestinal self-renewal, and *Lgr5* has been identified as an intestinal Wnt/β-catenin target gene. *Lgr5* is expressed only in proliferating, slender cells at the base of the intestinal crypt in between Paneth cells and below the +4 cells, termed crypt base columnar (CBC) cells. Lineage tracing experiments demonstrate that Lgr5 expressing cells are both actively self-renewing and pluripotent, suggesting they could be the intestinal stem cell [43]. However, more recently, Bulavin and co-workers have argued that all lineage tracing studies in the intestine are complicated by the finding that the dose of tamoxifen required to activate CreER also kills the +4 cells. In this model, the committed progenitor CBC/*Lgr5* + cell may be recruited to become a +4 stem cell after radiation or tamoxifen damage to the +4 cells. This would lead to marking of +4 cells by *Lgr5*-driven Cre [16]. Thus, studies with *Lgr5:CreER* and tamoxifen may in fact also be studying the biology of the +4 cell.

The identification of Lgr5 as a marker for the crypt base columnar cells allowed for more in depth study on their properties and has revealed that these cells represent a rapidly proliferating population that does not possess the protective mechanisms for their genome as found in the +4 cell [37, 43]. The Lgr5+ crypt base columnar cells are actively cycling, as evidenced by their labeling kinetics and expression of proliferation markers Ki67 and phospho-histone H3. These cells are more radiation resistant compared to the label-retaining cells in the +4 position [43]. Further studies have shown that unlike the +4 position cells, Lgr5+ cells are not label-retaining and undergo symmetrical division with random segregation of chromosomes, implying that the "immortal strand" hypothesis is not a protective mechanism in these cells [44, 45]. This may be more consistent with Lgr5 cells being committed long-term progenitor cells, rather than immortal stem cells.

One theory of stem cell homeostasis is that the stem cells divide asymmetrically, producing one stem daughter cell, and one differentiating daughter cell (also known as the transit amplifying cell) to self-renew and produce differentiated progeny with each division. In the Lgr5 + putative stem cells however, homeostasis appears to be controlled by neutral drift dynamics instead. In the neutral drift theory, during symmetrical division the parent stem cell produces two identical daughter cells with potential to follow either fate resulting in two stem cells, two differentiated transit-amplifying cells or one of each. This being so, the regulation of the crypt follows a stochastic model in which the stem cells adopt fates depending on their environment such as the loss of neighboring stem cells or overcrowding within the niche [12, 13, 46]. This model implicates the stem cell niche as a key regulator of stem cell homeostasis instead of simply intrinsic properties of the stem cell itself.

Based on the gene expression signature of the Lgr5+ cells, other proposed stem cell markers have been identified, including various Wnt target genes such as *Ascl2* (Achaete scute-like 2) [47], *Tnfrsf19* (TNF receptor superfamily member 19), Ring finger nuclease 43 (*Rnf43*)/Zinc and ring finger 3 (*Znrf3*) [48] as well as *Olfm4* [49]. Ascl2 is a basic helix loop helix transcription factor that together with β-catenin and Tcf4 regulates the expression of various genes including *Lgr5* [50]. Ectopic expression of *Ascl2* in the intestinal epithelium induces hyperproliferation. Rnf43 and Znrf3 are E3 ubiquitin ligases that negatively regulate the Wnt/β-catenin pathway by ubiquitinating Wnt receptors Frizzled and LRP6 on the cell surface. This ubiquitination targets the Wnt receptors for internalization and lysosomal degradation. Demonstrating their importance in regulation of crypt homeostasis, deletion of both *Rnf43* and *Znrf3* genes in mice results in greater numbers of proliferating cells, increased levels of β-catenin in these hyperproliferative cells, enlarged crypts and adenoma formation [48, 51]. Furthermore, loss of function mutations in *RNF43* are found in several human cancers [52].

2.5 Other Stem Cell Specific Markers

Lrig1 (Leucine rich repeats and immunoglobulin like domains 1) is a transmembrane protein and a negative feedback inhibitor of ErbB signaling. *Lrig1* positive cells are located primarily at positions 1–5, thus it was initially reported as a marker for the crypt base columnar stem cells [53]. However, a study using *Lrig1* reporter (*Lrig1-IRES-CreER;ROSA-LacZ*) mice and Lrig1-specific antibody shows these cells to be distinct from the Lgr5+ population [54]. Lineage tracing studies also demonstrate that *Lrig1+* cells can repopulate the crypt and the villus and generate all the differentiated progeny. Knock out of *APC* in the *Lrig1*-expressing cells also leads to adenoma formation, demonstrating their stem cell potential (keeping in mind the caveat regarding the effects of tamoxifen on stem cell dynamics). Moreover, BrdU labeling and Ki67 staining shows this cell population to be slowly cycling and less proliferative than the *Lgr5+* cells, but more proliferative than the *Bmi1* and *mTert* expressing cells. The authors propose that these cells serve as an intermediate population between the stem cell zone and the +4 position models [54].

Musashi-1, an RNA-binding protein, is expressed in both the +1 and +4 cells of the intestinal crypts and may therefore serve as a general marker of the intestinal stem cell [55–57].

Evidence has emerged of extensive plasticity of cell populations in the intestine and of interactions between these different putative stem cell populations [58]. As described above, long-term lineage tracing studies have demonstrated that *mTert, Bmi1* and *Hopx* are bona-fide intestinal stem cell markers and that the Bmi1+ and mTert+ cells reside at +4 position whereas Lgr5 marks the mitotically active stem cells that are distinct from the mTert, Bmi1 and Hopx+ population. Lgr5+ cells are sensitive to Wnt perturbations, ablated by irradiation and contribute to homeostatic regeneration. The finding that Lgr5 marks a distinctive, highly proliferative population of the small intestinal and colonic stem cells has challenged the existence of quiescent stem cells. Specific elimination of *Lgr5* expressing cells by knocking a human diphtheria toxin receptor gene into the *Lgr5* locus does not change intestinal epithelial homeostasis, implying either that this cell population is not essential for intestinal function or that other cell populations are capable of compensating for its loss [59]. The only notable difference upon ablation of Lgr5+ cells is the increase of enteroendocrine cells in the crypts [35]. Following cessation of diphtheria treatment, the Lgr5+ cells rapidly regenerate in the intestinal crypts in vivo and in organoid cultures. However, depletion of Lgr5+ cells during radiation induced damage or Wnt pathway inhibition leads to the complete loss of intestinal architecture [59, 60]. This implies that these cells are not essential for normal intestinal homeostasis but are required for regeneration of intestinal epithelium following damage.

Tian et al. demonstrate that Bmi1+ cells can give rise to Lgr5+ cells in the small intestinal crypts after ablation of the Lgr5+ population under both normal conditions and during post-injury regeneration [35]. An independent study also demonstrates that *Bmi1* and *Lgr5* mark two functionally distinct intestinal stem cells in vivo [61]. Interconversion between *Hopx* expressing cells and *Lgr5* expressing cells occurs as demonstrated by gene profiling and single cell organoid cultures [40, 61]. The *Hopx* positive +4 cells represent the quiescent population of reserve intestinal stem cells that are resistant to irradiation. Consistent with this model, a single *Hopx* expressing cell is able to generate rapidly proliferating *Lgr5*+ cells. Conversely, isolated single *Lgr5* expressing cells from *Lgr-EGFP-ERCre;Hopx:LacZ* mice, cultured ex vivo to form organoids, are initially negative for *Hopx*, but express *Hopx* after 7 days. β-Gal expression (marking Hopx+ cells) also overlaps with GFP expression (derived from the Lgr5 locus) in the organoids derived from crypts of these mice [40]. Further in vivo fate mapping studies of the Lgr5 cells with *Lgr-EGFP-ERCre;Hopx:LacZ;R26mTmG/+* mice provide evidence that the slow cycling intestinal stem cells at +4 position dynamically interconvert with the rapidly cycling Lgr5+ cells in the crypt base.

Following intestinal damage, certain non-stem cell populations are also able to regain stem-like properties. For example, Dll1+ secretory precursor cells normally produce short-lived secretory clones but are capable of reverting to organoid-producing Lgr5+ stem cells in vitro upon Wnt stimulation, as well as reverting to Lgr5+ stem cells in vivo upon irradiation [62]. Similarly, label-retaining Paneth or enteroendocrine precursors are also capable of replenishing the stem cell population and differentiating into multiple lineages under regenerative and post-injury conditions [63].

Taken together, these studies indicate that the intestinal epithelium possesses a highly complex signaling network to ensure maintenance of multiple cell populations with differing proliferative capabilities and differentiation states, and to allow the transition between these populations in response to insults or damage. This functional redundancy complicates the study of the normal physiology of the intestine because experimental techniques can be biased towards or against specific populations. For example, as mentioned above, +4 position stem cells are killed preferentially by tamoxifen in lineage tracing experiments causing them to be replaced by the Lgr5+ stem cells, which thus become over-represented. A question that is also raised by these discoveries is what signaling pathways regulate this plasticity and which components of the stem cell niche are responsible.

3 Lineage Specification of Intestinal Stem Cells

Intestinal homeostasis requires appropriate lineage specification of the intestinal stem cells. As mentioned earlier, intestinal stem cells differentiate into four major cell types that populate the intestinal epithelium: the absorptive enterocytes and the three secretory cell types—enteroendocrine cells, goblet cells and Paneth cells. The fate of these cell types is determined by various molecular signals.

Enterocytes, also termed columnar cells, constitute more than 80 % of the intestinal epithelium. Caudal-related homeobox transcription factor (*Cdx1*), thyroid hormone and Kruppel-like factor (*Klf4*) regulate the differentiation of enterocytes [64–68]. Notch signaling plays an important role in regulating the differentiation of secretory versus absorptive cell lineages. A basic helix-loop-helix transcription factor, Math1/ATOH1, which is regulated by the Delta-Notch signaling pathway regulates the development of a common secretory cell progenitor [69, 70]. Further differentiation of secretory precursors to enteroendocrine cells involves two other basic helix-loop-helix transcription factors Neurogenin 3 and NeuroD, as well as a pancreatic-duodenal homeobox 1 gene (*Pdx1*) [71, 72]. *Spdef*, an Ets-domain transcription factor and *Sox9*, an HMG-box transcription factor, are both Wnt target genes and promote terminal differentiation to goblet and Paneth cells [73–75].

4 The Intestinal Stem Cell Niche

The intestinal stem cell niche is essential for the maintenance of intestinal homeostasis by providing a suitable microenvironment and signaling milieu for the self-renewal and differentiation of stem cells [76–79]. The stem cell niche has multiple components, which may be divided into two separate parts: a specialized and a non-specialized niche. The specialized niche consists of the basement membrane and one or a few epithelial cell types that lie next to and locally regulate the stem cells. The non-specialized niche is comprised of the mesenchymal cells that lie in the lamina propria underneath the basement membrane and provide broader regulation of the stem cells. These include the mesenchymal stem cells, fibroblasts, myofibroblasts, vascular endothelium, lymphatic vessels, adipocytes, neurons and blood cells [80]. Two key signaling pathways controlling the intestinal stem cell niche are Wnt and Notch, while additional signaling pathways such as Bmp and Hedgehog are also involved.

4.1 Wnt Signaling

Wnt/β-catenin signaling is integral to normal intestinal homeostasis and is essential for the proliferation of the epithelial cells in the crypts [81]. Wnts are autocrine or paracrine signaling proteins essential for embryonic development, cell proliferation and differentiation. They are highly conserved across species and can stimulate multiple downstream pathways including the Wnt/β-catenin pathway, planar cell polarity pathway and Wnt/calcium pathway. Wnts act as ligands for the Frizzled family of receptors and also interact with co-receptors such as lipoprotein receptor-related proteins, (LRP5/6) and receptor tyrosine kinases (Ryk and Ror and Tyrosine-protein kinase-like 7 (PTK7)). Binding of Wnts to their receptors leads to the recruitment of Disheveled to the Frizzleds and this plays a crucial role in determining which downstream pathway is activated. In the Wnt/β-catenin pathway in the absence of Wnt ligand, a degradation complex composed of an Axin and

APC scaffold facilitates the phosphorylation of β-catenin by GSK3 and CK1, which then targets β-catenin for ubiquitination by β-TrCP and proteosomal degradation [82]. Binding of Wnts to their receptors inhibits GSK3 and prevents the degradation of β-catenin leading to its cytoplasmic accumulation and eventual translocation to the nucleus. In the nucleus, β-catenin acts as a co-activator of transcription factors, TCF and LEF, triggering the transcription of numerous Wnt target genes. These target genes are then responsible for the migration and proliferation of the intestinal stem cell compartment [83, 84]. It should be noted that alternative pathways downstream of Wnt have been proposed [84].

Functional studies have demonstrated the importance of the Wnt/β-catenin pathway in regulating proliferation and differentiation of intestinal stem cells. Knockout of the downstream β-catenin effector Tcf-4 in mice prevents the proliferation of cells in the inter-villus region of the small intestine resulting in lethality within 24 h of birth [85]. Knockdown of β-catenin itself results in a similar loss of intestinal architecture and function [86, 87]. Inhibition of the Wnt pathway upstream using a Wnt inhibitor, Dkk1, confirms these results. Dkk1 interacts with LRP5/6 causing its internalization, thus inhibiting the interaction of Wnts with Frizzleds and LRP5/6 on the cell surface. Homozygous intestinal epithelium-specific expression of Dkk1 results in the development of grossly abnormal intestines with shorter and fewer villi as compared to non-transgenic controls [88]. A dramatic reduction in the size and number of crypts, goblet cells, enteroendocrine cells and Paneth cells, demonstrates the importance of Wnts for both proliferation and differentiation of the secretory lineage [88]. Transient adenoviral expression of Dkk1 in adult mice produces a similar phenotype of reduced proliferation in the small intestine and colon and progressive loss of crypts, villi and glands [89]. Finally, small molecule inhibitors of PORCN, which block Wnt secretion, also produce a lack of proliferation in the small intestine [60]. Thus, Wnt production and Wnt signaling is essential for the proliferation of the intestinal stem cell compartment.

Conversely, too much Wnt/β-catenin activity is detrimental. Activation of the Wnt signaling pathway by overexpressing Wnt agonist R-spondin 1 results in massive proliferation of intestinal crypts [28]. $Apc^{Min/+}$ mice carrying a mutation in one allele of *Apc* spontaneously develop multiple adenomas in the intestinal epithelium, mimicking the human disease, familial adenomatous polyposis, caused by the truncation of APC. A key difference between the human disease and murine model is that the polyps are predominantly colonic in humans but present in the small intestine in mice [90]. Further study of these $Apc^{Min/+}$ mice has also shown that different APC mutations result in different levels of activation of the Wnt pathway and hence different degrees of polyposis, indicating the fine control of Wnt signaling on phenotype [91, 92]. Activating the Wnt signaling pathway while simultaneously inhibiting the BMP pathway in normal human intestinal epithelial crypt cells increases their proliferation and induces a gene expression profile similar to that of crypt-base columnar cells [93]. Conversely, stimulation of Wnt signaling by expression of a Lef1/β-catenin fusion protein in progenitor cells in the small intestine of a chimeric mouse results in apoptotic cell death only of the cells expressing Lef-1/β-catenin. Cells without Lef1/β-catenin expression show normal apoptosis, and the intestine as a whole is morphologically and histologically normal [94].

Wnts are also essential for intestinal proliferation ex vivo. R-spondin1, a Wnt sensitizer, is an essential component for culturing intestinal organoids [26, 28]. R-spondins (RSPOs) are a family of four proteins containing thrombospondin repeats that enhance Wnt signaling by binding to LGR5 and its paralogs, LGR4 and LGR6, together with the E3 ubiquitin ligases RNF43/ZNRF3, thus inhibiting the activity of the latter [51, 95, 96]. This causes the accumulation of Frizzled receptors on the cell surface, and hence an increased sensitivity to Wnts. Chromosomal translocations resulting in increased expression of RSPO2 and RPSO3 have been identified in a number of human cancers including colorectal cancers. Consistent with their importance in the Wnt/β-catenin pathway, RSPO fusions are only found in colorectal cancers that do not have APC or β-catenin mutations [97]. While the role of Wnts in regulating intestinal homeostasis is well established, the cells making R-spondins and Wnts in the stem cell niche are still being defined [60].

One important conclusion from these studies is that the level of Wnt signaling in the small intestinal crypts is carefully modulated in normal homeostasis such that only cells which receive optimal amounts of Wnt will be able to survive and proliferate [98]. Excessive Wnt signaling results in unrestrained proliferation and neoplastic growth, indicating the existence of a complex regulatory network in vivo that maintains precise levels of Wnt signaling and therefore normal activity of the intestinal stem cells.

4.2 Notch Signaling

The Notch pathway plays a central role in cell fate decisions and differentiation of the intestinal epithelium. Notch is a single transmembrane receptor that undergoes proteolytic cleavage by γ-secretase upon ligand binding, freeing an intracellular domain (NICD) that translocates to the nucleus. The NICD then binds to the transcription factor CLS (or CBF1) to regulate transcription. There are four Notch receptors and several ligands, such as Delta-like and Jagged in mammals.

Notch signaling mostly works at very short distances, such as through contact of adjacent cells or by expression of ligand and receptor on the same cell [99, 100]. Studies using lineage tracing have confirmed the endogenous expression of Notch-1 and Notch-2 receptors specifically in the crypt stem cells, at both the +4 position and crypt base (Fig. 1b). Notch signaling is also active in the intestinal crypt stem cells and progenitors, but not in any of the three differentiated secretory cell types [101]. Similar to the Wnt pathway, Notch signaling is essential for maintaining the undifferentiated and proliferative state of the crypts. Labeling of all the cell types in lineage tracing experiments demonstrates activation of the Notch signaling in the adult intestinal stem cells [102].

The Notch pathway is also a key regulator of absorptive versus secretory cell fate decisions in the intestine. Notch signaling stimulates the expression of Hairy/Enhancer of Split (Hes1), which then inhibits the function of several basic helix-loop-helix transcription factors including Math1, which is critical for differentiation of the secretory lineage [103, 104]. Constitutive activation of the Notch1 receptor in the intestinal epithelium using the *Villin* promoter results in postnatal lethality, 3 days after birth. The mice have grossly abnormal intestinal architecture with impaired differentiation of the secretory lineage. Their intestines lack goblet cells and have reduced enteroendocrine and Paneth cells, but increased numbers of proliferating intestinal progenitors. This is accompanied by upregulation of Hes1 and downregulation of Math1 and neurogenin-3, while the components of the Wnt pathway such as β-catenin nuclear translocation and levels of Tcf4 or Lef1 are not affected [100].

Consistent with the importance of Notch signaling in promoting absorptive cell differentiation, inhibition of the Notch signaling pathway either through conditional knock out of the common downstream transcription factor *CSL*, or by using a γ-secretase inhibitor, causes a phenotype opposite to the constitutive Notch1 knock-in. In this case, proliferative crypt cells terminally differentiate to goblet cells, and proliferation of the intestinal epithelium ceases. The Paneth cell and enteroendocrine cell numbers and location remain normal and Wnt signaling remains active [105]. Likewise, conditional inactivation of both the *Notch1* and *Notch2* receptors also results in complete conversion of the intestinal epithelial cells to goblet cells [106]. The *Hes1*$^{-/-}$ mice also have more goblet cells and less enterocytes compared to wild-type controls, but show no difference in the proliferation of the intestinal precursors [103]. Inhibition of the Notch pathway by deletion of both Notch ligands *Dll1* and *Dll4* confirms these findings, with the complete differentiation of progenitors into goblet cells and loss of the proliferative crypt compartment [106]. Staining for *Olfm4* using in situ hybridization also shows the absence of crypt base columnar stem cells in these mice. Taken together, these studies indicate that the Notch pathway is essential for the balance between proliferation and appropriate differentiation in the intestine crypt.

4.3 Bone Morphogenic Proteins (BMP)

BMPs, originally discovered to induce bone formation, belong to a family of growth factors that are integral to the normal development of various tissues. BMP4 is expressed in the intravillus mesenchyme of adult mice. The BMP receptor *Bmpr1a* is expressed highly in +4 position stem cells and in an increasing gradient along the crypt-villus axis but not in the proliferating cell zone [107] (Fig. 1a). This BMP signaling axis is believed to inhibit intestinal stem cell proliferation and self-renewal, thus maintaining the highly proliferative stem cell population only at the base of the crypts and promoting differentiation as cells move up the crypt. Consistent with this theory, expression of the BMP antagonist *Noggin* under the

Villin promoter leads to increased proliferation causing the development of ectopic epithelial invaginations containing proliferating cells that later develop into crypt-villus units. These crypt-villus units are grossly normal, with all terminally differentiated cell types. After several months, these mice develop intestinal polyps characterized by branched villi with dilated cysts similar to the human disease, juvenile polyposis [108]. Studies inhibiting BMP signaling by conditionally inactivating the BMP receptor *Bmpr1a* confirm this phenotype and suggest cross-talk with the Wnt pathway by inhibition of β-catenin activity [107, 109]. Likewise *Smad4*$^{+/-}$ mice also develop inflammatory polyposis lesions albeit at a later stage [110]. Studies in human colonic epithelium concur with the findings of BMP activity and interactions with the Wnt pathway [111, 112]. The requirement for Noggin to culture organoids ex vivo also reinforces the importance of inhibiting BMP signaling for self renewal and proliferation of intestinal stem cells [26].

4.4 Hedgehog

In the Hedgehog pathway, the binding of Hedgehog (Hh) ligands to Patched receptor relieves the inhibition of smoothened (*SMO*), leading to activation of *Gli* transcription factors, which then accumulate in the nucleus and control transcription of Hh target genes. The Hedgehog pathway, acting through Sonic (Shh) and Indian (Ihh) hedgehog proteins, is required for morphogenesis and embryonic development in a multitude of tissues. In the mouse, evidence of its role in the limbs, central nervous system [113] and foregut [114], among others, has been previously described. Ramalho-Santos and coworkers demonstrate that in the small intestine, both *Shh* (at very low levels) and *Ihh* are expressed at the base of the villi. *Ihh* is expressed throughout the epithelium in the colon and *Shh* is expressed mostly in the crypts (Fig. 1b). Importantly, they also demonstrate that hedgehog signaling is integral to anterior-posterior patterning, radial patterning, as well as proliferation and differentiation of the epithelial stem cells in the gastrointestinal tract. The intestines of *Ihh*$^{-/-}$ and *Shh*$^{-/-}$ mice show numerous intestinal abnormalities both gross and microscopic. *Ihh*$^{-/-}$ mice have smaller villi and less proliferation in the stem cell compartment, whereas *Shh*$^{-/-}$ mice show overgrown duodenal villi [115]. Inhibition of the Hedgehog pathway by expression of a pan-hedgehog inhibitor, Hhip, in the epithelial cells using the Villin promoter results in hyper-proliferation of the epithelium with formation of ectopic crypt-like structures and mislocalization of cells in the underlying stroma. This is accompanied by abnormally high and ectopic Wnt signaling [116]. This interaction of the Hedgehog pathway with the Wnt pathway is also demonstrated in rat colons. Expression of Wnt target genes is inhibited by ectopic expression of Ihh in vitro, and is restricted to the colon crypt base by Hedgehog signaling in vivo [117]. Besides inhibition of the Wnt pathway and a concomitant decrease in epithelial precursors, activation of Hedgehog signaling in the colon also increases epithelial Bmp signaling [118].

These signaling pathways are extremely complex and involve numerous interconnections and regulatory feedback loops. Numerous other signaling molecules and transcription factors have also been implicated including Forkhead [119], Yes-associated protein [120], Epithelial growth factor [26], Glucagon-like peptide-2 [121] and many more.

5 The Stem Cell Niche

5.1 Paneth Cells

In the small intestine, the Paneth cells are located at the base of the crypts, interspersed with the putative crypt base columnar stem cells just below the +4 position (Fig. 1a). The Paneth cell population is unique in that firstly, it is the only differentiated cell type which migrates down into the crypt instead of up into the villi, and secondly it has a much slower cycling time of around 60 days compared to the other differentiated cells with cycling times of 3–5 days [122]. The Paneth cells play a non-essential role in the physiology of the intestine. They secrete antimicrobial peptides that control gut flora as well as factors important for the development of the villus microvasculature [123]. Paneth cells also express components of the signaling pathways such as Wnt3, Wnt11, EGF, Tgf-α and Dll4, and therefore were originally hypothesized to be essential for intestinal proliferation and stem cell maintenance. The ability of the Paneth cells to support epithelial proliferation ex vivo is demonstrated by their enhancement of organoid forming efficiency when combined with Lgr5+ crypt base columnar stem cells compared to single Lgr5+ cells. Addition of exogenous Wnt3A in the absence of Paneth cells can also recapitulate this increase in organoid forming efficiency [124]. The adjacent location of the Notch ligand-expressing Paneth cells and the Notch receptor-expressing Lgr5+ crypt base columnar cells has been cited in support of a role for Paneth cells in Notch signaling [101].

Despite the role of Paneth cells ex vivo, the importance of their role in the intestinal stem cell niche in vivo has been called into question. Several different approaches to ablating the Paneth cells from the intestinal epithelium have not compromised intestinal proliferation. For example, even after more than 95 % of Paneth cells are killed by the expression of an attenuated diphtheria toxin gene under the Paneth cell-specific cryptdin-2 gene, no changes in tissue architecture, proliferation or differentiation are seen [125]. Growth factor independent 1 (*Gfi1*) deficient mice have no apparent Paneth cells, fewer goblet cells and more enteroendocrine cells and show normal crypt-villus structure and proliferation as assessed by both Ki67 staining and BrdU incorporation [126]. Deletion of *Sox9* in the intestinal epithelium also results in the absence of differentiated Paneth cells in the intestinal crypts but these crypts are larger and full of proliferating cells. *Sox9*$^{-/-}$ mice have normal body weight for up to 1 year [75]. The role of *Sox9* in the development of Paneth cells and goblet cells was confirmed by an independent study [74].

Counter arguments to the redundancy of Paneth cells claim that the ablation of Paneth cells in many of the above techniques is incomplete with enough Paneth cells remaining to sustain the normal function of the intestine. For instance, *Gfi1*$^{-/-}$ mice produced in two independent studies show Paneth cells are still present albeit in reduced numbers [124, 127]. Similarly, with cryptdin-controlled toxin expression and Sox9 deletion, depletion of Paneth cells is incomplete (95 % ablation) or temporary [124]. A more recent reassessment of this question utilizes an intestinal epithelial specific *Atoh1* (*Math1*) knockout. *Atoh1* is essential for differentiation of all secretory lineages upstream in their pathways, and also for Paneth cell survival. Therefore, this experiment maintains a complete and permanent absence of Paneth cells in the intestine. In crypts where *Atoh1* is deleted, no Paneth cells are present, and Lgr5+ crypt base columnar cells occupy the whole crypt base with increased proliferation, normal differentiation and intact Wnt signaling [128]. These conclusions were replicated in a second study that additionally demonstrates that while loss of Paneth cells in vivo produces no phenotype, *Math1*-deficient crypts could not be cultured as organoids without exogenous Wnt supplementation [129].

These studies, taken as a whole, show that ex vivo Paneth cells supply important signaling factors such as Wnts, but that in vivo they are fully dispensable and not needed to sustain intestinal stem cells. This implies either functional redundancy, or that in vivo there is a different source of Wnts and other key factors that support the intestinal stem cell niche. This source has been proposed to be the underlying mesenchyme surrounding the niche [60].

5.2 Mesenchyme Provides Signals to the Intestinal Stem Cell Niche

The intestinal mesenchyme contains many different cell types that perform functions ranging from immune regulation to maintenance of proliferation and differentiation [130]. Recently, numerous studies have highlighted the role of the mesenchyme in regulating various signaling pathways essential for intestinal homeostasis.

Wnt signaling is required for the maintenance of intestinal homeostasis and various studies provide data to support the essential role of the stroma as the source of Wnts (Fig. 1b). Organoid cultures are one avenue of studying the role of the intestinal mesenchyme on stem cell function by allowing the isolation of components in an in vitro system. As described before, the Paneth cells produce Wnts and are therefore essential for the ex vivo culture of intestinal crypts, when supplemented with significant quantities of RSPO1. However, in the presence of mesenchymal cells, Wnt3 produced by Paneth cells is not required for ex vivo culture [131]. Furthermore, murine and human subepithelial myofibroblasts can support human intestinal organoid formation as well as increase the duration of ex vivo survival of organoids [132]. Epithelial and myofibroblast co-cultures implanted subcutaneously into mice form enteroids while crypt mono-cultures cannot, [132, 133] demonstrating that myofibroblasts can also provide support to isolated intestinal epithelium in vivo without supply of exogenous factors. This effect of myofibroblast co-culture holds true with whole crypts as well as single Lgr5+ stem cells.

In addition to Wnts, the stroma is also an important source of the Wnt agonists, R-spondins (Fig. 1b). The subepithelial myofibroblasts express high levels of R-spondins that are sufficient to sustain organoid growth even in the absence of exogenously supplied R-spondins [60]. Supplementing organoid or organoid/myofibroblast co-cultures with exogenous R-spondin does not enhance the enteroid forming efficiency, but leads to formation of larger and more complex enteroids in the co-cultures [133]. Similarly, co-culture of colonic crypts with immortalized colonic myofibroblasts also results in a significantly higher efficiency of colonoid formation than crypts alone or in co-culture with L cells or a cell line from a young adult mouse colon (YAMC) [134].

PORCN is a membrane bound O-acyl transferase that post-translationally palmitoleates all mammalian Wnts at a conserved serine residue, and is essential for the secretion and binding of Wnts to the Frizzled receptors [135–137]. Complete inhibition of Wnt secretion from the epithelium in the $Porcn^{flox/flox}/Villin^{Cre}$ mice prevents the formation of organoids from the isolated crypts ex vivo unless supplied with exogenous Wnt3A, consistent with the critical role of Wnt signaling [60]. Importantly, intestinal proliferation, homeostasis and regenerative response to radiation damage are not affected in $Porcn^{flox/flox}/Villin^{Cre}$ mice. This study provides evidence that the Wnts from the stroma are sufficient for maintaining the intestinal epithelium.

Mesenchymal cells are also an important source of BMPs. *BMP4* is expressed in the intestinal intra-villus mesenchyme and phosphorylated SMAD1, 5 and 8 are observed in the nuclei of the villus epithelial cells indicative of paracrine BMP signaling to the adjacent villus epithelium [108]. This expression extends to the intercrypt mesenchymal cells, including those next to the +4 position putative stem cells (Fig. 1b). The stromal BMP signal appears to geographically restrict the crypt-forming region, since the crypts appear de novo at any place in the epithelium when BMP signaling is blocked. Noggin, a BMP antagonist, is expressed mostly in the submucosal layer at the crypt base and in lesser amounts near the +4 position [107] (Fig. 1b). Other BMP antagonists such as gremlin1/2 and chordin-like 1 are secreted by the myofibroblasts located near the crypts of the human colon [111]. Thus, the intestinal mesenchyme produces both activating and inhibitory signals that set up a gradient along the crypt-villus axis, limiting BMP activity at the crypt base and promoting proliferation of the epithelial stem cells (Fig. 1a).

Components of the Hedgehog signaling pathway are also expressed in both the epithelial and mesenchymal layers of the intestine. Specifically, *Shh* and *Ihh* are expressed in the epithelium while their receptors *Ptch1* and *Ptch2* as well as three downstream Gli transcription factors and target genes are expressed in the mesenchyme [116, 138] (Fig. 1b). This strongly suggests that Hedgehog signaling is paracrine from the epithelium to the mesenchyme. Furthermore, expression of a Hedgehog inhibitor in the intestinal epithelium results in mislocalization of the subepithelial myofibroblasts to the villus tips, near ectopic pre-crypt structures and proliferating epithelium [116]. Deletion of epithelial-specific *Ihh* also results in the absence of subepithelial myofibroblasts at the crypt base accompanied by abnormal proliferation and ectopic crypt-like structures of the epithelium. This phenotype is not observed when the Hedgehog pathway is inhibited in the epithelium itself by the deletion of *Smoothened*. This implies that the loss of myofibroblasts and their paracrine signaling is responsible for the abnormalities observed instead of autocrine Hedgehog signaling [139]. How this paracrine signal is transmitted from the producing mesenchymal cells to the receiving epithelial cells is still not fully understood but actin-based signaling filopodia called cytonemes have been proposed as the mechanism [140]. While more thoroughly studied in Drosophila systems, these cytonemes have been identified in vertebral development as well and may explain the precise spatial and temporal control of signaling in the intestinal stem cell niche [141].

5.3 An Integrated Model of the Intestinal Stem Cell Niche

The data in aggregate supports the conclusion that there are at least two different intestinal stem cell populations, quiescent and rapidly proliferating, which maintain normal intestinal homeostasis. Upon damage to the intestinal epithelium, these stem cells are capable of interconverting to replace the lost populations. The bidirectional signaling between the intestinal stroma and epithelium plays a key role in regulating the proliferation, differentiation and plasticity of the intestinal stem cells. As detailed above, subepithelial myofibroblasts are the main source of Wnt and R-spondins and are sufficient to sustain intestinal homeostasis even in the absence of epithelial Wnts. Receptors of the Wnt ligands (FZDs) and their agonists (LGRs), on the other hand, are abundant on the epithelial cells. Hedgehog signaling is also paracrine from the epithelium, which secretes Hedgehog, to the myofibroblasts bearing the receptors Patched. Hedgehog signaling regulates the localization of the subepithelial myofibroblasts at the base of the crypts and its inhibition leads to mislocalization of the subepithelial myofibroblasts to the tips of the villi. Additionally, BMP signaling, which allows for the differentiation of the epithelial cells by regulating Wnt signaling, is also bidirectional in an increasing gradient to the tip of the villi. Mesenchymal cells secrete BMPs, while their receptors are expressed by the +4 position cells. Taken together, subepithelial myofibroblasts and other stromal cells in the intestinal stem cell niche are source of the signals that regulate the proliferation, differentiation and plasticity of these intestinal stem cells (Fig. 1b).

5.4 Relevance of the Stem Cell Niche to Disease States and Therapeutics

Dysregulation of signaling in the stem cells or their niche can result in abnormal proliferation of the intestinal epithelium, which can then progress to cancer. Consistent with the role of Wnt signaling in intestinal homeostasis, one of the most common mutations found in human colon cancer is the inactivation of the APC gene, which results in stabilization of β-catenin and activation of additional pathways [142]. The loss of APC results in abnormal proliferation and neoplastic formation in the colon and is therefore integral to the development of familial adenomatous polyposis [143]. Dysregulation of the BMP signaling pathway caused by mutations of BMPR1A [144] and SMAD4 [145] results in juvenile polyposis, characterized by the formation of multiple polyps in the gastrointestinal tract and an increased risk of gastrointestinal adenocarcinomas. The cancer stem cell niche may also be integral for the maintenance of the cancer phenotype or the process of metastasis by cytokine and growth factor regulation, as well as by activation of inflammatory pathways [146, 147].

A profound understanding of the stem cell niche and its signaling pathways has permitted the long term ex vivo organoid culture of intestinal stem cells. Beyond their use in experimental analyses, these organoid cultures can be used to develop disease models from human patient biopsies for in depth study of disease pathophysiology as well as testing of novel therapeutics [29]. The ability to generate tissue-engineered small intestine for re-implantation into patients with short bowel syndrome or other intestinal diseases has also been explored. These tissue-engineered small intestines are created by attaching clusters of epithelial and mesenchymal cells isolated from the intestine onto scaffolds, and then implanting these loaded scaffolds into rats [148]. Human derived tissue-engineered small intestines have been produced by implantation into immunodeficient mice [149, 150]. In-depth knowledge of the intestinal stem cell niche can help improve the efficacy of these techniques [151].

6 Conclusion

The model of the intestinal stem cell and its niche is constantly evolving, as new pieces of the puzzle are discovered and fit into place. It appears that there are two different stem cell populations in the intestinal epithelium that control normal homeostasis—a rapidly proliferating population at the base of the crypt expressing markers like *Lgr5, Ascl2, Olfm4, Rnf43* and *Znrf3* and a quiescent population at the +4 position expressing *Bmi1, Hopx* and *mTert.* A great deal of plasticity exists within the intestine as these two stem cell populations are able to interconvert, and other intestinal epithelial cell types also have the potential to regain stemness post-injury. A complex array of signaling pathways involving both the epithelium and its underlying mesenchyme are involved in regulation of the stem cell niche. Crosstalk between different pathways, as well as spatial control and redundancy of signaling factors in the mesenchyme and epithelium allow for fine regulation of development and homeostasis. An improved understanding of the intestinal stem cell niche has also led to progress in the development of an in vitro model system of the intestine allowing for more physiologically relevant study of disease pathophysiology and testing of novel therapeutics. An integral part of understanding the behavior of these intestinal stem cells is characterizing the surrounding niche. Further studies to identify the source of the signaling molecules in the niche will provide an insight into how the sub-compartmentalization of the intestinal stem cells is maintained.

Acknowledgments Studies in the Virshup lab at the Duke-NUS Graduate Medical School are supported in part by the National Research Foundation Singapore and administered by the Singapore Ministry of Health's National Medical Research Council under the STAR Award Program to D.M.V.

References

1. Lipkin M. Cell Replication in the gastrointestinal tract of man. Gastroenterology. 1965;48:616–24.
2. Heath JP. Epithelial cell migration in the intestine. Cell Biol Int. 1996;20(2):139–46.
3. Potten CS, Kovacs L, Hamilton E. Continuous labelling studies on mouse skin and intestine. Cell Tissue Kinet. 1974;7(3):271–83.
4. Halprin KM. Epidermal "turnover time"—a re-examination. Br J Dermatol. 1972;86(1):14–9.
5. Bergmann O, Bhardwaj RD, Bernard S, Zdunek S, Barnabé-Heider F, Walsh S, et al. Evidence for cardiomyocyte renewal in humans. Science. 2009;324(5923):98–102.
6. Kajstura J, Gurusamy N, Ogórek B, Goichberg P, Clavo-Rondon C, Hosoda T, et al. Myocyte turnover in the aging human heart. Circ Res. 2010;107(11):1374–86.
7. Cheng H, Leblond CP. Origin, differentiation and renewal of the four main epithelial cell types in the mouse small intestine. V. Unitarian theory of the origin of the four epithelial cell types. Am J Anat. 1974;141(4):537–61.
8. Bjerknes M, Cheng H. The stem-cell zone of the small intestinal epithelium. I. Evidence from Paneth cells in the adult mouse. Am J Anat. 1981;160(1):51–63.
9. Rothenberg ME, Nusse Y, Kalisky T, Lee JJ, Dalerba P, Scheeren F, et al. Identification of a cKit(+) colonic crypt base secretory cell that supports Lgr5(+) stem cells in mice. Gastroenterology. 2012;142(5):1195–205.
10. Novelli MR, Williamson JA, Tomlinson IP, Elia G, Hodgson SV, Talbot IC, et al. Polyclonal origin of colonic adenomas in an XO/XY patient with FAP. Science. 1996;272(5265):1187–90.
11. Novelli M, Cossu A, Oukrif D, Quaglia A, Lakhani S, Poulsom R, et al. X-inactivation patch size in human female tissue confounds the assessment of tumor clonality. Proc Natl Acad Sci U S A. 2003;100(6):3311–4.
12. Ritsma L, Ellenbroek SIJ, Zomer A, Snippert HJ, de Sauvage FJ, Simons BD, et al. Intestinal crypt homeostasis revealed at single-stem-cell level by in vivo live imaging. Nature. 2014;507(7492):362–5.

13. Snippert HJ, van der Flier LG, Sato T, van Es JH, van den Born M, Kroon-Veenboer C, et al. Intestinal crypt homeostasis results from neutral competition between symmetrically dividing Lgr5 stem cells. Cell. 2010;143(1):134–44.
14. Gonzalez MA, Bernad A. Characteristics of adult stem cells. Adv Exp Med Biol. 2012;741:103–20.
15. Potten CS. Extreme sensitivity of some intestinal crypt cells to X and gamma irradiation. Nature. 1977;269(5628):518–21.
16. Zhu Y, Huang Y-F, Kek C, Bulavin DV. Apoptosis differently affects lineage tracing of Lgr5 and Bmi1 intestinal stem cell populations. Cell Stem Cell. 2013;12(3):298–303.
17. Bjerknes M, Cheng H. The stem-cell zone of the small intestinal epithelium. II. Evidence from Paneth cells in the newborn mouse. Am J Anat. 1981;160(1):65–75.
18. Fuchs E. The tortoise and the hair: slow-cycling cells in the stem cell race. Cell. 2009;137(5):811–9.
19. Li L, Clevers H. Coexistence of quiescent and active adult stem cells in mammals. Science. 2010;327(5965):542–5.
20. Kretzschmar K, Watt FM. Lineage tracing. Cell. 2012;148(1–2):33–45.
21. Sauer B, Henderson N. Targeted insertion of exogenous DNA into the eukaryotic genome by the Cre recombinase. New Biol. 1990;2(5):441–9.
22. Utomo AR, Nikitin AY, Lee WH. Temporal, spatial, and cell type-specific control of Cre-mediated DNA recombination in transgenic mice. Nat Biotechnol. 1999;17(11):1091–6.
23. Feil R, Brocard J, Mascrez B, LeMeur M, Metzger D, Chambon P. Ligand-activated site-specific recombination in mice. Proc Natl Acad Sci U S A. 1996;93(20):10887–90.
24. Yeh KY, Chopra DP. Epithelial cell cultures from the colon of the suckling rat. Vitro. 1980;16(11):976–86.
25. Kaeffer B. Mammalian intestinal epithelial cells in primary culture: a mini-review. Vitro Cell Dev Biol Anim. 2002;38(3):123–34.
26. Sato T, Vries RG, Snippert HJ, van de Wetering M, Barker N, Stange DE, et al. Single Lgr5 stem cells build crypt-villus structures in vitro without a mesenchymal niche. Nature. 2009;459(7244):262–5.
27. Yin X, Farin HF, van Es JH, Clevers H, Langer R, Karp JM. Niche-independent high-purity cultures of Lgr5+ intestinal stem cells and their progeny. Nat Methods. 2014;11(1):106–12.
28. Ootani A, Li X, Sangiorgi E, Ho QT, Ueno H, Toda S, et al. Sustained in vitro intestinal epithelial culture within a Wnt-dependent stem cell niche. Nat Med. 2009;15(6):701–6.
29. Sato T, Clevers H. Growing self-organizing mini-guts from a single intestinal stem cell: mechanism and applications. Science. 2013;340(6137):1190–4.
30. Yui S, Nakamura T, Sato T, Nemoto Y, Mizutani T, Zheng X, et al. Functional engraftment of colon epithelium expanded in vitro from a single adult $Lgr5^+$ stem cell. Nat Med. 2012;18(4):618–23.
31. Potten CS. Stem cells in gastrointestinal epithelium: numbers, characteristics and death. Philos Trans R Soc Lond B Biol Sci. 1998;353(1370):821–30.
32. Roth S, Fodde R. Quiescent stem cells in intestinal homeostasis and cancer. Cell Commun Adhes. 2011;18(3):33–44.
33. Rando TA. The immortal strand hypothesis: segregation and reconstruction. Cell. 2007;129(7):1239–43.
34. Potten CS, Owen G, Booth D. Intestinal stem cells protect their genome by selective segregation of template DNA strands. J Cell Sci. 2002;115(Pt 11):2381–8.
35. Tian H, Biehs B, Warming S, Leong KG, Rangell L, Klein OD, et al. A reserve stem cell population in small intestine renders Lgr5-positive cells dispensable. Nature. 2011;478(7368):255–9.
36. Sangiorgi E, Capecchi MR. Bmi1 is expressed in vivo in intestinal stem cells. Nat Genet. 2008;40(7):915–20.
37. May R, Sureban SM, Hoang N, Riehl TE, Lightfoot SA, Ramanujam R, et al. Doublecortin and CaM kinase-like-1 and leucine-rich-repeat-containing G-protein-coupled receptor mark quiescent and cycling intestinal stem cells, respectively. Stem Cells. 2009;27(10):2571–9.

38. Gerbe F, Brulin B, Makrini L, Legraverend C, Jay P. DCAMKL-1 expression identifies Tuft cells rather than stem cells in the adult mouse intestinal epithelium. Gastroenterology. 2009;137(6):2179–80.
39. Gagliardi G, Moroz K, Bellows CF. Immunolocalization of DCAMKL-1, a putative intestinal stem cell marker, in normal colonic tissue. Pathol Res Pract. 2012;208(8):475–9.
40. Takeda N, Jain R, LeBoeuf MR, Wang Q, Lu MM, Epstein JA. Interconversion between intestinal stem cell populations in distinct niches. Science. 2011;334(6061):1420–4.
41. Montgomery RK, Carlone DL, Richmond CA, Farilla L, Kranendonk ME, Henderson DE, et al. Mouse telomerase reverse transcriptase (mTert) expression marks slowly cycling intestinal stem cells. Proc Natl Acad Sci U S A. 2011;108(1):179–84.
42. Hughes KR, Gândara RMC, Javkar T, Sablitzky F, Hock H, Potten CS, et al. Heterogeneity in histone 2B-green fluorescent protein-retaining putative small intestinal stem cells at cell position 4 and their absence in the colon. Am J Physiol Gastrointest Liver Physiol. 2012;303 (11):G1188–201.
43. Barker N, van Es JH, Kuipers J, Kujala P, van den Born M, Cozijnsen M, et al. Identification of stem cells in small intestine and colon by marker gene Lgr5. Nature. 2007;449 (7165):1003–7.
44. Escobar M, Nicolas P, Sangar F, Laurent-Chabalier S, Clair P, Joubert D, et al. Intestinal epithelial stem cells do not protect their genome by asymmetric chromosome segregation. Nat Commun. 2011;29(2):258.
45. Schepers AG, Vries R, van den Born M, van de Wetering M, Clevers H. Lgr5 intestinal stem cells have high telomerase activity and randomly segregate their chromosomes. EMBO J. 2011;30(6):1104–9.
46. Lopez-Garcia C, Klein AM, Simons BD, Winton DJ. Intestinal stem cell replacement follows a pattern of neutral drift. Science. 2010;330(6005):822–5.
47. van der Flier LG, van Gijn ME, Hatzis P, Kujala P, Haegebarth A, Stange DE, et al. Transcription factor achaete scute-like 2 controls intestinal stem cell fate. Cell. 2009;136 (5):903–12.
48. Koo BK, Spit M, Jordens I, Low TY, Stange DE, van de Wetering M, et al. Tumour suppressor RNF43 is a stem-cell E3 ligase that induces endocytosis of Wnt receptors. Nature. 2012;488(7413):665–9.
49. van der Flier LG, Haegebarth A, Stange DE, van de Wetering M, Clevers H. OLFM4 is a robust marker for stem cells in human intestine and marks a subset of colorectal cancer cells. Gastroenterology. 2009;137(1):15–7.
50. Schuijers J, Junker JP, Mokry M, Hatzis P, Koo BK, Sasselli V, et al. Ascl2 acts as an R-spondin/Wnt-responsive switch to control stemness in intestinal crypts. Cell Stem Cell. 2015;16(2):158–70.
51. Hao HX, Xie Y, Zhang Y, Charlat O, Oster E, Avello M, et al. ZNRF3 promotes Wnt receptor turnover in an R-spondin-sensitive manner. Nature. 2012;485(7397):195–200.
52. Giannakis M, Hodis E, Mu XJ, Yamauchi M, Rosenbluh J, Cibulskis K, et al. RNF43 is frequently mutated in colorectal and endometrial cancers. Nat Genet. 2014;46(12):1264–6.
53. Wong VW, Stange DE, Page ME, Buczacki S, Wabik A, Itami S, et al. Lrig1 controls intestinal stem-cell homeostasis by negative regulation of ErbB signalling. Nat Cell Biol. 2012;14(4):401–8.
54. Powell AE, Wang Y, Li Y, Poulin EJ, Means AL, Washington MK, et al. The pan-ErbB negative regulator Lrig1 is an intestinal stem cell marker that functions as a tumor suppressor. Cell. 2012;149(1):146–58.
55. Kayahara T, Sawada M, Takaishi S, Fukui H, Seno H, Fukuzawa H, et al. Candidate markers for stem and early progenitor cells, Musashi-1 and Hes1, are expressed in crypt base columnar cells of mouse small intestine. FEBS Lett. 2003;535(1–3):131–5.
56. Potten CS, Booth C, Tudor GL, Booth D, Brady G, Hurley P, et al. Identification of a putative intestinal stem cell and early lineage marker; musashi-1. Differentiation. 2003;71 (1):28–41.

57. Maria Cambuli F, Rezza A, Nadjar J, Plateroti M. Brief report: musashi1-eGFP mice, a new tool for differential isolation of the intestinal stem cell populations. Stem Cells. 2013;31 (10):2273–8.
58. Philpott A, Winton DJ. Lineage selection and plasticity in the intestinal crypt. Curr Opin Cell Biol. 2014;31:39–45.
59. Metcalfe C, Kljavin NM, Ybarra R, de Sauvage FJ. Lgr5+ stem cells are indispensable for radiation-induced intestinal regeneration. Cell Stem Cell. 2014;14(2):149–59.
60. Kabiri Z, Greicius G, Madan B, Biechele S, Zhong Z, Zaribafzadeh H, et al. Stroma provides an intestinal stem cell niche in the absence of epithelial Wnts. Development. 2014;141 (11):2206–15.
61. Yan KS, Chia LA, Li X, Ootani A, Su J, Lee JY, et al. The intestinal stem cell markers Bmi1 and Lgr5 identify two functionally distinct populations. Proc Natl Acad Sci U S A. 2012;109 (2):466–71.
62. van Es JH, Sato T, van de Wetering M, Lyubimova A, Nee AN, Gregorieff A, et al. Dll1 +secretory progenitor cells revert to stem cells upon crypt damage. Nat Cell Biol. 2012;14 (10):1099–104.
63. Buczacki SJ, Zecchini HI, Nicholson AM, Russell R, Vermeulen L, Kemp R, et al. Intestinal label-retaining cells are secretory precursors expressing Lgr5. Nature. 2013;495(7439):65–9.
64. Alkhoury F, Malo MS, Mozumder M, Mostafa G, Hodin RA. Differential regulation of intestinal alkaline phosphatase gene expression by Cdx1 and Cdx2. Am J Physiol Gastrointest Liver Physiol. 2005;289(2):G285–90.
65. Soubeyran P, André F, Lissitzky JC, Mallo GV, Moucadel V, Roccabianca M, et al. Cdx1 promotes differentiation in a rat intestinal epithelial cell line. Gastroenterology. 1999;117 (6):1326–38.
66. Li IC, Chan CT, Lu YF, Wu YT, Chen YC, Li GB, et al. Zebrafish krüppel-like factor 4a represses intestinal cell proliferation and promotes differentiation of intestinal cell lineages. PLoS ONE. 2011;6(6):e20974.
67. Hinnebusch BF, Siddique A, Henderson JW, Malo MS, Zhang W, Athaide CP, et al. Enterocyte differentiation marker intestinal alkaline phosphatase is a target gene of the gut-enriched Kruppel-like factor. Am J Physiol Gastrointest Liver Physiol. 2004;286(1): G23–30.
68. Malo MS, Zhang W, Alkhoury F, Pushpakaran P, Abedrapo MA, Mozumder M, et al. Thyroid hormone positively regulates the enterocyte differentiation marker intestinal alkaline phosphatase gene via an atypical response element. Mol Endocrinol. 2004;18(8):1941–62.
69. Yang Q, Bermingham NA, Finegold MJ, Zoghbi HY. Requirement of Math1 for secretory cell lineage commitment in the mouse intestine. Science. 2001;294(5549):2155–8.
70. VanDussen KL, Samuelson LC. Mouse atonal homolog 1 directs intestinal progenitors to secretory cell rather than absorptive cell fate. Dev Biol. 2010;346(2):215–23.
71. Li HJ, Ray SK, Singh NK, Johnston B, Leiter AB. Basic helix-loop-helix transcription factors and enteroendocrine cell differentiation. Diabetes Obes Metab. 2011;13(Suppl 1):5–12.
72. Yamada S, Kojima H, Fujimiya M, Nakamura T, Kashiwagi A, Kikkawa R. Differentiation of immature enterocytes into enteroendocrine cells by Pdx1 overexpression. Am J Physiol Gastrointest Liver Physiol. 2001;281(1):G229–36.
73. Gregorieff A, Stange DE, Kujala P, Begthel H, van den Born M, Korving J, et al. The ets-domain transcription factor Spdef promotes maturation of goblet and Paneth cells in the intestinal epithelium. Gastroenterology. 2009;137(4):1333–45.e1–3.
74. Bastide P, Darido C, Pannequin J, Kist R, Robine S, Marty-Double C, et al. Sox9 regulates cell proliferation and is required for Paneth cell differentiation in the intestinal epithelium. J Cell Biol. 2007;178(4):635–48.
75. Mori-Akiyama Y, van den Born M, van Es JH, Hamilton SR, Adams HP, Zhang J, et al. SOX9 is required for the differentiation of Paneth cells in the intestinal epithelium. Gastroenterology. 2007;133(2):539–46.
76. Fuchs E, Tumbar T, Guasch G. Socializing with the neighbors: stem cells and their niche. Cell. 2004;116(6):769–78.

77. Mikkers H, Frisén J. Deconstructing stemness. EMBO J. 2005;24(15):2715–9.
78. Schofield R. The relationship between the spleen colony-forming cell and the haemopoietic stem cell. Blood Cells. 1978;4(1–2):7–25.
79. Schofield R. The stem cell system. Biomed Pharmacother. 1983;37(8):375–80.
80. Ema H, Suda T. Two anatomically distinct niches regulate stem cell activity. Blood. 2012;120(11):2174–81.
81. Clarke AR. Wnt signalling in the mouse intestine. Oncogene. 2006;25(57):7512–21.
82. Holland JD, Klaus A, Garratt AN, Birchmeier W. Wnt signaling in stem and cancer stem cells. Curr Opin Cell Biol. 2013;25(2):254–64.
83. Holmberg J, Genander M, Halford MM, Annerén C, Sondell M, Chumley MJ, et al. EphB receptors coordinate migration and proliferation in the intestinal stem cell niche. Cell. 2006;125(6):1151–63.
84. Yu J, Virshup DM. Updating the Wnt pathways. Biosci Rep. 2014;34(5):e00142.
85. Korinek V, Barker N, Moerer P, van Donselaar E, Huls G, Peters PJ, et al. Depletion of epithelial stem-cell compartments in the small intestine of mice lacking Tcf-4. Nat Genet. 1998;19(4):379–83.
86. Ireland H, Kemp R, Houghton C, Howard L, Clarke AR, Sansom OJ, et al. Inducible Cre-mediated control of gene expression in the murine gastrointestinal tract: effect of loss of beta-catenin. Gastroenterology. 2004;126(5):1236–46.
87. Fevr T, Robine S, Louvard D, Huelsken J. Wnt/beta-catenin is essential for intestinal homeostasis and maintenance of intestinal stem cells. Mol Cell Biol. 2007;27(21):7551–9.
88. Pinto D, Gregorieff A, Begthel H, Clevers H. Canonical Wnt signals are essential for homeostasis of the intestinal epithelium. Genes Dev. 2003;17(14):1709–13.
89. Kuhnert F, Davis CR, Wang HT, Chu P, Lee M, Yuan J, et al. Essential requirement for Wnt signaling in proliferation of adult small intestine and colon revealed by adenoviral expression of Dickkopf-1. Proc Natl Acad Sci U S A. 2004;101(1):266–71.
90. Su LK, Kinzler KW, Vogelstein B, Preisinger AC, Moser AR, Luongo C, et al. Multiple intestinal neoplasia caused by a mutation in the murine homolog of the APC gene. Science. 1992;256(5057):668–70.
91. Oshima M, Oshima H, Kitagawa K, Kobayashi M, Itakura C, Taketo M. Loss of Apc heterozygosity and abnormal tissue building in nascent intestinal polyps in mice carrying a truncated Apc gene. Proc Natl Acad Sci U S A. 1995;92(10):4482–6.
92. Li Q, Ishikawa TO, Oshima M, Taketo MM. The threshold level of adenomatous polyposis coli protein for mouse intestinal tumorigenesis. Cancer Res. 2005;65(19):8622–7.
93. Guezguez A, Paré F, Benoit YD, Basora N, Beaulieu JF. Modulation of stemness in a human normal intestinal epithelial crypt cell line by activation of the WNT signaling pathway. Exp Cell Res. 2014;322(2):355–64.
94. Wong MH, Huelsken J, Birchmeier W, Gordon JI. Selection of multipotent stem cells during morphogenesis of small intestinal crypts of Lieberkuhn is perturbed by stimulation of Lef-1/beta-catenin signaling. J Biol Chem. 2002;277(18):15843–50.
95. de Lau W, Barker N, Low TY, Koo BK, Li VS, Teunissen H, et al. Lgr5 homologues associate with Wnt receptors and mediate R-spondin signalling. Nature. 2011;476(7360):293–7.
96. Carmon KS, Gong X, Lin Q, Thomas A, Liu Q. R-spondins function as ligands of the orphan receptors LGR4 and LGR5 to regulate Wnt/beta-catenin signaling. Proc Natl Acad Sci U S A. 2011;108(28):11452–7.
97. Seshagiri S, Stawiski EW, Durinck S, Modrusan Z, Storm EE, Conboy CB, et al. Recurrent R-spondin fusions in colon cancer. Nature. 2012;488(7413):660–4.
98. Wong MH. Regulation of intestinal stem cells. J Investig Dermatol Symp Proc. 2004;9(3):224–8.
99. Artavanis-Tsakonas S, Rand MD, Lake RJ. Notch signaling: cell fate control and signal integration in development. Science. 1999;284(5415):770–6.
100. Fre S, Huyghe M, Mourikis P, Robine S, Louvard D, Artavanis-Tsakonas S. Notch signals control the fate of immature progenitor cells in the intestine. Nature. 2005;435(7044):964–8.

101. Fre S, Hannezo E, Sale S, Huyghe M, Lafkas D, Kissel H, et al. Notch lineages and activity in intestinal stem cells determined by a new set of knock-in mice. PLoS ONE. 2011;6(10): e25785.
102. Vooijs M, Ong CT, Hadland B, Huppert S, Liu Z, Korving J, et al. Mapping the consequence of Notch1 proteolysis in vivo with NIP-CRE. Development. 2007;134(3):535–44.
103. Jensen J, Pedersen EE, Galante P, Hald J, Heller RS, Ishibashi M, et al. Control of endodermal endocrine development by Hes-1. Nat Genet. 2000;24(1):36–44.
104. Ueo T, Imayoshi I, Kobayashi T, Ohtsuka T, Seno H, Nakase H, et al. The role of Hes genes in intestinal development, homeostasis and tumor formation. Development. 2012;139 (6):1071–82.
105. van Es JH, van Gijn ME, Riccio O, van den Born M, Vooijs M, Begthel H, et al. Notch/gamma-secretase inhibition turns proliferative cells in intestinal crypts and adenomas into goblet cells. Nature. 2005;435(7044):959–63.
106. Pellegrinet L, Rodilla V, Liu Z, Chen S, Koch U, Espinosa L, et al. Dll1- and dll4-mediated notch signaling are required for homeostasis of intestinal stem cells. Gastroenterology. 2011;140(4):1230–1240.e1–7.
107. He XC, Zhang J, Tong WG, Tawfik O, Ross J, Scoville DH, et al. BMP signaling inhibits intestinal stem cell self-renewal through suppression of Wnt-beta-catenin signaling. Nat Genet. 2004;36(10):1117–21.
108. Haramis AP, Begthel H, van den Born M, van Es J, Jonkheer S, Offerhaus GJ, et al. De novo crypt formation and juvenile polyposis on BMP inhibition in mouse intestine. Science. 2004;303(5664):1684–6.
109. Tian Q, He XC, Hood L, Li L. Bridging the BMP and Wnt pathways by PI3 kinase/Akt and 14-3-3zeta. Cell Cycle. 2005;4(2):215–6.
110. Takaku K, Oshima M, Miyoshi H, Matsui M, Seldin MF, Taketo MM. Intestinal tumorigenesis in compound mutant mice of both Dpc4 (Smad4) and Apc genes. Cell. 1998;92(5):645–56.
111. Kosinski C, Li VS, Chan AS, Zhang J, Ho C, Tsui WY, et al. Gene expression patterns of human colon tops and basal crypts and BMP antagonists as intestinal stem cell niche factors. Proc Natl Acad Sci U S A. 2007;104(39):15418–23.
112. Reynolds A, Wharton N, Parris A, Mitchell E, Sobolewski A, Kam C, et al. Canonical Wnt signals combined with suppressed TGFβ/BMP pathways promote renewal of the native human colonic epithelium. Gut. 2014;63(4):610–21.
113. Chiang C, Litingtung Y, Lee E, Young KE, Corden JL, Westphal H, et al. Cyclopia and defective axial patterning in mice lacking Sonic hedgehog gene function. Nature. 1996;383 (6599):407–13.
114. Litingtung Y, Lei L, Westphal H, Chiang C. Sonic hedgehog is essential to foregut development. Nat Genet. 1998;20(1):58–61.
115. Ramalho-Santos M, Melton DA, McMahon AP. Hedgehog signals regulate multiple aspects of gastrointestinal development. Development. 2000;127(12):2763–72.
116. Madison BB, Braunstein K, Kuizon E, Portman K, Qiao XT, Gumucio DL. Epithelial hedgehog signals pattern the intestinal crypt-villus axis. Development. 2005;132(2):279–89.
117. van den Brink GR, Bleuming SA, Hardwick JC, Schepman BL, Offerhaus GJ, Keller JJ, et al. Indian Hedgehog is an antagonist of Wnt signaling in colonic epithelial cell differentiation. Nat Genet. 2004;36(3):277–82.
118. van Dop WA, Uhmann A, Wijgerde M, Sleddens-Linkels E, Heijmans J, Offerhaus GJ, et al. Depletion of the colonic epithelial precursor cell compartment upon conditional activation of the hedgehog pathway. Gastroenterology. 2009;136(7):2195–2203.e1–7.
119. Kaestner KH, Silberg DG, Traber PG, Schütz G. The mesenchymal winged helix transcription factor Fkh6 is required for the control of gastrointestinal proliferation and differentiation. Genes Dev. 1997;11(12):1583–95.
120. Barry ER, Morikawa T, Butler BL, Shrestha K, de la Rosa R, Yan KS, et al. Restriction of intestinal stem cell expansion and the regenerative response by YAP. Nature. 2013;493 (7430):106–10.

121. Rowland KJ, Brubaker PL. The "cryptic" mechanism of action of glucagon-like peptide-2. Am J Physiol Gastrointest Liver Physiol. 2011;301(1):G1–8.
122. Ireland H, Houghton C, Howard L, Winton DJ. Cellular inheritance of a Cre-activated reporter gene to determine Paneth cell longevity in the murine small intestine. Dev Dyn. 2005;233(4):1332–6.
123. Stappenbeck TS, Hooper LV, Gordon JI. Developmental regulation of intestinal angiogenesis by indigenous microbes via Paneth cells. Proc Natl Acad Sci U S A. 2002;99(24):15451–5.
124. Sato T, van Es JH, Snippert HJ, Stange DE, Vries RG, van den Born M, et al. Paneth cells constitute the niche for Lgr5 stem cells in intestinal crypts. Nature. 2011;469(7330):415–8.
125. Garabedian EM, Roberts LJ, McNevin MS, Gordon JI. Examining the role of Paneth cells in the small intestine by lineage ablation in transgenic mice. J Biol Chem. 1997;272(38):23729–40.
126. Shroyer NF, Wallis D, Venken KJ, Bellen HJ, Zoghbi HY. Gfi1 functions downstream of Math1 to control intestinal secretory cell subtype allocation and differentiation. Genes Dev. 2005;19(20):2412–7.
127. Bjerknes M, Cheng H. Cell lineage metastability in Gfi1-deficient mouse intestinal epithelium. Dev Biol. 2010;345(1):49–63.
128. Kim TH, Escudero S, Shivdasani RA. Intact function of Lgr5 receptor-expressing intestinal stem cells in the absence of Paneth cells. Proc Natl Acad Sci U S A. 2012;109(10):3932–7.
129. Durand A, Donahue B, Peignon G, Letourneur F, Cagnard N, Slomianny C, et al. Functional intestinal stem cells after Paneth cell ablation induced by the loss of transcription factor Math1 (Atoh1). Proc Natl Acad Sci U S A. 2012;109(23):8965–70.
130. Powell DW, Pinchuk IV, Saada JI, Chen X, Mifflin RC. Mesenchymal cells of the intestinal lamina propria. Annu Rev Physiol. 2011;73:213–37.
131. Farin HF, van Es JH, Clevers H. Redundant sources of Wnt regulate intestinal stem cells and promote formation of Paneth cells. Gastroenterology. 2012;143(6):1518–1529.e7.
132. Lahar N, Lei NY, Wang J, Jabaji Z, Tung SC, Joshi V, et al. Intestinal subepithelial myofibroblasts support in vitro and in vivo growth of human small intestinal epithelium. PLoS ONE. 2011;6(11):e26898.
133. Lei NY, Jabaji Z, Wang J, Joshi VS, Brinkley GJ, Khalil H, et al. Intestinal subepithelial myofibroblasts support the growth of intestinal epithelial stem cells. PLoS ONE. 2014;9(1): e84651.
134. Hirokawa Y, Yip KH, Tan CW, Burgess AW. Colonic myofibroblast cell line stimulates colonoid formation. Am J Physiol Gastrointest Liver Physiol. 2014;306(7):G547–56.
135. Proffitt KD, Virshup DM. Precise regulation of porcupine activity is required for physiological Wnt signaling. J Biol Chem. 2012;287(41):34167–78.
136. Biechele S, Cox BJ, Rossant J. Porcupine homolog is required for canonical Wnt signaling and gastrulation in mouse embryos. Dev Biol. 2011;355(2):275–85.
137. Janda CY, Waghray D, Levin AM, Thomas C, Garcia KC. Structural basis of Wnt recognition by Frizzled. Science. 2012;337(6090):59–64.
138. Wang LC, Nassir F, Liu ZY, Ling L, Kuo F, Crowell T, et al. Disruption of hedgehog signaling reveals a novel role in intestinal morphogenesis and intestinal-specific lipid metabolism in mice. Gastroenterology. 2002;122(2):469–82.
139. Kosinski C, Stange DE, Xu C, Chan AS, Ho C, Yuen ST, et al. Indian hedgehog regulates intestinal stem cell fate through epithelial-mesenchymal interactions during development. Gastroenterology. 2010;139(3):893–903.
140. Gradilla AC, Guerrero I. Cytoneme-mediated cell-to-cell signaling during development. Cell Tissue Res. 2013;352(1):59–66.
141. Stanganello E, Hagemann AI, Mattes B, Sinner C, Meyen D, Weber S, et al. Filopodia-based Wnt transport during vertebrate tissue patterning. Nat Commun. 2015;5(6):5846.
142. Phelps RA, Chidester S, Dehghanizadeh S, Phelps J, Sandoval IT, Rai K, et al. A two-step model for colon adenoma initiation and progression caused by APC loss. Cell. 2009;137 (4):623–34.

143. Fodde R. The APC gene in colorectal cancer. Eur J Cancer. 2002;38(7):867–71.
144. Howe JR, Bair JL, Sayed MG, Anderson ME, Mitros FA, Petersen GM, et al. Germline mutations of the gene encoding bone morphogenetic protein receptor 1A in juvenile polyposis. Nat Genet. 2001;28(2):184–7.
145. Houlston R, Bevan S, Williams A, Young J, Dunlop M, Rozen P, et al. Mutations in DPC4 (SMAD4) cause juvenile polyposis syndrome, but only account for a minority of cases. Hum Mol Genet. 1998;7(12):1907–12.
146. Ernst M, Ramsay RG. Colorectal cancer mouse models: integrating inflammation and the stroma. J Gastroenterol Hepatol. 2012;27(1):39–50.
147. Yeung TM, Buskens C, Wang LM, Mortensen NJ, Bodmer WF. Myofibroblast activation in colorectal cancer lymph node metastases. Br J Cancer. 2013;108(10):2106–15.
148. Vacanti JP, Morse MA, Saltzman WM, Domb AJ, Perez-Atayde A, Langer R. Selective cell transplantation using bioabsorbable artificial polymers as matrices. J Pediatr Surg. 1988;23(1 Pt 2):3–9.
149. Barthel ER, Levin DE, Speer AL, Sala FG, Torashima Y, Hou X, et al. Human tissue-engineered colon forms from postnatal progenitor cells: an in vivo murine model. Regen Med. 2012;7(6):807–18.
150. Levin DE, Barthel ER, Speer AL, Sala FG, Hou X, Torashima Y, et al. Human tissue-engineered small intestine forms from postnatal progenitor cells. J Pediatr Surg. 2013;48(1):129–37.
151. Spurrier RG, Grikscheit TC. Tissue engineering the small intestine. Clin Gastroenterol Hepatol. 2013;11(4):354–8.

Dental Pulp Stem Cell Niche

Jinhua Yu, Mohamed Jamal, Franklin Garcia-Godoy and George T.-J. Huang

Abbreviations

DPSCs	Dental pulp stem cells
SHED	Stem cells from exfoliated deciduous teeth
PDLSCs	Periodontal ligament stem cells
DFPCs	Dental follicle precursor cells
SCAP	Stem cells from apical papilla
MSCs	Mesenchymal stromal/stem cells
BMMSCs	Bone marrow stromal/stem cells
ECM	Extracellular matrix
HA/TCP	Hydroxyapatite/tricalcium phosphate
LAB	Living autologous fibrous bone
SBP/DPSCs	Stromal bone producing DPSCs
MDPSCs	Mobilized DPSCs
PLLA	Poly-L-lactic acid
HA-PCL	Hydroxyapatite-polycaprolactone
nHA	Nano-hydroxyapatite
DSPP	Dentin sialophosphoprotein
DSP	Dentin sialoprotein
DPP	Dentin phosphoprotein
DMP-1	Dentin matrix protein-1
BMPs	Bone morphogenetic proteins
FGFs	Fibroblast growth factors

J. Yu
Key Laboratory of Oral Diseases of Jiangsu Province, Institute of Stomatology, Nanjing Medical University, Nanjing, Jiangsu 210029, China

M. Jamal
Department of Endodontics, Henry M. Goldman School of Dental Medicine, Boston University, Boston, MA 02118, USA

F. Garcia-Godoy · G.T.-J. Huang (✉)
Department of Bioscience Research, College of Dentistry, University of Tennessee Health Science Center, Memphis, TN 38163, USA
e-mail: gtjhuang@uthsc.edu

K. Turksen (ed.), *Tissue-Specific Stem Cell Niche*,
Stem Cell Biology and Regenerative Medicine,
DOI 10.1007/978-3-319-21705-5_8

VEGF	Vascular endothelial growth factor
IGFs	Insulin-like growth factors
TGF-β	Transforming growth factor
PDGF	Platelet-derived growth factor
EGF	Epidermal growth factor
FKBP1A	FK506 binding protein 1A
OPN	Osteopontin
MAPK	Mitogen-activated protein kinases
ERK	Extracellular signal-regulated kinases
JNK	c-Jun-N-terminal kinase
SP	Side population

1 Introduction

The first type of dental stem cell was isolated from the human pulp tissue of permanent teeth termed postnatal dental pulp stem cells (DPSCs) [1]. Subsequently, stem cells from exfoliated deciduous teeth (SHED) [2], periodontal ligament stem cells (PDLSCs) [3], dental follicle precursor cells (DFPCs) [4], and stem cells from apical papilla (SCAP) [5, 6] were isolated and characterized. These dental tissue-derived stem cells are mesenchymal stem/stromal cell (MSC)-like populations, therefore, they have the characteristics of MSCs defined by the International Society for Cell Therapy as those cells that can (i) adhere to plastic in standard culture conditions; (ii) have the capacity for in vitro trilineage differentiation to adipogenic, chondrogenic, and osteogenic cells; and (iii) show specific expression profile (express CD73, CD90, and CD105 and lack the expression of CD34, CD45, HLA-DR, CD14 or CD11b, CD79a, or CD19) [7, 8]. This chapter will mainly review the properties of DPSCs, their subpoulations and their niches. The issue of identifying their niches is complicated by the lack of specific markers for MSCs. There is a long list of markers that are used to detect in situ MSCs in tissues, however, most of these markers are also expressed by other cell types. Therefore, the interpretation of the identified MSC niches in tissues should be validated with in vitro studies by culturing those putative MSCs and subjecting them to various analysis as well as testing their tissue regeneration capacities using in vivo study models. The regulation of DPSC niches has been studied including the interactions between the extracellular matrix (ECM), growth factors and stem cell receptors within the niche. The intracellular signaling as the response to these interactions plays a pivotal role in the status of the stem cell in the niche. This chapter provides an updated review of such a regulation in the DPSC niche.

2 The Discovery of Stem Cells in Pulp and Apical Papilla

2.1 Dental Pulp Stem Cells

In 2000, Gronthos et al. isolated the first MSC-like cells from human dental pulp tissue of permanent teeth and termed them dental pulp stem cells (DPSCs) [1]. These cells were obtained by enzymatic digestion of the pulp tissue. DPSCs have typical fibroblast-like morphology, are clonogenic and can maintain their high proliferation rate even after extensive subculturing. There is no specific biomarker to identify the DPSCs, however, DPSCs express several markers including the mesenchymal and bone marrow stem cell markers, STRO-1 and CD146 [1, 8].

Culturing DPSCs under various differentiation conditions can guide these cells towards osteo-dentinogenic, adipogenic, neurogenic, chondrogenic and myogenic lineages [9, 10]. Following their transplantation in animal models, DPSCs mixed with hydroxyapatite/tricalcium phosphate (HA/TCP) particles were able to form pulp-dentin complex-like tissues with odontoblast-like cells [1, 9, 11]. In contrast to DPSCs, bone marrow stromal/stem cells (BMMSCs) do not form pulp-dentin complex, instead, they form bone and marrow. This indicates that under the same osteoinductive stimulus such as HA/TCP, DPSCs are inherently programmed to regenerate dentin- and pulp-like tissues [12].

2.2 Stem Cells from Apical Papilla

In 2006, Sonoyama et al. reported a new population of dental stem cells from the apical papilla, termed SCAP [5]. SCAP are clonogenic fibroblast-like cells, having a higher proliferation rate than DPSCs. SCAP express mesenchymal surface markers, STRO-1 and CD146. SCAP also express CD24, which could be a unique marker for this cell population [5, 6]. The capacity of SCAP to differentiate into functional dentinogenic cells has been verified by the same approaches as for the DPSCs. SCAP can undergo osteogenic, adipogenic, chondrogenic and neurogenic differentiation when they are cultured in the appropriate inductive medium. As in the case of DPSCs, when SCAP mixed with HA/TCP were transplanted into immunocompromised mice, a typical dentin-/pulp-like structure was formed, with odontoblast-like cells [5, 6]. Using a tooth root fragment model, root canal space was filled with a scaffold seeded with SCAP and resulted in regenerating new vascularized pulp-like tissue with deposition of a new layer of dentin-like tissue, demonstrating the regenerative potential of SCAP [13]. The discovery of SCAP was considered a paradigm shift in the field of endodontics, as its anatomical location and odontogenic potential could provide a biological explanation for the regenerative endodontic procedures of immature necrotic pulp with periapical periodontitis [14, 15].

3 Dental Pulp Stem Cell Niche

Stem cell niches are considered to be tissue specific with different local microenvironments and serve two purposes i) to maintain tissue homeostasis and ii) to regenerate damaged tissues [16–18]. There are coordinated interactions between the stem cell and its local microenvironment in this niche. The players involved in these interactive activities include, but not limited to, extracellular matrix, adjacent differentiated cells, secreted and cell surface molecules, mechanical signals, spatial arrangements and certain metabolic conditions. MSCs have been reported to be associated with vasculature which is considered as their niche [19, 20].

3.1 DPSC Niche

To localize the DPSC niche in human dental pulp, an effort was made to detect the expression of MSC markers STRO-1 and CD146, and a pericyte marker 3G5. As shown in Fig. 1, these makers are in the perivascular or perineural sheath regions [21]. For the STRO-1/CD146 double staining, one can observe that STRO-1 and CD146 seem to represent two populations in the vasculature. Selected STRO-1$^+$, CD146$^+$ or 3G5$^+$subpopulation of DPSCs form up to 20-fold more colony-forming unit-fibroblasts (CFU-F) than the non-selected population. CD146$^+$ DPSCs form ectopic pulp-dentin complex in vivo when mixed with HA/TCP powder [21]. Interestingly, STRO-1 is expressed in the neurosheath layer as well. Therefore, there are non-vascular DPSCs niches. Since cultured DPSCs express neural markers in high percentage (nestin: 92.7 %, NF-H: 42.3 %, GalC: 92.3 %, and βIII-tubulin: 95.9 %) [22], they were used to locate stem cell niches in the pulp of young permanent teeth. Nestin was found to be expressed throughout the tissue and βIII-tubulin was in the subodontoblast zone [23]. However, no further investigation was made to verify whether nestin$^+$ or βIII-tubulin$^+$ in pulp possess MSC characteristics in vitro or tissue regeneration capacity in animal study models, these neural markers are yet validated as markers for locating DPSCs niches in pulp. Other stem cell markers that have been used to verify MSCs are pluripotent stem cell markers such as OCT4, NANOG and SOX2 [24, 25]. OCT4 expression was detected in pulp of human deciduous teeth either in the vascular or non-vascular niches [26]. These pluripotent markers are expressed in cultured DPSCs and their expression levels can be enhanced with small molecule treatments [27]. Another potential marker that may locate DPSC niche is CXCR4, the receptor of SDF-1. Jiang et al. detected weak CXCR4 expression in healthy human pulp on the blood vessel walls. The expression is stronger in inflamed pulp and located in the non-vascular niche in addition to the blood vessel walls [28, 29]. Many non-vascular associated CXCR4$^+$ cells in inflamed pulp are likely inflammatory cells. NOTCH has been also considered as a marker for identifying DPSCs niches since NOTCH3 protein has been detected in pericytes of the injured pulp of rat molars [30, 31].

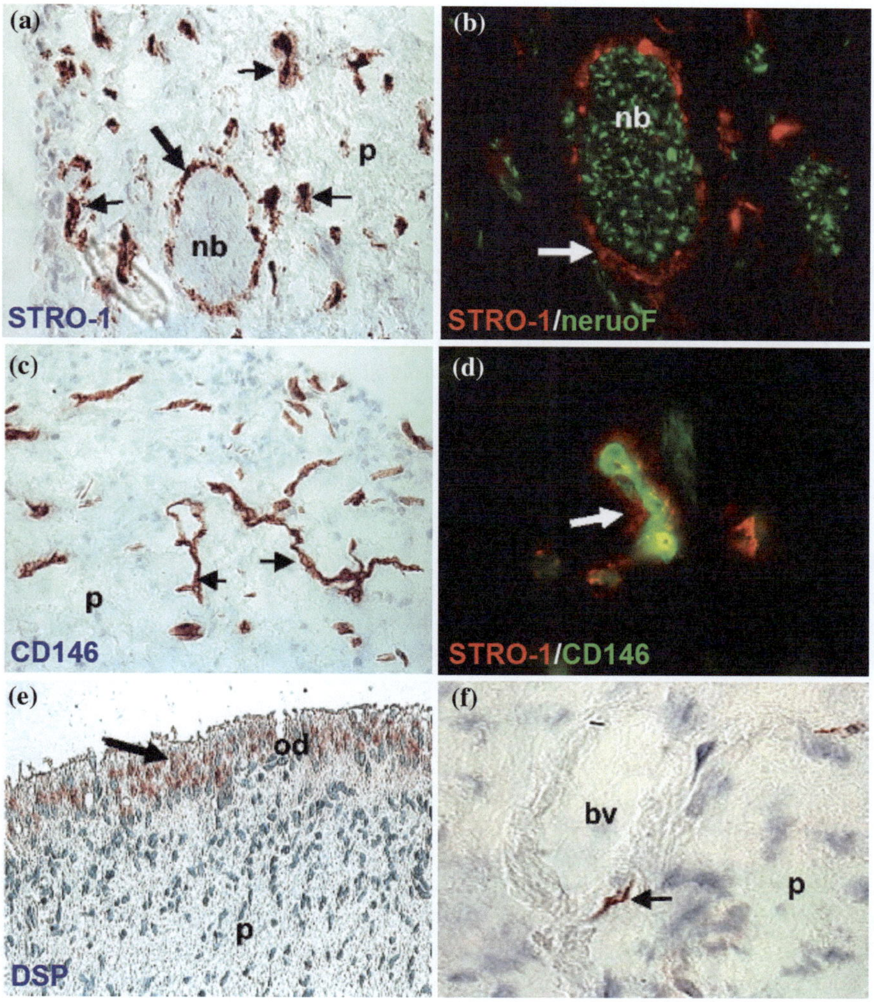

Fig. 1 Reactivity of perivascular makers in human dental pulp. **a** Immunolocalization of the STRO-1 antigen on blood vessels (*small arrows*) in human dental pulp (p) and around perineurium (*large arrow*) surrounding a nerve bundle (nb), 20x. **b** Dual immunofluorescence staining showing reactivity of the STRO-1 antibody labeled with Texas red to dental pulp perineurium (*arrow*) in combination with an antineurofilament antibody labeled with fluorescein isothiosyanate staining the inner nerve bundle (nb), 40x. **c** Immunolocalization of the CD146 antigen to blood vessel walls in human dental pulp tissue, 20x. **d** Dual immunofluorescence staining showing reactivity of the STRO-1 antibody labeled with Texas red to a blood vessel and the CC9 antibody labeled with fluorescein isothiosyanate. **e** Immunohistochemical staining of pulp tissue with a rabbit polyclonal anti-DSP antibody (*arrow*) to the odontoblast outer layer (od), 20x. **f** 3G5 reactivity to a single pericyte (*arrow*) in a blood vessel (bv) wall, 40x. Tissue sections were counterstained with hematoxylin. Adapted from [21] with permission

Recently, in a mouse model it was shown that a significant population of MSCs during development are derived from peripheral nerve-associated glia. Glial cells generate multipotent MSCs that produce pulp cells and odontoblasts indicating another non-vascular DPSC niche [32]. One caveat is that rodent incisors have a very different apical structure than its molar and human teeth, therefore, it remains to be tested whether glial cells in human pulps are also a stem cell source for odontoblasts. Using a mouse Ng2-Cre/Rosa26R model, pericytes on blood vessels are tracked and confirmed as the DPSC niches in molar teeth [33]. Although this Rosa26/Cre mouse model also identified non-pericyte DPSC niche close to cervical loop region in the pulp of incisor which is different from human pulp, it is likely that human pulp also uses dual origin of DPSC niches (vascular vs non-vascular) for the homeostasis and rapid repair from damage of the pulp [33].

3.2 SCAP Niche

Limited studies on MSC niches in apical papilla are reported. STRO-1 is detected mainly in the perivascular region of apical papilla and other regions scattered in the tissue [5]. CD105 expression in the apical papilla also appears to be associated with blood vessels [34]. NOTCH3 and CXCR4 were also found to be expressed paravascularly [35, 36].

4 Subpopulations of Dental Pulp Stem Cells

As mentioned above, DPSCs isolated via described methods are heterogeneous populations of stem/progenitor cells. Although lack of specific markers for DPSCs, a number of markers were used by researchers to select subpopulations of DPSCs (Table 1).

Table 1 DPSC subpopulations

Subpopulation	Isolation method	Tissue formation in vivo[a]	Reference
$STRO1^+$ $CD146^+$	MACS and FACS	Pulp-dentin like complex	[21]
$CD34^+$c-$kit^+$$CD45^+$or SBP/DPSCs	FACS	Bone-like	[39]
SP DPSCs	FACS	Pulp-dentin like	[43]
$CD31^-$$CD146^-$ SP	FACS	Enhance host angiogenesis	[44] [45]
$CD105^+$	FACS	Pulp-dentin like	[46]
MDPSCs	G-CSF mobilization	Pulp-dentin like	[48]
SSEA-4^+	FACS	dentin-like	[50]

[a]Some authors use "osteodentin" to describe "dentin-like" tissue formation in vivo

4.1 STRO-1 and CD146

Upon DPSC isolation and characterization, several subsets of cells were found to have different surface markers, proliferation rates and differentiation potential. These findings indicate the heterogeneous nature of DPSCs and a possible hierarchy among the cells [1, 21]. Using magnetic-activated cell sorting (MACS) or fluorescence activated cells sorting (FACS), Shi et al., isolated STRO-1^+ and CD146^+ subpopulations from unfractioned heterogeneous DPSCs. They found that 82 % and 96 % of the colony-forming cells were represented in the STRO-1 and CD146 subpopulations respectively, suggesting that DPSCs may reside within these subpopulations [21]. In rats, Yang et al., confirmed the feasibility of using FACS to isolate subpopulations of DPSCs and they were able to successfully isolate STRO-1^+cells from rat DPSCs. They found that STRO-1^+ rat DPSCs had better odontogenic potential than STRO-1^- subpopulation [37, 38].

4.2 $CD34^+c$-Kit^+CD45^-

Papaccio and colleagues isolated CD34^+c-kit$^+$CD45^- subpopulation of DPSCs [39]. The rationale for their selection criteria is based on the findings that (i) CD34 and c-kit are markers for stromal cells, (ii) c-kit is expressed in neural crest-derived cells, and (iii) CD45 is a marker for hematopoietic progenitors. Therefore CD34^+c-kit$^+$CD45^- would be most probably a population of stromal stem cells of neural crest origin. Although CD34 has not been considered to be expressed by MSC type, studies have been shown that CD34^+ fraction exists in MSC population [40, 41]. This CD34^+c-kit$^+$CD45^- subpopulation from DPSCs was found to be clonogenic, able to differentiate into pre-osteoblasts which in turn can generate osteoblasts capable of producing living autologous fibrous bone (LAB) tissue in vitro. They termed this population as stromal bone producing DPSCs (SBP/DPSCs). Interestingly, when they transplanted LAB in vivo, LAB formed lamellar bone with osteocytes [39]. The caveat is that the bone regenerated by DPSCs lacks the key feature of the bona fide bone, i.e., absence of bone marrow formation. The same group isolated a similar subpopulation from dental pulp cells of deciduous teeth and found similar findings [42].

4.3 $CD31^-CD146^-$ SP and $CD105^+$

Nakashima's group used several methods to isolate different subpopulations of DPSCs. In 2006, they isolated a side population (SP) fraction from human, bovine, canine and porcine dental pulp cells based on the exclusion of the DNA binding dye

Hoechst 33342 by flow cytometry [43]. The SP cells showed self-renewal properties with longer proliferative lifespan than non-SP cells. These cells demonstrated potential for trilineage differentiation. Interestingly, the SP cells expressed higher levels of CD146, CD31 and FLK-1 than the non-SP cells suggesting their possible perivascular/pericyte origin. The autogenous transplantation of BMP2-treated SP cells on amputated pulp stimulated reparative dentin formation suggesting that the SP subfraction of DPSCs can be used for cell-based dentin regeneration [43]. The research team also isolated a subpopulation of SP subfraction based on the expression of CD31, CD146 and CD105. $CD31^{-}CD146^{-}$ SP cells and $CD105^{+}$ cells were highly proliferative and showed multilineage differentiation potential [44]. In a rat cerebral ischemia model, local transplantation of porcine $CD31^{-}CD146^{-}$ SP cells and $CD105^{+}$ cells resulted in acceleration of neovascularization of the ischemic zone [45]. Transplantation of $CD105^{+}$ pulp cells with SDF-1 induced complete pulp regeneration in adult dog teeth [46]. Interestingly, in the study model the transplanted cells were found to express pro-angiogenic factors VEFG-A and were in the proximity of the newly formed vasculature, implying a possible trophic effect of transplanted cells on enhancing vascularization [47]. Recently the same group developed a method for cell isolation with an aim to isolate clinical-grade pulp stem cells based on good manufacturing practice [48]. The method is based on the ability of granulocyte colony-stimulating factor (G-CSF) to mobilize a subpopulation of DPSCs and they termed this subpopulation as mobilized DPSCs (MDPSCs) [48, 49]. The isolated MDPSCs were found to be clonal, proliferative and express stem cells markers (such as CD105, CXCR4 and SOX2) more than unfractionated pulp cells. They also showed multidifferentiation potential into endothelial, neuronal, adipose and odontoblast-like cells. When they transplanted this subpopulation of cells with G-CSF into mature dog teeth after pulpectomy, they found pulp-like tissue with vasculature and odontoblast-like cells attached to the dentinal walls 14 days after transplantation [48].

4.4 Stage-Specific Embryonic Antigen (SEEA)-4

Kawanabe and colleagues used SEEA-4 to isolate a subpopulation of human DPSCs. They found 45.5 % of the dental pulp cells expressed SSEA-4, and the SSEA-4^{+} dental pulp cells are clonogenic, showed multilineage differentiation potential and had the capacity to form ectopic pulp-dentin-like complex in vivo. Interestingly SSEA-4^{+} cells lacked adipogenic differentiation potential [50].

5 Regulation of DPSC Niche

5.1 Extracellular Matrix in DPSC Niche

5.1.1 Effects of Extracellular Matrix on the Homeostasis of Stem Cell Niche

Extracellular matrix (ECM) mainly consists of four families, i.e., collagens, glycoproteins, proteoglycans and elastins [51]. The collagens are the main component of ECM, which exists in almost all tissues of organism. The glycoprotein contains fibronectin and laminin. The proteoglycans and elastins distribute on the cell surface or are present in the matrix, some of which can combine with cell receptors.

Stem cells reside in a dynamic and specialized niche that provides extracellular cues to allow the survival and to maintain a balance between self-renewal and differentiation of stem cells [52]. ECM proteins are the key components shaping the niche and maintaining stem cell homoeostasis [53]. ECM can organize a platform for molecular complex assembly, ultimately leading to integrate signals emanated from soluble and matrix-bound factors and from cell-matrix interactions as well [54].

Differential expression of EMC components in the niche can define tissue specificity. Subtle variations in ECM due to altered protein biosynthesis, post-synthetic modifications or imbalanced proteolytic degradation of ECM components, can promote a self-amplifying process leading to a progressive weakening of cell-ECM interactions and subsequent self-renewal capability [53]. The reduction of instructive cues in injured and senescent ECM affects the dynamics of stem cell niche. A reciprocal crosstalk also exists, as stem cells can support their viability and tissue specificity by secreting ECM proteins [55]. In fact, stem cell lineage selection is strictly dependent on compliant ECM protein compositions. The typical hierarchy of tissue rigidity depends on both ECM and constituent cells. Evidence has shown that soft matrices, similar to those found in the brain, drive stem cells into neurogenic lineages. In contrast, myogenic and osteogenic lineage specificity depends on stiffer matrices [56].

5.1.2 Effects of ECM on DPSC Proliferation

The chemical composition of ECM, free- and ECM-linked bioactive molecules, and mechanic and adhesive forces generated in the dynamic space of the niche are key players in stem cell proliferation and migration. To date, various scaffolds have been used to mimic ECM environment around the DPSC niche. Stem cells grown on an electrospun poly-L-lactic acid (PLLA) scaffolds have a higher proliferation rate than on 2-dimensional conditions [57]. The hydroxyapatite-polycaprolactone (HA-PCL) composite scaffolds can also promote the proliferation of both human BMMSCs and DPSCs [58]. Porous silicon scaffolds with 36 nm pore size can offer

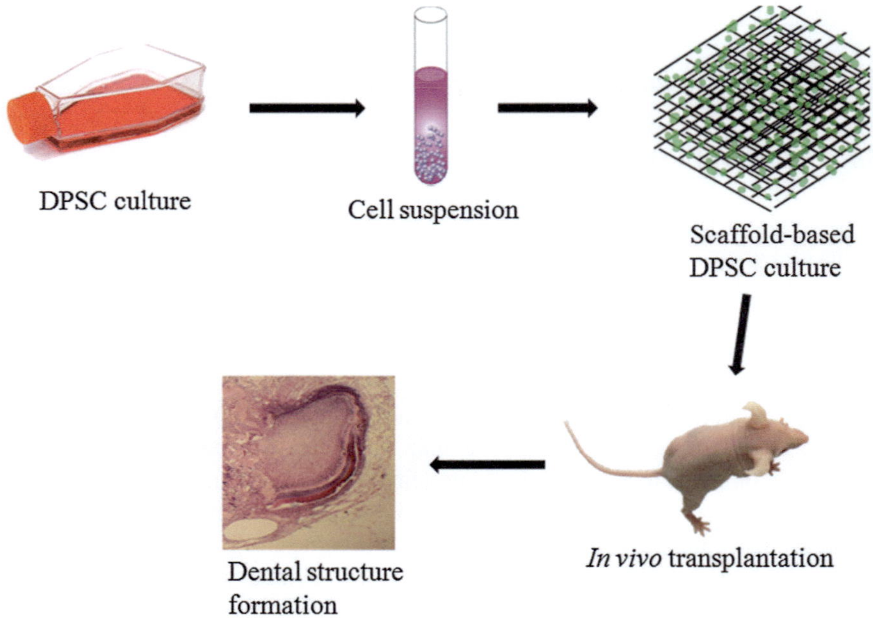

Fig. 2 Scaffolds as DPSC carrier for establishment of ECM environment in vivo. DPSCs are seeded onto the 3D scaffolds and transplanted into animals to form dental structures in vivo

the best adhesion and fastest growth rate for DPSCs [59]. DPSCs also showed some proliferation and mineralization characteristics on electrospun poly (epsilon-caprolactone) (PCL)/gelatin scaffolds with or without the addition of nano-hydroxyapatite [60].

However, scaffold-based engineering for dental tissue regeneration (Fig. 2) has achieved only limited success due to following complications. Firstly, the existence of scaffolds negatively affects the sufficient cell-cell interactions in DPSCs niche. Secondly, intrinsic positional information in DPSC niche is interrupted to some extent in the presence of these artificial scaffolds. Thirdly, the acidic side-products of some scaffolds such as polyglycolic acid (PGA), poly(lactic-co-glycolic acid) (PLGA) and PLLA, may exert an adverse effect on the proliferation of stem cells in vivo. Finally, the nutrient delivery and metabolic waste removal inside the central scaffolds are often limited following the dentinogenesis, which may negatively affect the growth and differentiation of DPSCs [61].

To avoid the obstacles of biodegradable scaffolds, cell pellet/aggregate engineering has been developed and provided some significant advantages in dental tissue engineering (Fig. 3). Cell-cell and cell-matrix interactions in cell pellet/aggregate engineering are more sufficient and natural than those in scaffold-based tissue regeneration. Although native intercellular ECM disappears during primary cell culturing, re-aggregated DPSCs can produce new ECM that acts

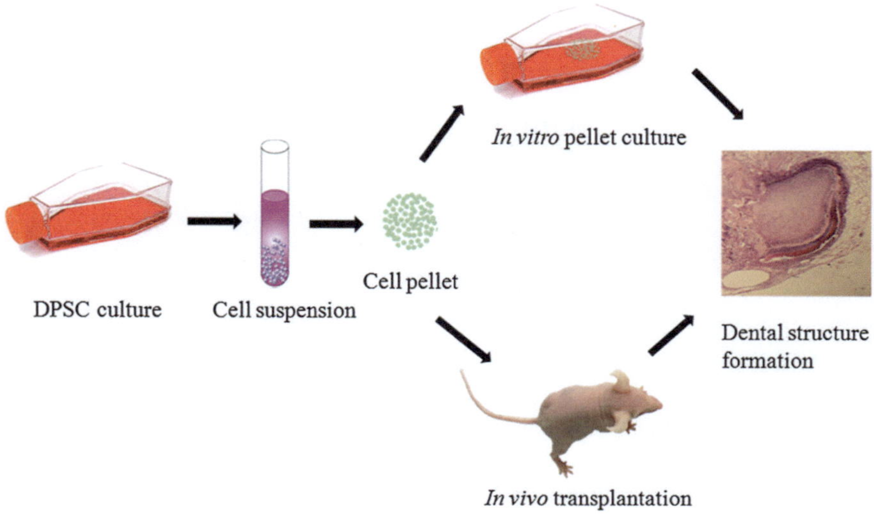

Fig. 3 Cell pellet tissue engineering. DPSCs are harvested as pellets and cultured in vitro or directly transplanted into animals to generate dental tissues in vivo

as a natural scaffold and generate new growth factors necessary for cell proliferation and adhesion.

5.1.3 Effects of ECM on DPSC Differentiation

Some reports have shown that natural MSCs are sensitive to ECM elasticity. They can differentiate specifically into a particular lineage with the changes of ECM elasticity index and produce a corresponding cell phenotype. According to the matrix elasticity, tissues can be classified into 3 distinct groups: (i) soft tissues, such as brain; (ii) stiff tissues, such as muscles; and (iii) rigid tissues, such as bone. It is thought that the great variation in tissues or ECM microenvironments provides specific conditions that can drive the cell differentiation into different lineages [56]. Both matrix stiffness and soluble factors can modulate MSC-specific lineage commitment via RhoA signaling and Rho-kinase (ROCK) activity [62].

Many studies have demonstrated that the use of biodegradable scaffolds can mimic the functions of ECM to facilitate the differentiation of DPSCs. Collagen scaffolds can induce DPSCs to produce an abundant deposition of mineralized ECM, in which type I collagen is considered to offer the initiation sites for calcification [63]. Moreover, DPSCs can differentiate into the osteo/odontoblastic phenotype by means of composite HA-PCL scaffolds. The osteo/odontogenic differentiation is mainly affected by calcium phosphate mineral components in scaffolds and then sol–gel reactions allow calcium phosphate nanoparticles to be finely dispersed into a PCL matrix [58]. Nanotopography and hyaluronic acid

Table 2 Effects of growth factors on DPSCs

Growth factors	Functions		
	Proliferation	Differentiation	Morphogenesis
BMPs	Positive	Positive	Positive
TGF-β	Positive	Positive	Positive
FGFs	N/A	Positive	Positive
VEGF	Positive	Positive	Positive
IGFs	Positive	Positive	Positive
NGF	N/A	Positive	N/A

N/A: not available

provide important chemical cues to promote the chondrogenic differentiation of DPSCs [64].

As a mineralized connective tissue, dentin is composed of organic components of collagen and non-collagen proteins which contain some important ECM components. Among them, dentin sialophosphoprotein (DSPP), dentin sialoprotein (DSP), dentin phosphoprotein (DPP) and dentin matrix protein-1 (DMP-1) are considered to play an important role in DPSC differentiation [65].

5.2 *Growth Factors in DPSCs Niche*

Many growth factors can modulate the microenvironment of DPSCs including bone morphogenetic proteins (BMPs), fibroblast growth factors (FGFs), vascular endothelial growth factor (VEGF) and insulin-like growth factors (IGFs), transforming growth factor (TGF) β, platelet-derived growth factor (PDGF), epidermal growth factor (EGF), etc. (Table 2). These growth factors function synergistically and/or antagonistically to organize and pattern dental tissues and organs [66].

5.2.1 Effects of Growth Factors on DPSC Proliferation and Differentiation

(i) BMPs

Bone morphogenetic proteins (BMPs)-2, -4, and -7 are key signals that participate in the epithelial-mesenchymal interactions during tooth development. During the root formation, pre-odontoblasts and pulp cells can express BMP-4, while BMP-2 and -7 are expressed in the early odontoblasts. As odontoblasts become mature and start to secrete dentin matrix, BMP-4 expression is significantly downregulated. In contrast to BMP-2, -4, and -7, BMP-3 works as a BMP antagonist, which has an important role in the maintenance of the pulp [67].

Recombinant human BMP-2 (rhBMP-2) has been reported to induce both bovine and human adult pulp cells to differentiate into odontoblast lineages [68]. rhBMP-2, -4, and -7 are capable of stimulating the differentiation of DPSCs and form the regenerative dentin [69–71]. When reconstituted in a collagen gel drop and stimulated by microencapsulated and control-released BMP4, BMP7 and Wnt3a, postnatal dental epithelial and mesenchymal cells can orchestrate de novo formation of enamel- and dentin-like tissues in two integrated layers in vivo [72, 73].

(ii) TGF-β

TGF-β family comprises a group of diverse growth factors including TGF-βs, BMPs, growth differentiation factors (GDFs), activins, inhibins, anti-Müllerian hormone (AMH), and nodal growth differentiation factor (NODAL) [74]. In humans, three TGF-β isoforms, TGF-β1, TGF-β2 and TGF-β3 are expressed. TGF-β1 is the most abundant isoform, which is released mainly from platelets, macrophages and bone. TGF-β1 has been shown to increase the proliferation, migration and production of the ECM of DPSCs. In a study investigating effects of short-term versus prolonged administration of TGF-β, longer duration of TGF-β treatment can produce more efficient mineralization in MSCs. FK506 binding protein 1A (FKBP1A) interacts with TGF-BR1 and upregulates the odontoblast differentiation of dental pulp cells. Clearly, a complex series of interactions involving TGF-β family members are associated with the mineralization process in DPSCs [75].

(iii) FGFs

FGFs are a family of multifunctional polypeptide growth factors that regulate the proliferation, differentiation, survival and motility of cells of mesenchymal, epithelial and neuroectodermal origin. These properties of FGFs are essential in embryonic development. In adults, FGFs are mainly involved in inflammatory processes, wound healing and angiogenesis. The family comprises two prototypic members, i.e., acidic FGF (aFGF) and basic FGF (bFGF), as well as 21 additionally related polypeptide growth factors [76–79].

bFGF can exert a significant effect on the proliferation of human DPSCs. It promotes DPSC self-renewal and neuronal differentiation but inhibits their mineralization under osteogenic conditions. In addition, bFGF-based neuronal differentiation of DPSCs occurs via the FGFR and PLCγ intracellular transduction pathways [80]. bFGF increases the stemness gene expression and proliferation of SCAP [81].

Many in vivo studies have demonstrated that bFGF is a potent inducer of committed progenitor differentiation in hard-tissue regeneration. The ectopic transplantation of dental pulp cells/FGF-2 in vivo increases the formation of dentin-like structures. bFGF dramatically induces the mRNA expression of DSPP and bone sialoprotein (BSP) in immature dental pulp cells. Dental pulp cells pre-treated with bFGF2 show the increased ALP activity and calcified nodule formation [82]. bFGF combined with TGFβ1 induces the differentiation of dental pulp cells into odontoblast-like cells, and synergistically upregulates the effects of TGFβ1 during odontoblast differentiation [83, 84].

(iv) VEGF

VEGF is the most potent and specific angiogenic factor. VEGF family currently includes six known members: VEGF-A, B, C, D, E, and placenta growth factor. Three tyrosine kinases, VEGFR-1, VEGFR-2 and VEGFR-3 have been identified as VEGF receptors. VEGFs generated by DPSCs act directly on themselves in an autocrine manner to induce their proliferation and differentiation. This action is regulated by VEGFR-2 and partly by AP-1 signaling via the c-FOS protein [85]. When cultured in a 3-D fibrin mesh, VEGF can induce DPSCs to acquire endothelial cell-like features, displaying focal organization into capillary-like structures [86]. In addition, VEGF can enhance the differentiation of immature DPSCs into vascular endothelial cells and promote their neovascularization [87].

(v) IGFs

IGFs, including IGF-1 and IGF-2, are evolutionarily conserved peptide structurally related to insulin. IGFs induce a variety of cellular responses, including cell proliferation, differentiation, migration, and survival. IGF-1 and two other proteins, IGF-1-binding-protein-3 (IGFBP-3) and IGF-1 receptor (IGF-1R), serve as regulators of cellular proliferation. There is growing evidence that IGF system plays a very important role in the development, homeostasis and regeneration of dental tissues. IGFs regulate tooth morphogenesis and repair by controlling cell proliferation and differentiation. Treatment of human DPSCs with IGF-1 increases cell proliferation, ALP activity, and DNA synthesis. IGFBP-3 may inhibit cell proliferation through competitive binding to IGF-1, and also affect cell proliferation irrespective of the presence of IGF-1 [88]. IGF-1 with PDGF-BB has a synergistic effect on the proliferation of dental pulp cells in vitro [89]. Recent studies have revealed that IGF-1 promotes osteogenic differentiation of dental stem cells via the MAPK pathway in vitro and enhances their osteogenic mineralization in vivo [90]. IGF-1 also triggers the osteogenic differentiation of human DPSCs through mTOR signaling pathway [91]. Both Torc 1 and Torc 2 play a role in the modulation of DPSCs in which Torc 1 is proved to be essential.

(vi) NGF

Nerve growth factor (NGF), from the family of neurotrophins, is a small signaling protein which is known to be necessary to promote the growth and survival of sympathetic fibers and sensory nerves. NGF and its receptor tropomyosin-related kinase A (TrkA) play a crucial role in the development and function of the nervous system. DPSCs exert an intriguing regenerative potential in the damaged central nervous system of the rodent, that is attributed in part to their multiple-differentiation ability to replace lost neurons, but mainly to the production of neurotrophic factors including NGF and brain derived neurotrophic factor (BDNF) that promote neuron survival and axon guidance [92, 93]. NGF is involved in the guidance of trigeminal axons in embryonic teeth. Expression of NGF mRNAs in postnatal teeth is correlated with trigeminal axon growth, indicating that NGF plays a paramount role in the final innervation pattern of dental pulp and dentin. Neurotrophins can speed up the mRNA expressions of DSPP, ALP, osteopontin (OPN), type I collagen and BMP-2 as well as the formation of calcified substances in DPSCs [94].

In general, growth factors can affect both the proliferation and differentiation of DPSCs. They work synergistically and/or antagonistically as a complicated network to regulate the DPSC niche. Their efficiency is also modulated by DPSC themselves and other components inside or outside the niche.

Table 3 Signaling pathways in DPSC Niche

Function/effect	Homeostasis	Proliferation		Differentiation	
	Maintenance	Positive	Negative	Positive	Negative
Signaling pathways	NOTCH	NF-κB PI3 K/Akt MAPK NOTCH ITGA5	PI3 K/Akt mTOR	NF-κB	NF-κB
				MAPK	NOTCH
				NOTCH Id1/BMP2 Smad	WNT/β-catenin ITGA5
				FGFR	
				PLCγ	

5.3 Signaling Pathways in DPSC Niche

A complex network of signaling pathways, cytokines, growth factors, and chemokines, participates in maintaining and regulating the homeostasis of DPSC niche as well as the self-renewal and multiple differentiation of DPSCs (Table 3).

5.3.1 NOTCH Signaling Pathway

NOTCH signaling is an important signaling pathway for niche regulation, controlling stem cell maintenance as well as cell fate decision. Although NOTCH receptors are absent in adult rat pulp tissue, their expression is reactivated after dental injury [30]. Generation of daughter cells from asymmetric stem cell division is closely associated with NOTCH signaling [95]. NOTCH ligand Delta1 is known to influence the proliferation and differentiation of many types of tissue specific stem cells. NOTCH-Delta1 signaling is expressed in human DPSCs and can enhance the proliferation of DPSCs [96]. The activation of NOTCH signaling by either Jagged1 or N1ICD inhibits odontoblast differentiation of DPSCs without affecting their proliferation [97]. Therefore, NOTCH signaling pathway plays an important role in maintaining the correct balance between proliferation and differentiation of DPSCs [96].

5.3.2 MAPK Signaling Pathway

Mitogen-activated protein kinases (MAPK) signaling pathway is crucial in balancing cell apoptosis, survival, migration, proliferation, differentiation and other cellular processes. Three main MAPK families (extracellular signal-regulated kinases (ERK), c-Jun-N-terminal kinase (JNK) and p38) are distinctly involved in these processes [98].

2-hydroxyethyl methacrylate (HEMA), a kind of resin-based dental materials, can inhibit the migration of DPSCs by phosphorylation of p38 or JNK MAPK pathways [99]. P38 MAPK and IGF-1R are responsible for the mitotic quiescence of DPSCs. There is a cross talk between IGF-1R and p38 MAPK signaling pathways in DPSCs, and the signals provided by these pathways converge at STAT3 and inversely regulate its activity to maintain the quiescence or to promote self-renewal and differentiation of these cells [100]. Mechanical stretch increases the proliferation while inhibiting the osteo/odontogenic differentiation of DPSCs. The stretch-induced proliferation of DPSCs is mediated by Akt, ERK1/2 and P38 MAPK signaling pathways, while down-regulation of osteo/odontogenic differentiation was associated with PI3 K/Akt and ERK pathways [101]. Natural mineralized scaffolds (DDM and CBB) can also induce the osteo/odontogenic differentiation of DPSCs via MAPK signaling pathways [102].

5.3.3 NF-κB Signaling Pathway

NF-κB is a type of transcription factor that is activated in most cell types after stimulation by a variety of factors, including cytokines, growth factors, and hormones. It regulates the expression of a large number of genes, and it is usually activated in cells after RANKL, IL-1, or TNF binds to their respective receptors [103]. NF-κB pathway plays an important role in the osteo/odontogenic differentiation of DPSCs. TNF-α (an activator of NF-κB pathway) enhances the odontoblastic phenotype of DPSCs [104]. 17β-estradiol promotes the osteo/odontogenic potency of human DPSCs by activation of NF-κB signaling pathway [105]. Moreover, DPSCs from the injured pulp present a lower proliferative capacity, upregulated osteogenic potential and weakened odontogenic capacity than healthy DPSCs and this process is also associated with NF-κB signaling pathway[106].

5.3.4 TGF-β Signaling Pathway

Transforming growth factor-β1 (TGF-β1), a multi-functional cytokine, is important for the homeostasis of dental tissues. TGF-β1, TGF-β2, and TGF-β3 mRNA are expressed in human DPSCs [107]. TGF-β2 possibly regulates the differentiation of dental pulp cells at specific stages synergistically with other factors via multiple signaling pathways, especially the ALK/Smad2/3-signal transduction pathways [107]. TGF-β signaling also participates in the NGF regulation during pulp tissue repair and can up-regulate NGF levels in human dental pulp cells via p38 and JNK MAPK pathways [108]. Additionally, TGF-β signaling controls the odontoblast differentiation and dentin formation during tooth morphogenesis [109].

5.3.5 WNT/β-Catenin Signaling Pathway

Canonical WNT signaling plays a pivotal role in tooth development and stem cell self-renewal through β-catenin. Over-expression of β-catenin can sufficiently suppress the differentiation and mineralization of DPSCs [110]. Zinc-bioglass (ZnBG) incorporated within calcium phosphate cements can activate the odontogenic differentiation and promote the angiogenesis of dental pulp cells in vitro. ZnBG up-regulates integrins α1, α2, β1, and β3 and activates the downstream signaling pathways including the canonical and non-canonical WNT signaling pathways [111].

5.3.6 Eph/Ephrin Signaling Pathway

The Eph family of receptor tyrosine kinases and their ligands, the ephrin molecules, are reported to play an essential role in the migration of neural crest cells during tooth development and stem cell niche maintenance [112]. DPSCs stimulated by EphB2-Fc and EphB1-Fc which interfere the EphB/EphrinB signaling exhibit a significant rounder and smaller morphology. This suggests that EphB/EphrinB interactions can mediate cell attachment, spreading and migration in the DPSC niche [113].

5.3.7 Signaling Synergism and Crosstalks

Signaling pathways can affect each other synergistically in balancing cellular activities of DPSCs. A large number of regulatory genes in odontogenic and osteogenic differentiation interact or crosstalk via NOTCH, WNT, TGF-β/BMP, and cadherin signaling pathways [114]. Extracellular phosphate (Pi) can regulate BMP-2 expression via cAMP/protein kinase A and ERK1/2 MAPK pathways in human DPSCs [115]. TGF-β1 can down-regulate the differentiation ability of human DPSCs through ALK5/SMAD2/3 signaling pathways [116]. P38 MAPK pathway is implicated in regulating ALP activity in human DPSCs and may interact with SMAD pathways [117]. AMPK, AKT and mTOR signaling pathways are associated with each other in the process of human DPSC differentiation [118]. LPS can induce the over-expression of WNT5α signaling in human DPSCs while it is critically associated with other signaling pathways. LPS-induced WNT5a expression is mediated through the TLR4/MyD88/PI3-kinase/AKT pathway, which then initiates NF-κB activation in human DPSCs [119].

5.4 Interactions Between Different Components in DPSC Niche

5.4.1 Interactions Between ECM and Growth Factors

In the stem cell microenvironment, ECM components are capable of trapping growth factors and regulate their local concentrations and availability [120]. Researchers have studied ECM and growth factors in different combinations and sequences to direct the differentiation of stem cells in vitro to specific cell phenotypes by precisely controlling the biochemical composition of the cell microenvironment [121].

(i) Protective effects of ECM on growth factors

Generally, ECM is a reservoir of growth factors. The bioavailability of growth factors can be regulated by ECM proteins via establishing stable gradients of growth factors. ECM proteins (e.g., fibronectin, vitronectin, collagens and proteoglycans) themselves or in combination with heparin and heparin sulphate, actively integrate many growth factors, such as FGFs, HGF, and VEGFs. Similarly, type II and IV collagens can act as an insoluble and localized reservoir for morphogens to bind BMPs and TGF-β through conserved structure modules [53].

(ii) ECM regulates the activity and expression levels of growth factors

Interactions between ECM proteins and growth factors can increase integrin-growth factor receptor crosstalks and cellular responses [53]. Likewise, ECM is able to regulate the activity of TGF-βs via controlling the proteolytic activation of latent transforming growth factor-β-binding protein-1 (LTBP1) [122]. As a type of ECM molecule, tenascin C (TNC) is an extrinsic regulator of neural stem cell behavior. TNC and other ECM molecules present in germinal zones of CNS can play a paramount role in limiting or magnifying growth factor expressions in neural stem cells. TNC regulates the responses to two growth factors (FGF2 and BMP4), thus modulating EGF receptor expression so as to promote EGF receptor acquisition [123].

5.4.2 Interactions Between ECM and Signaling Molecules

ECM contains abundant signaling molecules that modulate cell signaling through the interactions with their respective receptors [53]. ECM relays complicated signals to stem cells in the niche. During the formation of mineralized tissues such as bone and dentin, several intracellular signaling pathways are embedded in these biological processes, representing an interconnected network of proteins to decipher clues from ECM [124].

The importance for cellular attachment to an adhesive ECM potentially acts either to improve cell flattening and the development of tension, which has been indicated to be essential for a productive mitogenic response to growth factors in a variety of cell types, or by ligation of integrin receptors, which appears to synergize with growth factors so as to co-stimulate the MAPK signaling pathway [125].

5.4.3 Growth Factors and Signaling Pathways

Many factors have been described as MSC differentiation inducers including platelet-derived growth factor (PDGF), bFGF, TGF β, EGF, and IGF [126]. These growth factors are able to interplay with distinct cell receptors via activating different intracellular signaling pathways.

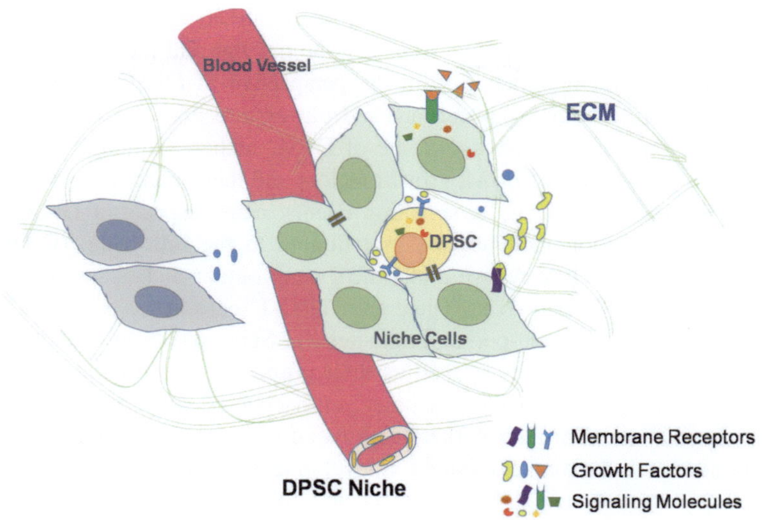

Fig. 4 Hypothetic model of regulation in DPSC niche. A perivascular DPSC niche containing a DPSC and the niche supporting cells, ECM, soluble factors, receptors and intracellular signaling factors

The members of TGF-β family may have a beneficial effect for neuronal survival and promote neurogenesis [120]. The expression of TGF-β type I and Smad2 signaling by GlcN/oligo-GlcN is coincident with their promoting effects on the early osteogenic differentiation of DPSCs. Exogenous GlcN is able to promote the osteogenic differentiation of DPSCs via the increased expression of TGF-β type I. Furthermore, GlcN can activate the Smad2 signaling pathway via increasing TGF-β type I, and modulate the phosphorylation of Runx2/Cbfa1 during initial osteogenic differentiation of human DPSCs [127].

In summary, the regulation of DPSC niche involves ECM, growth factors, bioactive molecules, signaling pathways as well as the interactions between each other. These factors act antagonistically or synergistically as a network to regulate the status of DPSCs in the niche (Fig. 4).

6 Conclusion and Prospects

DPSCs are versatile stem cells and hold tremendous potential for regenerative medicine. There have been a significant amount of studies accumulated since their discovery. This review highlighted the localization of DPSC niche in the pulp, the subpopulations and most importantly the regulation of the niche at the molecular levels. Other dental stem cells such as SCAP are relatively less studied in terms of these areas mentioned above. Understanding the cellular and molecular regulation of DPSC niche is of importance from the perspective of DPSC biology. As we outlined in depth, a complex interconnected players are involved in such a regulation that determines the status and fate of DPSC niches. Although efforts have been put into the studies of ECM, growth factors and their interactions with DPSC receptors leading to intracellular signaling events, many aspects of DPSC niches still await investigation including but not limited to the following: (i) Different locations of DPSC niches. As we underlined, perivascular niches are not the only location where DPSCs may exist. There are STRO-1^+ cells at the perineural sheath as well as possible other non-perivascular sites. (ii) Subpopulations of DPSCs, their in situ location and functions as well as their properties in vitro. More extensive studies are needed to further identify other yet discovered subpopulations, if exist, and the distribution of surface markers within and between the subpopulations. We need to understand the marker expression level in association with other recognized stemness genes such as OCT4, NANOG and SOX2. (iii) The stem cell niche information and regulation specific to DPSC niche is still limited. More in situ microstructural studies are needed to understand specific DPSC niche microenvironment. Further defining DPSC niche properties may require improved and innovative 3D culture approaches that may be technological challenging. Combination of in vitro models such as using bioengineered scaffolds or cell pellet 3D cultures and in vivo animal models is likely needed to dissect the reconstituted artificial/regenerated DPSC niches. (iv) No information is available regarding DPSC niches in aging or diseased pulps. Acquiring such information mentioned above is critical not only to further understand the biology of DPSCs, but also for establishing stem cell-mediated therapeutic strategies [128].

Acknowledgments This work was supported in part by grants from the National Institutes of Health R01 DE019156 (G.T.-J.H.).

References

1. Gronthos S, Mankani M, Brahim J, Robey PG, Shi S. Postnatal human dental pulp stem cells (DPSCs) in vitro and in vivo. Proc Natl Acad Sci U S A. 2000;97(25):13625–30.
2. Miura M, Gronthos S, Zhao M, Lu B, Fisher LW, Robey PG, et al. SHED: stem cells from human exfoliated deciduous teeth. Proc Natl Acad Sci. 2003;100(10):5807–12.
3. Seo BM, Miura M, Gronthos S, Bartold PM, Batouli S, Brahim J, et al. Investigation of multipotent postnatal stem cells from human periodontal ligament. Lancet. 2004;364 (9429):149–55.
4. Morsczeck C, Gotz W, Schierholz J, Zeilhofer F, Kuhn U, Mohl C, et al. Isolation of precursor cells (PCs) from human dental follicle of wisdom teeth. Matrix Biol. 2005;24 (2):155–65.
5. Sonoyama W, Liu Y, Fang D, Yamaza T, Seo BM, Zhang C, et al. Mesenchymal stem cell-mediated functional tooth regeneration in Swine. PLoS One. 2006;1:e79.

6. Sonoyama W, Liu Y, Yamaza T, Tuan RS, Wang S, Shi S, et al. Characterization of the Apical papilla and its residing stem cells from human immature permanent teeth: a pilot study. J Endod. 2008;34(2):166–71.
7. Dominici M, Le Blanc K, Mueller I, Slaper-Cortenbach I, Marini F, Krause D, et al. Minimal criteria for defining multipotent mesenchymal stromal cells. The international society for cellular therapy position statement. Cytotherapy. 2006;8(4):315–7.
8. Huang GT, Gronthos S, Shi S. Mesenchymal stem cells derived from dental tissues vs. those from other sources: their biology and role in regenerative medicine. J Dent Res. 2009;88 (9):792–806.
9. Gronthos S, Brahim J, Li W, Fisher LW, Cherman N, Boyde A, et al. Stem cell properties of human dental pulp stem cells. J Dent Res. 2002;81(8):531–5.
10. Zhang W, Walboomers XF, Shi S, Fan M, Jansen JA. Multilineage differentiation potential of stem cells derived from human dental pulp after cryopreservation. Tissue Eng. 2006;12 (10):2813–23.
11. Batouli S, Miura M, Brahim J, Tsutsui TW, Fisher LW, Gronthos S, et al. Comparison of stem-cell-mediated osteogenesis and dentinogenesis. J Dent Res. 2003;82(12):976–81.
12. Gronthos S, Graves SE, Ohta S, Simmons PJ. The STRO-1+ fraction of adult human bone marrow contains the osteogenic precursors. Blood. 1994;84(12):4164–73.
13. Huang GTJ, Yamaza T, Shea LD, Djouad F, Kuhn NZ, Tuan RS, et al. Stem/progenitor cell–mediated de novo regeneration of dental pulp with newly deposited continuous layer of dentin in an in vivo model. Tissue Eng Part A. 2009;16(2):605–15.
14. Banchs F, Trope M. Revascularization of immature permanent teeth with apical periodontitis: new treatment protocol? J Endod. 2004;30(4):196–200.
15. Huang GTJ, Sonoyama W, Liu Y, Liu H, Wang S, Shi S. The hidden treasure in apical papilla: the potential role in pulp/dentin regeneration and bioroot engineering. J Endod. 2008;34(6):645–51.
16. Moore KA, Lemischka IR. Stem cells and their niches. Science. 2006;311(5769):1880–5.
17. Roeder I, Loeffler M, Glauche I. Towards a quantitative understanding of stem cell–niche interaction: Experiments, models, and technologies. Blood Cells Mol Dis. 2011;46(4):308–17.
18. Schofield R. The relationship between the spleen colony-forming cell and the haemopoietic stem cell. Blood Cells. 1978;4(1–2):7–25.
19. Crisan M, Yap S, Casteilla L, Chen C-W, Corselli M, Park TS, et al. A perivascular origin for mesenchymal stem cells in multiple human organs. Cell Stem Cell. 2008;3(3):301–13.
20. Ergun S, Tilki D, Klein D. Vascular wall as a reservoir for different types of stem and progenitor cells. Antioxid Redox Signal. 2011;15(4):981–95.
21. Shi S, Gronthos S. Perivascular niche of postnatal mesenchymal stem cells in human bone marrow and dental pulp. J Bone Miner Res. 2003;18(4):696–704.
22. Martens W, Wolfs E, Struys T, Politis C, Bronckaers A, Lambrichts I. Expression pattern of basal markers in human dental pulp stem cells and tissue. Cells Tissues Organs. 2012;196 (6):490–500.
23. Martel-Pelletier J, Lajeunesse D, Reboul P, Pelletier JP. Therapeutic role of dual inhibitors of 5-LOX and COX, selective and non-selective non-steroidal anti-inflammatory drugs. Ann Rheum Dis. 2003;62(6):501–9.
24. Greco SJ, Liu K, Rameshwar P. Functional similarities among genes regulated by OCT4 in human mesenchymal and embryonic stem cells. Stem Cells. 2007;25(12):3143–54.
25. Pierantozzi E, Gava B, Manini I, Roviello F, Marotta G, Chiavarelli M, et al. Pluripotency regulators in human mesenchymal stem cells: expression of NANOG but not of OCT-4 and SOX-2. Stem Cells Devel. 2011;20(5):915–23.
26. Lizier NF, Kerkis A, Gomes CM, Hebling J, Oliveira CF, Caplan AI, et al. Scaling-up of dental pulp stem cells isolated from multiple niches. PLoS ONE. 2012;7(6):e39885.
27. Al-Habib M, Yu Z, Huang GT. Small molecules affect human dental pulp stem cell properties via multiple signaling pathways. Stem Cells Dev. 2013;22(17):2402–13.

28. Jiang L, Zhu YQ, Du R, Gu YX, Xia L, Qin F, et al. The expression and role of stromal cell-derived factor-1alpha-CXCR4 axis in human dental pulp. J Endod. 2008;34(8):939–44.
29. Jiang H, Ling J, Gong Q. The expression of stromal cell-derived factor 1 (SDF-1) in inflamed human dental pulp. J Endod. 2008;34(11):1351–4.
30. Lovschall H, Mitsiadis TA, Poulsen K, Jensen KH, Kjeldsen AL. Coexpression of Notch3 and Rgs5 in the pericyte-vascular smooth muscle cell axis in response to pulp injury. Int J Dev Biol. 2007;51(8):715–21.
31. Mitsiadis TA, Feki A, Papaccio G, Caton J. Dental pulp stem cells, niches, and Notch signaling in tooth injury. Adv Dent Res. 2011;23(3):275–9.
32. Kaukua N, Shahidi MK, Konstantinidou C, Dyachuk V, Kaucka M, Furlan A, et al. Glial origin of mesenchymal stem cells in a tooth model system. Nature. 2014;513(7519):551–4.
33. Feng J, Mantesso A, De Bari C, Nishiyama A, Sharpe PT. Dual origin of mesenchymal stem cells contributing to organ growth and repair. Proc Natl Acad Sci U S A. 2011;108 (16):6503–8.
34. Abe S, Yamaguchi S, Amagasa T. Multilineage cells from apical pulp of human tooth with immature apex. Oral Sci Int. 2007;4(1):45–58.
35. Jamal M, Chogle SM, Karam SM, Huang GT. NOTCH3 is expressed in human apical papilla and in subpopulations of stem cells isolated from the tissue. Genes Dis. 2015 [E-pub].
36. Liu JY, Chen X, Yue L, Huang GT, Zou XY. CXC chemokine receptor 4 is expressed paravascularly in apical papilla and coordinates with stromal cell-derived factor-1alpha during transmigration of stem cells from apical papilla. J Endod. 2015 [E-pub].
37. Yang X, van den Dolder J, Walboomers XF, Zhang W, Bian Z, Fan M, et al. The odontogenic potential of STRO-1 sorted rat dental pulp stem cells in vitro. J Tissue Eng Regen Med. 2007;1(1):66–73.
38. Yang X, Zhang W, van den Dolder J, Walboomers XF, Bian Z, Fan M, et al. Multilineage potential of STRO-1+ rat dental pulp cells in vitro. J Tissue Eng Regen Med. 2007;1(2):128–35.
39. Laino G, d'Aquino R, Graziano A, Lanza V, Carinci F, Naro F, et al. A new population of human adult dental pulp stem cells: a useful source of living autologous fibrous bone tissue (LAB). J Bone Miner Res. 2005;20(8):1394–402.
40. Lin CS, Ning H, Lin G, Lue TF. Is CD34 truly a negative marker for mesenchymal stromal cells? Cytotherapy. 2012;14(10):1159–63.
41. Lin CS, Xin ZC, Dai J, Lue TF. Commonly used mesenchymal stem cell markers and tracking labels: limitations and challenges. Histol Histopathol. 2013;28(9):1109–16.
42. Laino G, Graziano A, d'Aquino R, Pirozzi G, Lanza V, Valiante S, et al. An approachable human adult stem cell source for hard-tissue engineering. J Cell Physiol. 2006;206(3):693–701.
43. Iohara K, Zheng L, Ito M, Tomokiyo A, Matsushita K, Nakashima M. Side population cells isolated from porcine dental pulp tissue with self-renewal and multipotency for dentinogenesis, chondrogenesis, adipogenesis, and neurogenesis. Stem Cells. 2006;24 (11):2493–503.
44. Iohara K, Zheng L, Wake H, Ito M, Nabekura J, Wakita H, et al. A novel stem cell source for vasculogenesis in ischemia: subfraction of side population cells from dental pulp. Stem Cells. 2008;26(9):2408–18.
45. Sugiyama M, Iohara K, Wakita H, Hattori H, Ueda M, Matsushita K, et al. Dental pulp-derived CD31/CD146 side population stem/progenitor cells enhance recovery of focal cerebral ischemia in rats. Tissue Eng Part A. 2011;17(9–10):1303–11.
46. Iohara K, Imabayashi K, Ishizaka R, Watanabe A, Nabekura J, Ito M, et al. Complete pulp regeneration after pulpectomy by transplantation of CD105+ stem cells with stromal cell-derived factor-1. Tissue Eng Part A. 2011;17(15–16):1911–20.
47. Iohara K, Zheng L, Ito M, Ishizaka R, Nakamura H, Into T, et al. Regeneration of dental pulp after pulpotomy by transplantation of CD31(-)/CD146(-) side population cells from a canine tooth. Regen Med. 2009;4(3):377–85.

48. Iohara K, Murakami M, Takeuchi N, Osako Y, Ito M, Ishizaka R, et al. A novel combinatorial therapy with pulp stem cells and granulocyte colony-stimulating factor for total pulp regeneration. Stem Cells Transl Med. 2013;2(7):521–33.
49. Nakashima M, Iohara K, Murakami M. Dental pulp stem cells and regeneration. Endod Top. 2013;28(1):38–50.
50. Kawanabe N, Murata S, Fukushima H, Ishihara Y, Yanagita T, Yanagita E, et al. Stage-specific embryonic antigen-4 identifies human dental pulp stem cells. Exp Cell Res. 2012;318(5):453–63.
51. Hynes RO. The extracellular matrix: not just pretty fibrils. Science. 2009;326(5957):1216–9.
52. Fuchs E, Tumbar T, Guasch G. Socializing with the neighbors: stem cells and their niche. Cell. 2004;116(6):769–78.
53. Brizzi MF, Tarone G, Defilippi P. Extracellular matrix, integrins, and growth factors as tailors of the stem cell niche. Curr Opin Cell Biol. 2012;24(5):645–51.
54. Discher DE, Mooney DJ, Zandstra PW. Growth factors, matrices, and forces combine and control stem cells. Science. 2009;324(5935):1673–7.
55. Kurtz A, Oh SJ. Age related changes of the extracellular matrix and stem cell maintenance. Prev Med 2012;54 Suppl:S50–S56.
56. Engler AJ, Sen S, Sweeney HL, Discher DE. Matrix elasticity directs stem cell lineage specification. Cell. 2006;126(4):677–89.
57. Alessandri M, Lizzo G, Gualandi C, Mangano C, Giuliani A, Focarete ML, et al. Influence of biological matrix and artificial electrospun scaffolds on proliferation, differentiation and trophic factor synthesis of rat embryonic stem cells. Matrix Biol. 2014;33:68–76.
58. D'Anto V, Raucci MG, Guarino V, Martina S, Valletta R, Ambrosio L. Behaviour of human mesenchymal stem cells on chemically synthesized HA-PCL scaffolds for hard tissue regeneration. J Tissue Eng Regen Med. 2013.
59. Collart-Dutilleul PY, Secret E, Panayotov I, Deville de Periere D, Martin-Palma RJ, Torres-Costa V, et al. Adhesion and proliferation of human mesenchymal stem cells from dental pulp on porous silicon scaffolds. ACS Appl Mater Interfaces. 2014;6(3):1719–28.
60. Yang X, Yang F, Walboomers XF, Bian Z, Fan M, Jansen JA. The performance of dental pulp stem cells on nanofibrous PCL/gelatin/nHA scaffolds. J Biomed Mater Res Part A. 2010;93(1):247–57.
61. Yu J, Shi J, Jin Y. Current approaches and challenges in making a bio-tooth. Tissue Eng Part B Rev. 2008;14(3):307–19.
62. Even-Ram S, Artym V, Yamada KM. Matrix control of stem cell fate. Cell. 2006;126 (4):645–7.
63. Zhang W, Walboomers XF, van Kuppevelt TH, Daamen WF, Bian Z, Jansen JA. The performance of human dental pulp stem cells on different three-dimensional scaffold materials. Biomaterials. 2006;27(33):5658–68.
64. Nemeth CL, Janebodin K, Yuan AE, Dennis JE, Reyes M, Kim DH. Enhanced chondrogenic differentiation of dental pulp stem cells using nanopatterned PEG-GelMA-HA hydrogels. Tissue Eng Part A. 2014;20(21–22):2817–29.
65. Huang B, Sun Y, Maciejewska I, Qin D, Peng T, McIntyre B, et al. Distribution of SIBLING proteins in the organic and inorganic phases of rat dentin and bone. Eur J Oral Sci. 2008;116 (2):104–12.
66. Zhang YD, Chen Z, Song YQ, Liu C, Chen YP. Making a tooth: growth factors, transcription factors, and stem cells. Cell Res. 2005;15(5):301–16.
67. Nagatomo KJ, Tompkins KA, Fong H, Zhang H, Foster BL, Chu EY, et al. Transgenic overexpression of gremlin results in developmental defects in enamel and dentin in mice. Connect Tissue Res. 2008;49(6):391–400.
68. Liu HC, E LL, Wang DS, Su F, Wu X, Shi ZP, et al. Reconstruction of alveolar bone defects using bone morphogenetic protein 2 mediated rabbit dental pulp stem cells seeded on nano-hydroxyapatite/collagen/poly(L-lactide). Tissue Eng Part A. 2011;17(19–20):2417–2433.

69. Nakashima M. Induction of dentin formation on canine amputated pulp by recombinant human bone morphogenetic proteins (BMP)-2 and -4. J Dent Res. 1994;73(9):1515–22.
70. Nakashima M. Induction of dentine in amputated pulp of dogs by recombinant human bone morphogenetic proteins-2 and -4 with collagen matrix. Arch Oral Biol. 1994;39(12):1085–9.
71. Rutherford RB, Gu K. Treatment of inflamed ferret dental pulps with recombinant bone morphogenetic protein-7. Eur J Oral Sci. 2000;108(3):202–6.
72. Jiang N, Zhou J, Chen M, Schiff MD, Lee CH, Kong K, et al. Postnatal epithelium and mesenchyme stem/progenitor cells in bioengineered amelogenesis and dentinogenesis. Biomaterials. 2014;35(7):2172–80.
73. Lin ZM, Qin W, Zhang NH, Xiao L, Ling JQ. Adenovirus-mediated recombinant human bone morphogenetic protein-7 expression promotes differentiation of human dental pulp cells. J Endod. 2007;33(8):930–5.
74. Massague J. Receptors for the TGF-beta family. Cell. 1992;69(7):1067–70.
75. Liu J, Jin T, Chang S, Ritchie HH, Smith AJ, Clarkson BH. Matrix and TGF-beta-related gene expression during human dental pulp stem cell (DPSC) mineralization. Vitro Cell Dev Biol Anim. 2007;43(3–4):120–8.
76. Nakatake Y, Hoshikawa M, Asaki T, Kassai Y, Itoh N. Identification of a novel fibroblast growth factor, FGF-22, preferentially expressed in the inner root sheath of the hair follicle. Biochim Biophys Acta. 2001;1517(3):460–3.
77. Nishimura T, Nakatake Y, Konishi M, Itoh N. Identification of a novel FGF, FGF-21, preferentially expressed in the liver. Biochim Biophys Acta. 2000;1492(1):203–6.
78. Powers CJ, McLeskey SW, Wellstein A. Fibroblast growth factors, their receptors and signaling. Endocr Relat Cancer. 2000;7(3):165–97.
79. Yamashita T, Yoshioka M, Itoh N. Identification of a novel fibroblast growth factor, FGF-23, preferentially expressed in the ventrolateral thalamic nucleus of the brain. Biochem Biophys Res Commun. 2000;277(2):494–8.
80. Osathanon T, Nowwarote N, Pavasant P. Basic fibroblast growth factor inhibits mineralization but induces neuronal differentiation by human dental pulp stem cells through a FGFR and PLC gamma signaling pathway. J Cell Biochem. 2011;112(7):1807–16.
81. Wu J, Huang GT, He W, Wang P, Tong Z, Jia Q, et al. Basic fibroblast growth factor enhances stemness of human stem cells from the apical papilla. J Endod. 2012;38(5):614–22.
82. Shimabukuro Y, Ueda M, Ozasa M, Anzai J, Takedachi M, Yanagita M, et al. Fibroblast growth factor-2 regulates the cell function of human dental pulp cells. J Endod. 2009;35 (11):1529–35.
83. He H, Yu J, Liu Y, Lu S, Liu H, Shi J, et al. Effects of FGF2 and TGFbeta1 on the differentiation of human dental pulp stem cells in vitro. Cell Biol Int. 2008;32(7):827–34.
84. Kim J, Park JC, Kim SH, Im GI, Kim BS, Lee JB, et al. Treatment of FGF-2 on stem cells from inflamed dental pulp tissue from human deciduous teeth. Oral Dis. 2014;20(2):191–204.
85. Grando Mattuella L, Westphalen Bento L, de Figueiredo JA, Nor JE, de Araujo FB, Fossati AC. Vascular endothelial growth factor and its relationship with the dental pulp. J Endod. 2007;33(5):524–30.
86. Marchionni C, Bonsi L, Alviano F, Lanzoni G, Di Tullio A, Costa R, et al. Angiogenic potential of human dental pulp stromal (stem) cells. Int J Immunopathol Pharmacol. 2009;22 (3):699–706.
87. Demarco FF, Conde MC, Cavalcanti BN, Casagrande L, Sakai VT, Nor JE. Dental pulp tissue engineering. Braz Dent J. 2011;22(1):3–13.
88. Magnucki G, Schenk U, Ahrens S, Navarrete Santos A, Gernhardt CR, Schaller HG, et al. Expression of the IGF-1, IGFBP-3 and IGF-1 receptors in dental pulp stem cells and impacted third molars. J Oral Sci. 2013;55(4):319–27.
89. Denholm IA, Moule AJ, Bartold PM. The behaviour and proliferation of human dental pulp cell strains in vitro, and their response to the application of platelet-derived growth factor-BB and insulin-like growth factor-1. Int Endod J. 1998;31(4):251–8.

90. Yu Y, Mu J, Fan Z, Lei G, Yan M, Wang S, et al. Insulin-like growth factor 1 enhances the proliferation and osteogenic differentiation of human periodontal ligament stem cells via ERK and JNK MAPK pathways. Histochem Cell Biol. 2012;137(4):513–25.
91. Feng X, Huang D, Lu X, Feng G, Xing J, Lu J, et al. Insulin-like growth factor 1 can promote proliferation and osteogenic differentiation of human dental pulp stem cells via mTOR pathway. Dev Growth Differ. 2014.
92. Leong WK, Henshall TL, Arthur A, Kremer KL, Lewis MD, Helps SC, et al. Human adult dental pulp stem cells enhance poststroke functional recovery through non-neural replacement mechanisms. Stem Cells Transl Med. 2012;1(3):177–87.
93. Mead B, Logan A, Berry M, Leadbeater W, Scheven BA. Intravitreally transplanted dental pulp stem cells promote neuroprotection and axon regeneration of retinal ganglion cells after optic nerve injury. Invest Ophthalmol Vis Sci. 2013;54(12):7544–56.
94. Esmaeili A, Alifarja S, Nourbakhsh N, Talebi A. Messenger RNA expression patterns of neurotrophins during transdifferentiation of stem cells from human-exfoliated deciduous teeth into neural-like cells. Avicenna J Med Biotechnol. 2014;6(1):21–6.
95. Artavanis-Tsakonas S, Muskavitch MA. Notch: the past, the present, and the future. Curr Top Dev Biol. 2010;92:1–29.
96. He F, Yang Z, Tan Y, Yu N, Wang X, Yao N, et al. Effects of Notch ligand Delta1 on the proliferation and differentiation of human dental pulp stem cells in vitro. Arch Oral Biol. 2009;54(3):216–22.
97. Zhang C, Chang J, Sonoyama W, Shi S, Wang CY. Inhibition of human dental pulp stem cell differentiation by Notch signaling. J Dent Res. 2008;87(3):250–5.
98. Johnson GL, Lapadat R. Mitogen-activated protein kinase pathways mediated by ERK, JNK, and p38 protein kinases. Science. 2002;298(5600):1911–2.
99. Williams DW, Wu H, Oh JE, Fakhar C, Kang MK, Shin KH, et al. 2-Hydroxyethyl methacrylate inhibits migration of dental pulp stem cells. J Endod. 2013;39(9):1156–60.
100. Vandomme J, Touil Y, Ostyn P, Olejnik C, Flamenco P, El Machhour R, et al. Insulin-like growth factor 1 receptor and p38 mitogen-activated protein kinase signals inversely regulate signal transducer and activator of transcription 3 activity to control human dental pulp stem cell quiescence, propagation, and differentiation. Stem Cells Devel. 2014;23(8):839–51.
101. Hata M, Naruse K, Ozawa S, Kobayashi Y, Nakamura N, Kojima N, et al. Mechanical stretch increases the proliferation while inhibiting the osteogenic differentiation in dental pulp stem cells. Tissue Eng Part A. 2013;19(5–6):625–33.
102. Zhang H, Liu S, Zhou Y, Tan J, Che H, Ning F, et al. Natural mineralized scaffolds promote the dentinogenic potential of dental pulp stem cells via the mitogen-activated protein kinase signaling pathway. Tissue Eng Part A. 2012;18(7–8):677–91.
103. Boyce BF, Yamashita T, Yao Z, Zhang Q, Li F, Xing L. Roles for NF-kappaB and c-Fos in osteoclasts. J Bone Miner Metab 2005;23 Suppl:11–15.
104. Paula-Silva FW, Ghosh A, Silva LA, Kapila YL. TNF-alpha promotes an odontoblastic phenotype in dental pulp cells. J Dent Res. 2009;88(4):339–44.
105. Wang Y, Zheng Y, Wang Z, Li J, Zhang G, Yu J. 10(-7) m 17beta-oestradiol enhances odonto/osteogenic potency of human dental pulp stem cells by activation of the NF-kappaB pathway. Cell Prolif. 2013;46(6):677–84.
106. Wang Y, Yan M, Wang Z, Wu J, Zheng Y, Yu J. Dental pulp stem cells from traumatically exposed pulps exhibited an enhanced osteogenic potential and weakened odontogenic capacity. Arch Oral Biol. 2013;58(11):1709–17.
107. Tai TF, Chan CP, Lin CC, Chen LI, Jeng JH, Chang MC. Transforming growth factor beta2 regulates growth and differentiation of pulp cells via ALK5/Smad2/3. J Endod. 2008;34 (4):427–32.
108. Yongchaitrakul T, Pavasant P. Transforming growth factor-beta 1 up-regulates the expression of nerve growth factor through mitogen-activated protein kinase signaling pathways in dental pulp cells. Eur J Oral Sci. 2007;115(1):57–63.

109. Oka S, Oka K, Xu X, Sasaki T, Bringas P Jr, Chai Y. Cell autonomous requirement for TGF-beta signaling during odontoblast differentiation and dentin matrix formation. Mech Dev. 2007;124(6):409–15.
110. Scheller EL, Chang J, Wang CY. Wnt/beta-catenin inhibits dental pulp stem cell differentiation. J Dent Res. 2008;87(2):126–30.
111. Zhang J, Park YD, Bae WJ, El-Fiqi A, Shin SH, Lee EJ, et al. Effects of bioactive cements incorporating zinc-bioglass nanoparticles on odontogenic and angiogenic potential of human dental pulp cells. J Biomater Appl. 2014.
112. Smith A, Robinson V, Patel K, Wilkinson DG. The EphA4 and EphB1 receptor tyrosine kinases and ephrin-B2 ligand regulate targeted migration of branchial neural crest cells. Curr Biol. 1997;7(8):561–70.
113. Stokowski A, Shi S, Sun T, Bartold PM, Koblar SA, Gronthos S. EphB/ephrin-B interaction mediates adult stem cell attachment, spreading, and migration: implications for dental tissue repair. Stem Cells. 2007;25(1):156–64.
114. Liu L, Ling J, Wei X, Wu L, Xiao Y. Stem cell regulatory gene expression in human adult dental pulp and periodontal ligament cells undergoing odontogenic/osteogenic differentiation. J Endod. 2009;35(10):1368–76.
115. Tada H, Nemoto E, Foster BL, Somerman MJ, Shimauchi H. Phosphate increases bone morphogenetic protein-2 expression through cAMP-dependent protein kinase and ERK1/2 pathways in human dental pulp cells. Bone. 2011;48(6):1409–16.
116. Lin PS, Chang MC, Chan CP, Lee SY, Lee JJ, Tsai YL, et al. Transforming growth factor beta1 down-regulates Runx-2 and alkaline phosphatase activity of human dental pulp cells via ALK5/Smad2/3 signaling. Oral Surg Oral Med Oral Pathol Oral Radiol Endod. 2011;111 (3):394–400.
117. Wang FM, Hu T, Zhou X. p38 mitogen-activated protein kinase and alkaline phosphatase in human dental pulp cells. Oral Surg Oral Med Oral Pathol Oral Radiol Endod. 2006;102 (1):114–8.
118. Pantovic A, Krstic A, Janjetovic K, Kocic J, Harhaji-Trajkovic L, Bugarski D, et al. Coordinated time-dependent modulation of AMPK/Akt/mTOR signaling and autophagy controls osteogenic differentiation of human mesenchymal stem cells. Bone. 2013;52 (1):524–31.
119. He WX, Wang ZH, Zhou ZY, Zhang YQ, Zhu QL, Wei KW, et al. Lipopolysaccharide Enhances Wnt5a expression through toll-like receptor 4, myeloid differentiating factor 88, Phosphatidylinositol 3-OH Kinase/AKT and nuclear factor kappa B pathways in human dental pulp stem cells. J Endod. 2014;40(1):69–75.
120. Solozobova V, Wyvekens N, Pruszak J. Lessons from the embryonic neural stem cell niche for neural lineage differentiation of pluripotent stem cells. Stem cell Rev. 2012;8(3):813–29.
121. Nair R, Ngangan AV, Kemp ML, McDevitt TC. Gene expression signatures of extracellular matrix and growth factors during embryonic stem cell differentiation. PLoS ONE. 2012;7 (10):e42580.
122. Dallas SL, Rosser JL, Mundy GR, Bonewald LF. Proteolysis of latent transforming growth factor-beta (TGF-beta)-binding protein-1 by osteoclasts. A cellular mechanism for release of TGF-beta from bone matrix. J Biol Chem. 2002;277(24):21352–60.
123. Garcion E, Halilagic A, Faissner A, ffrench-Constant C. Generation of an environmental niche for neural stem cell development by the extracellular matrix molecule tenascin C. Development. 2004;131(14):3423–3432.
124. Jadlowiec JA, Zhang X, Li J, Campbell PG, Sfeir C. Extracellular matrix-mediated signaling by dentin phosphophoryn involves activation of the Smad pathway independent of bone morphogenetic protein. J Biol Chem. 2006;281(9):5341–7.
125. Armstrong MT, Armstrong PB. Growth factor modulation of the extracellular matrix. Exp Cell Res. 2003;288(2):235–45.

126. Vidane AS, Zomer HD, Oliveira BM, Guimaraes CF, Fernandes CB, Perecin F, et al. Reproductive stem cell differentiation: extracellular matrix, tissue microenvironment, and growth factors direct the mesenchymal stem cell lineage commitment. Reprod Sci. 2013;20(10):1137–43.
127. Huang CH, Tseng WY, Yao CC, Jeng JH, Young TH, Chen YJ. Glucosamine promotes osteogenic differentiation of dental pulp stem cells through modulating the level of the transforming growth factor-beta type I receptor. J Cell Physiol. 2010;225(1):140–51.
128. Mayo V, Sawatari Y, Huang CY, Garcia-Godoy F. Neural crest-derived dental stem cells–where we are and where we are going. J Dent. 2014;42(9):1043–51.

Vascular Niche in HSC Development, Maintenance and Regulation

Süleyman Coşkun and Karen K. Hirschi

1 Introduction

A fertilized egg, as a single cell, has all the genetic information and potential to form an entire embryo. Through multiple cellular divisions and differentiation, newly forming cells start to acquire unique cellular functions and commit to specific cell lineages during development. This lineage specification results in the formation of three germ layers; ectoderm, mesoderm and endoderm, which then differentiate into specific cell types that form distinct tissues and organs in the body. "Stemness" potential, the ability to self-renew and give rise to multiple cell lineages, is not restricted to the developing embryo. Fortunately, a rare population of multipotent cells within most tissues (adult stem cells) is set aside in an undifferentiated state and, upon activation, can then provide a lifelong supply of cells that need to be replaced throughout life. During the lifespan of an organism, adult stem cells must be maintained in an undifferentiated, quiescent state within tissues, and be able to proliferate and differentiate quickly to repair the tissue when needed. Unlike embryonic stem cells (ESC), adult stem cells are largely quiescent and must maintain a balance between self-renewal and differentiation, which is tightly regulated by cellular microenvironment or niche in which they reside. In this review, we will focus on hematopoietic stem cells (HSC), specifically, and discuss their interactions with vascular endothelial cells during development, and provide an

---

S. Coşkun (✉) · K.K. Hirschi
Yale Cardiovascular Research Center, Department of Internal Medicine, Vascular Biology and Therapeutics Program, and Yale Stem Cell Center, Yale University School of Medicine, 300 George St, New Haven, CT 06511, USA
e-mail: suleyman.coskun@yale.edu

S. Coşkun
Department of Neurosurgery, Yale Program in Brain Tumor Research, Yale School of Medicine, The Anlyan Center, 300 Cedar St, New Haven, CT 06520, USA

K. Turksen (ed.), *Tissue-Specific Stem Cell Niche*,
Stem Cell Biology and Regenerative Medicine,
DOI 10.1007/978-3-319-21705-5_9

overview of the current knowledge of the role of the vascular niche in adult bone marrow in the regulation of HSC.

2 Hematopoietic Stem Cells

Efficient blood production and circulation is of vital importance for human survival. Starting at very early stages of development, blood cells perform the critical tasks of carrying oxygen and nutrients to all living cells, and removing metabolic waste products. Blood also contains immune cells that protect the body from infections and invading foreign bodies. Despite such critical tasks, blood cells have only limited lifespan therefore they need to be replaced often such that new blood cells enter the bloodstream continuously. Hematopoietic stem cells are one of the most widely studied among adult stem cells; they provide a lifelong supply of multi-lineage blood cell progenitors that give rise to short-lived blood cell types including myeloid cells, lymphocytes, megakaryocytes and red blood cells.

For proper long-term function of HSC, it is crucial to maintain an adequate number of HSC via self-renewal, and to provide a constant supply of blood cells via regulation of differentiation. Because of their unique localization within hematopoietic tissues, it was long proposed that HSC interact with their microenvironment (HSC niche), which ultimately controls their maintenance and function. The HSC niche concept was first proposed by Schofield in the late 1970s [1] and since then, researchers have defined the role of many niche components, cellular and non-cellular, in the regulation of HSC. In this section, we will summarize HSC development from their first emergence in the developing embryo and throughout their ontogeny within various hematopoietic tissues during embryogenesis. We will also discuss the regulation of HSC phenotype and function by vascular niche cells within fetal and adult bone marrow.

HSC Phenotype: The first evidence of the presence of HSC comes from studies in the early 1960s when it was discovered that adult mouse bone marrow contains a population of cells that can form myeloerythroid colonies in the spleen and rescue lethally irradiated mice, in which the hematopoietic repopulating activity had been ablated [2, 3]. Since then, multiple strategies have been developed for the identification and isolation of adult bone marrow HSC. HSC are characterized by expression of surface proteins Sca-1 and c-Kit and the lack of blood lineage (Lin) related protein expression (CD4, CD8, CD45R, Ter119, Gr-1, and Mac-1 except fetal HSC). Although the c-Kit^{+} Sca^{+} Lin^{-} (KSL) fraction of bone marrow cells contains all of the HSC activity [4], isolation of HSC using KSL markers yields a heterogeneous population that consists of more primitive and potent long-term HSC (LT-HSC) and their immediate progeny short-term HSC (ST-HSC). These multipotent cells represent only a small fraction of cells in any hematopoietic organ such that only 1 in 10,000–15,000 bone marrow cells are shown to have HSC properties. With recent developments in the fields of multicolor flow cytometry, high-speed cell sorting, inducible gene targeting and advanced imaging techniques, we have gained a wealth of knowledge about the cellular and molecular mechanisms underlying the development, self-renewal and differentiation of HSC. To further purify LT-HSC, per se, additional markers have been identified, such as thymus cell antigen-1 (Thy1.1), SLAM F1 (CD150), FMS-like tyrosine kinase-3 (Flt3), interleukin receptor α (IL-7Rα) and endoglin (CD105) [5–8]. Human LT-HSC are characterized phenotypically as $Thy1.1^{low/-}$ $CD34^{low/-}$ or $CD150^{+}$ $Flt3/Flk2^{-}$ KSL [4, 7, 9]. ST-HSC, on the other hand, are represented in the $CD34^{+}$ $Flt3^{-}$ KSL fraction. Another method to isolate adult marrow HSC involves the use of dye-exclusion properties, resulting in the sorting of a unique cell population referred to as the side population (SP) [10]. Combining both surface markers and the Hoechst dye exclusion function, so called SPKSL cells can be isolated, which represent HSC with higher purity compared to cells isolated with each technique individually [11].

When multi-lineage HSC and progenitors commit to lineage differentiation they either give rise to common lymphoid progenitors (CLP) that produces T- and B-lymphocytes or common myeloid progenitors (CMP) that produce erythroid, megakaryocyte, granulocyte and monocytes [12] (Fig. 1). Using flow cytometry and only four SLAM markers (CD150, CD48, CD229, and CD244) [13], HSC and

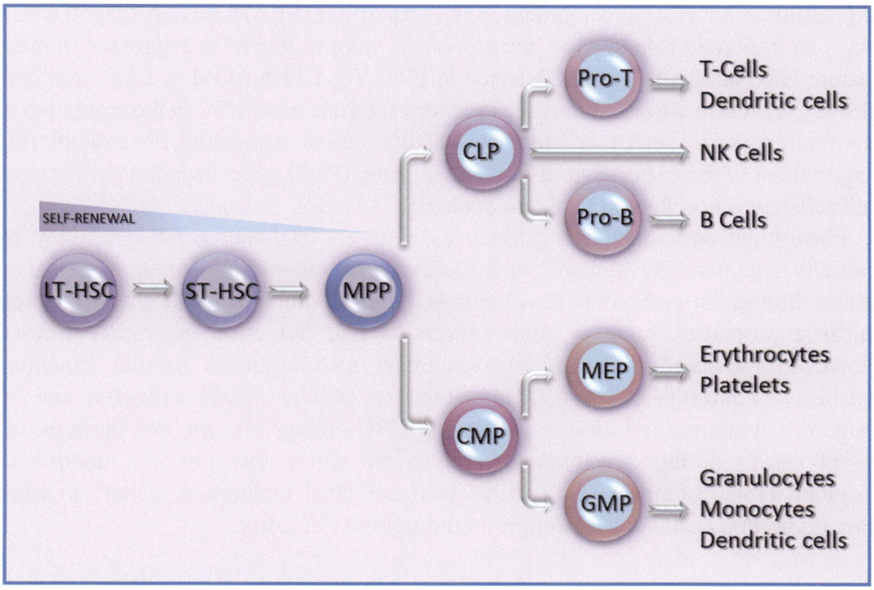

Fig. 1 Hematopoietic stem cell (HSC) lineage. HSC have the ability to self-renew and differentiate into all blood cell types throughout postnatal life. Long-term HSC (LT-HSC) are the most primitive HSC with higher self-renewal potential. Short-term HSC (ST-HSC) have all the potential of LT-HSC except they have limited self-renewal capacity. Multipotent progenitors (MPP) can differentiate into all types of hematopoietic cells but lack self-renewal ability. MPP give rise to common lymphoid progenitors (CLP) and common myeloid progenitors (CMP). CLP further differentiate into committed precursors of B and T lymphocytes, whereas CMP give rise to megakaryocyte/erythroid progenitors (MEP) and granulocyte/macrophage progenitors (GMP)

progenitors can be isolated into functionally distinct subpopulations with unique cell-cycle status, self-renewal potential, and repopulating abilities.

Hematopoietic stem and progenitor cells (HSPC) emerge within distinct tissues during development (discussed in next section) and exhibit phenotypic differences throughout their ontogeny [14–16]. CD41 is the earliest marker of HSPC, expressed by hematopoietic cells in the extraembryonic yolk sac (YS) as early as embryonic day (E)9.0 [17]. CD45 is another surface antigen that is expressed by hematopoietic cells in the YS [18], as well as the embryonic aorta-gonad-mesonephros (AGM) region and fetal liver [19]. Fetal liver HSC differ from adult bone marrow HSC in the expression of specific markers such as Mac-1, CD144, and AA4.1 [20–22], as well as in their general gene expression profiles [23–25]. Unlike adult bone marrow HSC, which are largely quiescent, 30 % of fetal HSC are actively cycling. Fetal liver HSC divide rapidly and give more robust and rapid reconstitution of irradiated recipients relative to adult HSC [21, 26]. Quiescent and activated fetal HSC also show differential cell surface protein expression. Quiescent HSC express CD38, which is downregulated upon activation concomitant with increased CD34 and Mac-1 expression [21, 27, 28]. Upon birth, most HSC become quiescent within one week postnatally [29]. In 7 week old mice, CD34 negative adult HSC emerge [30]. Consistent with these findings, cell surface markers CD41, CD45 and CD38 have been shown to be differentially expressed during HSC development [31]. Furthermore; SLAM protein CD150 which plays an important role in adult hematopoiesis were shown to be expressed in more mature HSC although it is not detected in E9.0 YS, E11.5 AGM or E12.5 placenta HSPC [31]. There are also differences between fetal and adult HSC in the regulation of stem cell properties such as self-renewal and differentiation potential. For example, the requirement of the expression of Polycomb group (PcG) genes, which regulate stem cell self-renewal, differs in fetal and adult HSC [32–34].

Phenotypic and functional differences between fetal and adult HSC may be partially explained by dynamic changes in the microenvironment in which they reside throughout embryonic development. Cellular function is determined by cell intrinsic properties such as gene expression and autocrine regulatory factors. However, the surrounding microenvironment also regulates cellular functions extrinsically through cell-to-cell or cell-to-extracellular matrix adhesion, and by paracrine cytokines and growth factors. As HSC emerge and migrate through different tissues during development, HSC niche within these tissues function to maintain HSC "stemness", as well as facilitate HSC maturation toward an adult phenotype that sustains hematopoiesis throughout adult life.

3 HSC Ontogeny

In mammals, HSC predominantly reside within bone marrow postnatally, but during development, HSPC emerge and migrate through different hematopoietic tissues in two distinct phases of blood cell production. The first phase of blood development is referred to as primitive hematopoiesis, during which only nucleated erythroblasts, macrophages and megakaryocytes are produced. In mice, this occurs within the extraembryonic YS at E7.25; in humans, primitive hematopoiesis similarly initiates within the YS at 18 days of gestation [35]. In contrast, during definitive hematopoiesis, all blood lineages are generated from HSPC. The first definitive HSPC in mice arise in the extraembryonic YS at E8.25 and then migrate to and/or emerge within the AGM at E9.5–E10.5, placenta at E9.5, fetal liver at E12.5, and finally within the fetal bone marrow before birth at E16.5 [36]. In human, adult repopulating HSC first appear in AGM specifically in dorsal aorta [37]. Hematopoietic stem and progenitor cells exhibit different phenotypes and functions within these early hematopoietic tissues, suggesting that microenvironment plays a critical role in regulating these properties.

Yolk Sac: Of all the microenvironments that generate and/or support HSPC throughout development, the extraembryonic YS is among the simplest tissue, composed of mesodermal and endodermal germ layers. Initially, at ~E7.0–7.5, the generation of primordial (unspecialized) endothelium and primitive erythroblasts occurs coincidentally within the mesoderm, forming structures referred to as blood islands [38]. One day later in development, the YS gives rise to definitive HSPC; thus, the contribution of the YS to hematopoiesis is of great interest.

It has been confirmed that progenitors arising within the YS migrate to the fetal liver and later colonize the fetal bone marrow. Furthermore, removal of the YS tissue results in failure of development of hematopoietic populations in the liver. In fact, initial studies suggested that the YS was not only the first, but also the sole, hematopoietic tissue where de novo HSC formation occurs [39, 40]. However, later studies proved this is not the case. That is, a seminal chicken–quail chimeric experiment in which quail intraembryonic compartments were grafted onto chicken YS before the onset of circulation demonstrated that only quail cells contributed to long-term multilineage hematopoiesis [41]. Similar results were later obtained by using Xenopus as a model organism [42]. Despite these experimental results, the true nature of the hematopoietic progenitors formed in the YS was still controversial due to their inability to repopulate lethally irradiated adult recipients in vivo [43, 44] even though YS multilineage HSPC are able to rescue lethally irradiated neonatal mice [45, 46]. However, when YS cells are co-cultured with AGM-derived stromal cells, they are able to repopulate irradiated adult mice, demonstrating that the AGM stroma is necessary and sufficient to induce adult repopulating ability among YS derived HSPC [47]. This study indicated that that both the YS and AGM contribute to definitive hematopoiesis, and emphasize the importance of microenvironment in the acquisition of HSPC functional properties.

Placenta: The placenta is another extraembryonic organ, derived from trophectoderm and mesoderm [48], that harbors HSC activity coincident with the emergence of HSC within the AGM region, and prior to HSC activity within the fetal liver. In vivo transplantation studies show that adult repopulating HSC are present in the placenta at E10.5 [49, 50]. Blood cells are known to be derived from mesoderm during embryogenesis, therefore placental HSC most likely arise from chorionic and allantoic mesoderm [51, 52]. Placental HSC are localized predominantly near vessels of the chorioallantoic mesenchyme and the fetal labyrinth [53].

Thus, similar to YS tissues, the vasculature in the placenta provides a unique microenvironment for HSC generation, expansion and maturation. The HSC pool in the placenta expands quite rapidly such that it becomes 15-fold larger than that within the AGM region. By E11, the HSC population in the placenta declines, as the HSC pool in the liver continues to grow, suggesting that the placenta may be a major source of the HSC that migrate to the liver [49].

AGM: The AGM is formed from embryonic splanchnopleural mesoderm, in close association with definitive endoderm that derives from the epiblast during gastrulation [54], and it is the first tissue within the embryo proper to generate HSC de novo with adult repopulating hematopoietic activity. As in the extraembryonic placenta, hematopoiesis is detected within the AGM after the onset of systemic circulation; therefore, it was unclear as to whether the HSPC detected within this tissue are generated in situ, or migrate in from elsewhere via blood circulation. However, further studies using organ explant cultures clearly showed that HSC with adult repopulating ability arise autonomously within the AGM region [44, 55], although the cellular origin of the AGM HSC has been debated, as described in the next section. Stromal cells within the AGM that support definitive hematopoiesis have been isolated and shown to exhibit some phenotypic similarities to vascular smooth muscle cells and bone marrow mesenchymal stem cells, although they lack their full multilineage potential [56, 57].

Fetal Liver: Multi-lineage HSPC generated within the YS, AGM and placenta migrate to the fetal liver at ~E11 [58]. Within 24 h, the number of HSC in the fetal liver increases, and continues to double from E12.5 to E14.5, until it starts to decrease at ~E15.5 [21]. This rapid expansion of HSC within the fetal liver suggests that this microenvironment provides mitogenic and self-renewal signals for HSC. The role of fetal liver as one of the main hematopoietic tissue for HSC expansion has been demonstrated with 38-fold increase in the presence of competitive repopulating units in fetal liver between E12 and E16 [59]. While the number of HSC increases dramatically due to the higher frequency of cycling HSC that undergo self-renewal, some also undergo rapid differentiation to give rise to hematopoietic progenitors [60]. Co-culture of fetal YS HSPC on fetal liver stromal cells promotes the development of adult repopulating ability therein [61].

Spleen: The spleen is another hematopoietic organ in the embryo in which the vascular niche is important for HSC maintenance. HSC has been shown to reside adjacent to sinusoids in the spleen [7]. Moreover when osteoblasts, the main HSC niche component in adult bone marrow, are ablated, most HSC migrate to spleen, which provides a vascular niche environment to maintain HSC activity as an extramedullary organ [62]. Interestingly, ablation of osteoblasts via ganciclovir (GV) treatment did not decrease HSC number as one would expect, instead, it took several weeks during which time bone marrow cellularity was severely reduced, implying that other cellular components are involved in HSC maintenance [63]. Although the spleen functions as a hematopoietic organ in developing embryo, in the normal adult, its function is limited to erythropoiesis [64–66].

Fetal Bone Marrow: Fetal bone marrow is the final destination for HSC within the developing embryo, and bone marrow serves as the main hematopoietic organ throughout postnatal life. Fetal bone marrow develops late in embryogenesis, at ~E16 when bone tissue is just forming. Earlier in development, bone structure is avascular cartilage differentiated from condensed mesenchyme. Upon vascularization of primordial bones, cartilage hypertrophy occurs, followed by mineralization and osteogenesis [67]. We found that long bones were still avascular cartilage at E15.5, and vascularization and LT-HSC emerged coincidentally at E16.5 [68]. Invading arteries branch out into capillaries that then form sinusoidal vessels throughout the bone marrow cavity. The first multi-lineage hematopoietic activity, which we found to be contained within the KSL population, was localized within the middle regions of fetal long bones that were vascularized first, and subsequently distributed toward the proximal and distal regions by E17.5, in a pattern that paralleled vascular expansion within bone tissue [68]. It was also shown that, after transplantation, HSC preferred to engraft in bone marrow vascular domains in mice [69]. These observations suggest that within fetal bone marrow, as in the YS, placenta and AGM (reviewed in [36]), the vasculature provides an initial niche to maintain HSC function as other components of the marrow niche emerge, particularly osteolineage cells, which we found to be required for fetal bone marrow HSC to acquire LT–HSC phenotype and function [68].

In summary, all developmental and postnatal hematopoietic sites provide unique microenvironment for HSC emergence, expansion and development. Within these hematopoietic sites, HSC are closely associated with vascular niche cells, in particular, vascular endothelial cells. In the next section, we will focus on HSC – vascular endothelial cell relationship both in terms of their developmental origins, and their interaction within bone marrow vascular niche.

4 Role of the Vascular Endothelium in Developmental Hematopoiesis

From their first emergence to their final destination in bone marrow, HSC reside in close proximity to and interact with the vascular endothelium. Given that vascular endothelium forms the inner linings of blood vessels, and HSC give rise to blood cells, the co-localization of both vascular endothelial cells and HSC within hematopoietic tissues seems anatomically natural. This mutually beneficial relationship between vascular endothelium and HSC is also critically important for HSC development, maintenance and function. Genetic studies in multiple model organisms have demonstrated that hematopoietic cells do not form in the absence of vasculature [70, 71]. This genetic evidence suggests a shared ontogeny, although the nature of which is still debated, and two main theories will be discussed herein.

Hemogenic Endothelium. Observations from as early as the 1920s [72] suggested that the hematopoietic cells generated during development within the YS and AGM bud from the endothelium. The specialized endothelial cells within the YS and AGM that exhibit blood-forming potential are referred to as hemogenic endothelial cells (reviewed in [73]). Early cell tracking studies helped us to visualize this event by labeling endothelial cells in vivo and monitoring their progeny [19, 74, 75]. Genetic tracing studies using the Cre/lox-based system revealed that HSC arise from VE-cadherin positive cells [76]. In addition, advanced dynamic in vivo imaging studies conducted in multiple model systems have further confirmed the generation of blood cells from aortic endothelium within the embryonic AGM [77–79]. Although similar cell clusters were seen both ventral and dorsal site of the artery, only the cells from ventral side of dorsal aorta showed definitive HSC initiation and expansion [80]. Using single cell tracing and genetic analysis, hemogenic endothelial cells have been shown to be committed to hematopoietic fate by losing endothelial potential and initiating hematopoietic program while they are still a part of the vessel wall, and HSC maturation is thought to occur over time [81, 82]. Moreover, it was demonstrated that the HSPC that are generated in the YS and AGM exhibit differences in hematopoietic potential. That is, early stage YS hemogenic endothelial cells are thought to only generate erythroid/myeloid progenitors; whereas, hemogenic endothelial cells from the later-stage AGM region give rise to multilineage HSC [83].

Although early phenotypic characterization studies suggested that both endothelial and hematopoietic cells express CD31 (PECAM-1), CD34 and Flk1 (VEGFR2) [84], later clonal analysis studies revealed that hemogenic endothelial cells exhibit a phenotype that is distinct from endothelial cells which do not exhibit blood forming potential. We defined the in vivo phenotype of YS and AGM hemogenic endothelial cells: they express the vascular endothelial growth factor receptor VEGFR2 (Flk-1), HSC marker c-Kit, and lack expression of blood cell lineage markers, including the pan-leukocyte marker CD45 [18, 85]. In addition, hemogenic endothelial cells exhibit Hoechst dye-efflux properties [18, 86] which are characteristic of bone marrow-derived HSC and other stem cell populations in adult tissues and collectively described as "side population" or SP cells [10, 87–90]. Hemogenic endothelial cells within the murine YS and AGM which demonstrate clonal multi-lineage hematopoietic potential are thus defined as Flk-1^{+}c-$Kit^{+}CD45^{-}$ SP cells [18]. Consistent with this, other studies also demonstrated that hemogenic cells can be identified by up-regulation of c-Kit, CD41 and CD45 during endothelial to hematopoietic transition [78, 91].

Hemangioblast. Although early observations suggested that specialized endothelial cells give rise to blood cells in the YS and AGM [72], in 1932, Murray suggested that a common bipotential progenitor, termed the "hemangioblast", gives rise to both endothelial and blood cells [92]. Later on, a clonal mesodermal precursor for blood and endothelial cells, the so-called blast colony-forming cell (BL-CFC), was identified in differentiating embryonic stem cell cultures [93, 94]. Using embryonic stem cell fate tracing approaches, distinct cell progeny, identified by their differential expression of Brachyury and fetal liver kinase 1 (Flk-1), were said to represent a developmental progression from mesodermal precursor to hemangioblast [95].

BL-CFC were considered to be the in vitro equivalent of in vivo hemangioblasts; however unlike describing a population of cells as previously conceptualized [92], the hemangioblast began to be defined as a clonal bipotent progenitor to blood and endothelium. Subsequently, the identification of very rare BL-CFC in the posterior region of the primitive streak of the E7.5 mouse embryo that expresses both the mesodermal marker Brachyury and Flk 1 was considered further evidence for the existence of the hemangioblasts in vivo [96]. Other studies found similar results, however these experiments relied on the isolation, culture, and/or manipulation of cells in vitro [97]. Thus, the presence of bipotential hemangioblasts in vivo has not been confirmed, to date.

Nonetheless, in vitro studies suggest that the hemangioblast and hemogenic endothelial cell models may not be mutually exclusive, and instead may represent different cellular states during hematovascular specification; that is, the hematopoietic cells may emerge from hemangioblasts via a hemogenic endothelium intermediate [98]. Other in vivo studies of murine YS development support the idea that the earliest blood and endothelial cell populations are specified independently, and that hemogenic endothelial cells first appear in the YS and produce definitive blood cells [99]. Thus, the earliest endothelial and blood cells are likely independently fated during gastrulation. Studies of HSPC derivation from embryonic stem cells and induced pluripotent stem cells further support the idea that definitive HSC are derived from hemogenic endothelial cells, and early blood cells are independently fated [100, 101]. Myers and Krieg present an elegant series of experiments, which suggest that the hemangioblast may be a state of competence rather than a bipotential progenitor state that exists in vivo [102].

5 Vascular Niches in Adult Bone Marrow

Not only are HSC derived from endothelial cells during development, but HSC also reside within a specialized vascular microenvironment within bone marrow postnatally. The so-called vascular niche(s) maintain and regulate HSC function via cell to cell contact and secreted factors (Fig. 2). Furthermore the complexity of vascular niches has been revealed by recent studies that suggest that there are three main vascular niche types; sinusoidal, arteriolar and perivascular niches. Herein, we will summarize the recent studies and the role of each vascular niche in adult bone marrow in the regulation of HSC maintenance and function.

Sinusoidal Niche: The sinusoidal vasculature of adult bone marrow is a unique network of tubular structures, made up of a single layer of endothelial cells that facilitate cellular exchange between the marrow cavity and blood circulation. Bone marrow sinusoids have a discontinuous vascular structure, as opposed to intact capillaries with fenestrae; thus, they function as a quick gateway for blood cells produced in bone marrow [103]. HSC labeling in tissue sections allowed researchers to visualize their close association with sinusoidal vessels in the bone marrow and spleen [7, 104, 105]. Using a combination of cell surface markers (SLAM proteins $CD150^{+}$ $CD48^{-}$ $CD41^{-}$), Kiel and coworkers showed that 60 % of SLAM-marked HSC were localized near the sinusoidal endothelial cells

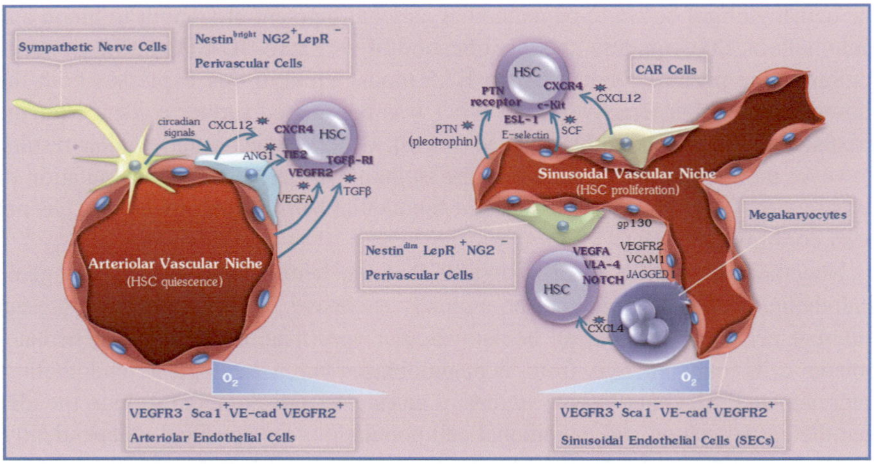

Fig. 2 Regulation of HSC maintenance and function within bone marrow vascular niche. Three main vascular niche types have been described within bone marrow; sinusoidal, arteriolar and perivascular niches. Self-renewal and differentiation properties of adult hematopoietic stem cells (HSC) are tightly regulated by each of these bone marrow vascular niche components via direct cell to cell interaction and/or soluble factors

(SEC) [106]. Phenotypic characterization of bone marrow SEC revealed that the majority of these cells in the bone marrow vascular niche are VEGFR3$^+$Sca1$^-$ VEcadherin$^+$VEGFR2$^+$ and they play critical role in engraftment and hematopoiesis [107], suggesting that SEC function as a niche for HSC. Three hematopoietic organs that contain sinusoidal vasculature are liver, spleen, and bone marrow and they all provide the necessary environment for maintenance and expansion of the HSC pool. Furthermore, both the liver and spleen become sites of extramedullary hematopoiesis in certain pathological conditions.

In bone marrow, SEC are distinguished from arterioles with the expression of VEGF-receptor-3 (VEGFR3) but the lack of Sca1 on their cell surface whereas arteriolar endothelium lacks VEGFR3 but express Sca1 and Tie2 cell surface proteins and associated with smooth muscle cells [108]. Sinusoidal and arteriolar endothelial cells both also express VEGFR2. Although deletion of VEGFR2 is dispensable for steady-state hematopoiesis in the adult; this results in failure of VEGFR3$^+$Sca1$^-$ SEC regeneration, as well as impaired HSPC production, following irradiation [107], suggesting that the SEC niche is indispensable for HSC engraftment and hematopoietic reconstitution within bone marrow. Furthermore, in a reciprocal transplantation study, it was shown that hematopoietic cells derived from acute graft-versus-host disease (GvHD) model mice were able to reconstitute hematopoiesis in healthy recipient mice, whereas, hematopoietic cells from healthy donor mice failed to reconstitute hematopoiesis in GvHD recipient mice which have a defective vascular niche [109]. Together, these studies indicated the importance of SEC in supporting HSC self-renewal and maintenance within the bone marrow niche.

In further support for this idea, SEC are also shown to express and secrete pleotrophin (PTN) in the bone marrow. PTN deficient mice (PTN-/-) exhibit significantly decreased HSC number in bone marrow. In contrast, mice deficient for protein tyrosine phosphatase receptor zeta, a receptor inhibited by PTN, have an increased number of HSC. PTN-/- mice also fail to reconstitute hematopoiesis after myelosuppression. Reciprocal transplantation studies demonstrate PTN plays an important in the vascular niche, rather than in HSC, via an intrinsic mechanism. In addition, inhibition of PTN by systemic administration of anti-PTN antibodies prevents HSC homing and retention in the bone marrow niche. Collectively, these studies reveal that SEC-derived PTN and regulates HSC self-renewal and hematopoietic reconstitution in vivo [110].

Arteriolar Niche: There is a growing body of evidence that HSC also reside close to non-sinusoidal vessels, particularly arteries and arterioles [111, 112]. Perhaps this is not surprising given that the first embryonic HSC with adult repopulating potential are detected at around E10.5 within the dorsal aorta of the AGM region and in the arteries that connect the dorsal aorta with the placenta [44, 113–116]. More recently, cerebrovascular endothelial cells in morphologically defined arterioles and arteries within the E10.5–E11.5 mouse embryo have also been shown to exhibit de novo hematopoietic potential [117].

In adult bone marrow, although 67 % of CD150^{+} CD48^{-} CD41^{-} Lin^{-} HSC are thought to reside close or adjacent to the sinusoids, 37 % reside close or adjacent to Sca-1^{hi} VEGFR2^{+} VEGFR3^{-} arterioles. Kunisaki and Frenette proposed a model which supports the existence of two distinct vascular niches, with different physiological and phenotypic properties [118]. According to this model, dormant HSC predominantly reside in arteriolar niche, whereas proliferating HSC are localized to sinusoidal niches [118]. In fact, sinusoidal vessels has been shown to express E-selectin which promotes HSC proliferation [119] and arteriolar vessels have been shown to maintain HSC quiescence [120], which together support the existence of two distinct vascular niche with unique regulatory function.

Perivascular Niche: Although the inner, luminal layer of blood vessels is composed of endothelial cells, they are usually invested by perivascular cells such as pericytes, vascular smooth muscle cells or reticular cells. In addition to these cells, small blood vessels can also be associated with mesenchymal stem cells, neuronal cell types, or megakaryocytes [7]. Perivascular cells that are in contact with sinusoidal and arterial endothelial cells are thought to be involved in HSC maintenance and regulation, although their specific contribution is still not clear.

For example, HSC localized close to sinusoidal vessels [106] are also thought to be associated with perivascular cells that exhibit high expression of CXCL12 [121]. CXCL12 abundant reticular cells (CAR) were also detected in the endosteal region of bone, which challenged the previously proposed separate niche hypothesis (vascular niche vs osteoblastic niche); rather, this finding suggested a unified niche theory in which HSC are simultaneously influenced by different stromal cell types that coordinately regulate their function.

Other studies in humans identified CD146-expressing perivascular mesenchymal progenitors with an ability to form ectopic hematopoietic bone when transplanted into mice [122]. These perivascular mesenchymal progenitors express angiopoietin and CXCL12, both of which have been shown to regulate HSC maintenance. In fact, further studies indicate that CD146-expressing perivascular cells are the human counterparts of CAR cells in mouse, which are mesenchymal stem/progenitor cells that express CXCL12 and SCF and localized adjacent to blood vessels in the bone marrow [123]. When CXCL12 and SCF were deleted from bone marrow in cell-specific manner, it was shown that their expression in endothelial and perivascular cells, but not other stromal or hematopoietic cell types, was necessary for HSC maintenance [124–126].

Further studies revealed that Nestin-GFP^{bright} cells (10 % of Nestin^{+} cells) and NestinGFP^{dim} cells (80 % of Nestin^{+} cells) represented different perivascular cell types, which are associated with arterioles and sinusoids, respectively [120]. Peri-arteriolar Nestin GFP^{bright} cells also express neuron-glial antigen 2 (NG2) but not Leptin Receptor (LEPR) and maintain dormant HSC, whereas peri-sinusoidal Nestin^{dim} cells express LEPR but not NG2 and maintain proliferating HSC [118]. Neurons of the sympathetic nervous system also reside in the perivascular region and are thought to regulate HSC within bone marrow. Interfering with the function of sympathetic nervous system via mutation or drug administration results in reduced progenitor mobilization from the bone marrow, suggesting that HSC migration and localization are regulated by the nervous system [127]. Cxcl12 expression is also tightly regulated by adrenergic innervation of the bone marrow [128], and sympathetic neuropathy drives malignant transformation of the HSC niche in hematopoietic diseases [129, 130].

Perhaps the most interesting perivascular niche component is megakaryocytes, as they are terminally differentiated hematopoietic cells derived from HSC. Megakaryocytes give rise to platelets, and are usually found incorporated into bone marrow sinusoids, extending their cytoplasmic protrusion where they deposit platelets. Defects in megakaryocyte and platelet development result in abnormalities in bone marrow hematopoiesis [131, 132]. Interestingly, bone marrow transplantation studies revealed that transplanted HSC are found near megakaryocytes [133], and when megakaryocytes are ablated, HSC engraftment is severely impaired [134]. Megakaryocytes haven been shown to regulate HSC proliferation through CXCL4 secretion [135] as well as by governing osteopontin availability within the niche [136].

Despite understanding that HSC are associated with, and regulated by, multiple cell types, the spatial and dynamic organization of specific niches within hematopoietic tissues are just now being explored with powerful lineage tracing and imaging tools [137]. For example, Wang and coworkers described bulged, cyst like HSC-containing pockets in bone marrow which they called as hemospheres [138]. These structures are composed of endothelial tubes and peripheral mesenchymal cells engulfing HSC clusters containing more than two CD150^{+} CD48^{-} cells. However, whether these structures provide a niche for maintenance and regulation of HSC, or just harbor transit amplifying HSC, is not clear and needs further investigation.

6 Vascular Niche-HSC Interactions

Vascular endothelial cells and hematopoietic cells share a common developmental origin and are regulated by similar signaling pathways. Moreover, the role of vasculature in HSC maintenance and regulation has been demonstrated in vitro and in vivo. However, studies of the identification of cellular phenotypes of vascular niche components, and of the molecular basis of vascular niche-HSC interactions, are still in their infancy. Here, we summarize the major regulatory mechanisms currently proposed to mediate endothelial cell-HSC interactions.

Cell-to-Cell Interactions. Co-culture studies employing adult and embryonic cell types have demonstrated that endothelial cells play a critical role in maintaining HSC potential. For example, $CD34^{+}$ human bone marrow cells were co-cultured with porcine microvascular endothelial cells, which resulted in ex vivo expansion of adult repopulating cells that successfully reconstitute SCID mice [139]. The repopulating ability of adult human bone marrow cells was also improved by co-culturing them with human brain endothelial cells [140]. Primary endothelial cells (Tie-2-GFP, $Flk\text{-}1^{+}$ and $CD41^{-}$) isolated from E9.5 YS and AGM also promote 9.4- and 11.4-fold increases, respectively, in hematopoietic progenitor production, as well as long-term repopulating ability [141]. Interestingly, endothelial cells isolated from different tissues have been shown to possess differing abilities to regulate HSC. For instance, Tie2-GFP endothelial cells were isolated from non-hematopoietic adult mouse tissues, including brain, heart, lung, liver and kidney, and co-cultured with adult HSC, and although the endothelial cells from the heart and liver maintained long-term repopulating ability, kidney endothelial cells were unsuccessful even in the presence of hematopoietic growth factors (SCF, TPO and IL6) [142]. Collectively, these co-culture studies demonstrate that HSC are regulated via direct contact with specific endothelial cell types; however, the molecular mechanisms by which endothelial cell contact regulates HSC are not clear.

Extrinsic signaling pathways mediated via cell-cell contact, such as Notch signaling, have been shown to be involved in the HSC niches niche regulation [143, 144]. Several Notch receptors (Notch1 and Notch4) and ligands (Delta-like 1, Delta-like 4 and Jagged-1) are expressed by both hematopoietic progenitors and endothelial cells [145–147], and Notch signaling is known to regulate the de novo formation of fetal HSC in the yolk sac and AGM region [85]. Notch1, specifically, has been shown to be required for definitive hematopoiesis in AGM [148, 149] and adult hematopoiesis in bone marrow [150, 151].

When hematopoietic stem and progenitor cells derived from umbilical cord were expanded in culture media supplemented with engineered Delta protein (Delta1ext-IgG), a Notch ligand, there was a 6-fold enrichment of repopulating HSC [152]. Bone marrow SEC also express Notch ligands and when they are co-cultured with HSC, there is a Notch-dependent increase in HSC numbers, suggesting that SEC regulate HSC via Notch signaling in vivo [153, 154]. In contrast, when VE-cadherin and VEGFR2 are inhibited in SEC with neutralizing antibodies, there is a significant downregulation of Notch ligands (Jagged1 and Jagged2), associated with decreased HSC self-renewal in vitro and repopulating ability in vivo [153]. Jagged-1 was also shown to regulate bone marrow hematopoiesis [155]. When Jagged-1 is conditionally deleted from endothelial cells, there is a decrease in hematopoiesis due to significant loss of the adult HSC pool [155].

Although Notch signaling is important for developmental and adult hematopoiesis, there are fundamental differences in the regulation via this pathway in fetal and adult HSC. Notch 1 signaling in fetal liver, for example, promotes HSC self-renewal and expansion without loss of stemness [156]; whereas, in adult bone marrow, Notch signaling induces HSC differentiation at the expense of self-renewal [157]. However, when Notch is activated upon bone marrow injury, it enhances HSC self-renewal and prevents depletion under stress condition [158]. This may be explained by cross-talk with other pathways during different stages of development, as well as changes in physiological conditions. Wnt signaling, for example, has been shown to regulate Notch signaling in vivo, such that soluble Wnt3a stimulates HSC expression of many Notch signaling components [143]. Other studies support a role for Wnt signaling in HSC regulation during fetal and adult development in a dose-dependent manner [144, 159]. That is, Wnt signaling was first reported to promote HSC self-renewal [144]; however, other reports reveal that an overabundance of Wnt signaling leads to exhaustion of the long-term HSC pool [160, 161].

Cell Adhesion Molecules and Cytokines. HSC are tethered to other niche cellular components through cell adhesion molecules (CAM), which regulate the localization and migration of HSC. Adult bone marrow endothelial cells have been shown to express E-selectin, P-selectin, vascular cell adhesion molecule 1 (VCAM-1) and intercellular adhesion molecule 1 (ICAM-1), all of which are not expressed by other endothelial cells under homeostatic conditions, but only during inflammation [162]. Rolling and extravasation of hematopoietic cells through blood vessels is dependent on interactions between VCAM-1 (CD106), ICAM-1 (CD54), and E- and P-selectin (CD62E and CD62P) expressed by bone marrow endothelial cells, and their binding partners VLA-4 (CD49d), LFA-1 (CD11a), and hyaluronan binding-cellular adhesion molecule (HCAM/CD44) expressed on HSC and their progeny [162].

ICAM and VCAM are calcium independent CAM that belong to an IgG superfamily. ICAM-1, binds to integrin alpha-L (LFA-1A) expressed by HSC. HSC also express VLA-4 and VLA-5, which form a heterodimer and bind to fibronectin and VCAM-1 on vascular endothelium [163, 164]. Fibronectin binding then promotes the expression of specific transcription factors such as c-Myb and GATA2 that have roles in stem cell maintenance [165].

Bone marrow vascular endothelial cells also express E-selectin and P-selectin, which are involved in hematopoietic cell adhesion [166, 167]. E-selectin is expressed mainly by bone marrow SEC, whereas P-selectin is present on endothelium of larger blood vessels [168]. Thus, it seems that these two endothelial selectins may be expressed in distinct, possibly nonoverlapping locations in the bone marrow and may have distinct functional roles in this tissue. E-selectin, specifically, is known to regulate HSC proliferation. That is, E-selectin deletion

(Sele−/−) or administration of an E-selectin antagonist, results in increased HSC quiescence [119]. Although glycoprotein ESL-1 and glycosphingolipids are proposed as possible ligands for E-selectin on HSC, the functional ligand(s) is currently unknown [119].

CXCL12 (SDF-1): Bone marrow stroma is relatively hypoxic and promotes constitutive expression of CXCL12, also known as SDF-1, via its hypoxia-inducible factor 1 (Hif-1)-inducible promoter [169]. The interaction between CXCL12 and its receptor CXC-chemokine receptor 4 (CXCR4) is important for HSC homing and mobility throughout HSC development [170]. CXCL12-CXCR4 interactions are thought to mediate HSC migration to, and retention by, niche cells, thereby regulating HSC quiescence [121]. Deletion of CXCR4 results in abnormal HSC proliferation, which eventually leads to decreased HSC numbers in the bone marrow [171]. Ablation of CXCL12-expressing cells also reduces HSC frequency in the bone marrow [123, 172]. In addition, deletion of CXCL12 from reticular cells around sinusoids causes HSC mobilization, followed by their entrance into the bloodstream [124], emphasizing the regulatory role of perivascular reticular cells around sinusoids in HSC retention and maintenance.

Sympathetic nerve cells, another perivascular cell type, also regulate CXCL12 expression via the circadian cycle [173] and, this then contributes to the regulation of HSC retention in the bone marrow [127, 128]. There are also the high levels of CXCL12 in peri-arteriolar nestin-GFP+ cells [120]; however, deletion of CXCL12 in peri-arteriolar cells with nestin-Cre targeting did not affect the HSC population. Therefore, peri-sinusoidal CXCL12 abundant reticular cells are likely to be the main source for CXCL12 during the regulation of HSC maintenance.

Stem Cell Factor (SCF): One of the first molecules identified to have a direct role in hematopoiesis is stem cell factor (SCF); its receptor is c-Kit [174]. Absence of SCF protein (the Sl mutation) or absence of cell surface c-Kit receptor localization (the W mutation) result in death in utero or in the perinatal period with severe macrocytic anemia, indicating SCF-c-Kit interactions are essential for hematopoiesis [175–177]. While c-Kit receptor appears to be predominantly expressed by HSC and their immediate progeny, SCF is expressed by many niche cellular components in bone marrow, such as mesenchymal stem cells, osteoblasts, fibroblasts, vascular endothelial cells and perivascular cells [123, 172, 178, 179].

The fact that multiple niche cell types express overlapping cytokines and growth factors, such as SCF, makes it is difficult to identify which cellular source is most critical for HSC regulation in vivo. To identify the critical cellular source of SCF in HSC regulation, Ding and coworkers [126] devised a strategy in which SCF is conditionally deleted from its known cellular sources in adult bone marrow: SEC, osteoblasts, perivascular stromal cells, or nestin-positive MSC [126]. These studies revealed that SCF produced and secreted by SEC and lepr-expressing perivascular stromal cells regulates HSC in the bone marrow niche, rather than osteoblasts or nestin-cre-or nestin-creER-expressing cells, as previously thought.

Vascular Endothelial Growth Factor (VEGF): Although a master regulator of angiogenesis, vascular endothelial growth factor (VEGF) signaling also regulates HSC homeostasis. There are two predominant receptor tyrosine kinases that mediate VEGF-A signaling in HSC and endothelial cells: VEGFR1 (or Flt-1) and VEGFR2 (or Flk-1 or KDR) [180] [181, 182]. Gene deletion studies revealed the importance of VEGF-A signaling in both blood and vascular cells; VEGF-A- and VEGFR2-deficient mice die early in development due to impaired hematopoietic and vascular development [71, 183, 184]. HSC have also been shown to produce VEGF-A, suggesting possible regulation via autocrine signaling. In fact, blocking intracellular and extracellular components of VEGF-A signaling with neutralizing antibodies diminishes HSC colony formation [185]. Addition of VEGFR2 inhibitor to HSC in culture also blocks colony formation; however, neutralizing VEGF-A antibodies have no effect. These data suggest that intracellular autocrine VEGF-A signaling also controls HSC behavior [185].

Transforming Growth Factor-beta (***TGF-beta***): TGF-β–SMAD signaling regulates HSC self-renewal and maintenance within the bone marrow niche [173]. Vascular endothelial cells produce TGF-β and its growth inhibitory properties have been shown to control HSC quiescence in culture systems [186–188]. Blocking TGF-β signaling with neutralizing antibodies causes HSC progenitors to enter cell cycle [189–191]. The growth inhibitory role of TGF-β has also been demonstrated in vivo; TGF-β injection into mice results in decrease in HSC progenitor proliferation in bone marrow [192]. In addition, human HSC with dominant-negative TGF-βRII exhibit increased proliferation and survival in vitro [193]. These results suggest that TGF-β secreted by niche endothelial cells, or by HSC, plays an important role in HSC regulation [194]. Recently, it was also shown that hypoxia and TGFβ pathways can converge on cell cycle regulation of HSC, which is a good example of combinatorial niche factors that are likely to involved in fine tuning in HSC maintenance and regulation [195].

Other Regulatory Factors. As vascular endothelial cells gain more attention as a main cellular component of the HSC niche, we gain more insight about regulatory factors that they secrete, which potentially regulate HSC maintenance. Vascular endothelial-derived molecule adrenomedullin has been identified through gene ontology studies of human brain endothelial cells' (HUBEC) transcriptome and is suggested to play a role in promoting HSC progenitors in vitro [196]. Zhang and coworkers, using microarray approach, identified insulin-like growth factor binding protein-2 (IGFB2) and angiopoietin-like protein 5 (Angptl-5) specifically expressed by the mouse fetal liver derived endothelial cells, and showed that including these molecules in serum free media supplemented with SCF, TPO and FGF1 (STF media) increased human cord blood-derived HSC by 20-fold, compared to serum free STF media only [197].

Glycoprotein 130 (***gp130***): Glycoprotein 130 (gp130) was first identified as β subunit of the IL-6R complex; however it is also found in other cytokine receptor complexes [198]. Conditional deletion of gp130 in endothelial and hematopoietic cells resulted in severe anemia in adult mice [199]. Reciprocal transplantation studies revealed that while gp130 deficient bone marrow reconstitute normal hematopoiesis in the wild type recipients, bone marrow cells derived from wild type donor repopulate gp130 deficient recipient mice but failed to rescue hematopoietic defects. The results indicated that gp130 on endothelial cells particularly around sinusoids are essential for normal hematopoiesis rather than gp130 on HSC [200].

Hypoxia: Local environmental factors are also important niche factors. One of the major local components of the niche is oxygen (O_2) concentration, as it directly affects the production of reactive oxygen species (ROS) in HSC. The superoxide anion radical, hydrogen peroxide, and the hydroxyl radical are important ROS. A low oxygen tension (hypoxia) zone is defined as preferred by multiple stem cell types probably due to their metabolic needs to maintain quiescence [201, 202]. It was demonstrated that HSC with long-term repopulation ability reside in hypoxic regions of the bone marrow [203]; quiescent HSC were known to be localized in the most hypoxic areas of the bone marrow [204]; however, hypoxic HSC are now known to be dispersed throughout bone marrow, not just in hypoxic endosteal compartments, as previously thought [201, 202, 205]. Initial studies relied on indirect methods to define hypoxic regions in the bone marrow. A hypoxic marker pimonidazol labelled HSC enriched in endosteal areas of bone marrow [206]; however, long-label retaining cells also stain positively with pimanidazol near vascular sinusoids [207]. Interestingly, cells adjacent to pimonidazol positive HSC are not stained, as expected, suggesting that hypoxia may be regulated in a cell autonomous manner, rather than being niche determined [111]. Oxygen levels within HSC are sensed via hypoxy-inducible factors (HIF)1α and HIF2α, which undergo degradation under normal oxygen conditions but are stabilized at lower oxygen levels and translocate into the nucleus to activate target genes [208]. One of the target genes in HSC is Meis 1, which regulates glycolysis and metabolic activity of LT-HSC [209, 210]. In endosteal niche, HIF1α also regulates HSC survival, quiescence and behavior via regulation of VEGF-A and CXCR4 [211, 212]. Inducible Hif1α deletion in HSC results in loss of quiescence and self-renewal properties [204, 213].

7 Summary

The potential of HSC in regenerative medicine relies upon removing them from their natural habitat, propagating them in culture, and placing them into a foreign tissue environment. Understanding the signals and mechanisms involved in maintaining stem cell identity and self-renewal capacity are crucial for harnessing their use for therapeutic strategies. HSC reside within specialized niches composed of supporting cells (mesenchymal cells, osteoblasts, vascular endothelial cells, perivascular cells, adipocytes, neural cells) and extracellular matrix. HSC niche(s) play significant roles in regulating the delicate balance between HSC maintenance and differentiation. Once outside their niche, "stemness" properties of HSC are lost in a short period of time; thus, limiting our ability to propagate them for cell therapies. From their first emergence to their final destination in bone marrow, HSC reside in close proximity to, and interact with vascular cells. There are three main vascular niche types; sinusoidal, arteriolar and perivascular. In recent years, vascular niches have gained attention as major regulators of HSC; however, more insight is needed to understand the precise molecular and cellular mechanisms that govern interactions between vascular niche and HSC. Gaining a better understanding of how the vascular niche maintains HSC in vivo is crucial for designing novel in vitro culture strategies for ex vivo propagation of HSC, enabling their clinical use for correcting hematopoietic and vascular disorders.

Acknowledgments KKH has been supported by NIH grants HL077675, HL096360, HL128064, EB005173, EB017103 and EB016629.

References

1. Schofield R. The relationship between the spleen colony-forming cell and the haemopoietic stem cell. Blood Cells. 1978;4(1–2):7–25.
2. Becker AJ, McCulloch EA, Till JE. Cytological demonstration of the clonal nature of spleen colonies derived from transplanted mouse marrow cells. Nature. 1963;197:452–4.
3. Till JE, McCulloch EA. A direct measurement of the radiation sensitivity of normal mouse bone marrow cells. Radiat Res. 1961;14:213–22.
4. Morrison SJ, Weissman IL. The long-term repopulating subset of hematopoietic stem cells is deterministic and isolatable by phenotype. Immunity. 1994;1(8):661–73.
5. Christensen JL, Weissman IL. Flk-2 is a marker in hematopoietic stem cell differentiation: a simple method to isolate long-term stem cells. Proc Natl Acad Sci USA. 2001;98(25):14541–6.
6. Kondo M, Weissman IL, Akashi K. Identification of clonogenic common lymphoid progenitors in mouse bone marrow. Cell. 1997;91(5):661–72.
7. Kiel MJ, Yilmaz OH, Iwashita T, Yilmaz OH, Terhorst C, Morrison SJ. SLAM family receptors distinguish hematopoietic stem and progenitor cells and reveal endothelial niches for stem cells. Cell. 2005;121(7):1109–21.
8. Chen CZ, Li M, de Graaf D, Monti S, Gottgens B, Sanchez MJ, et al. Identification of endoglin as a functional marker that defines long-term repopulating hematopoietic stem cells. Proc Natl Acad Sci USA. 2002;99(24):15468–73.
9. Osawa M, Hanada K, Hamada H, Nakauchi H. Long-term lymphohematopoietic reconstitution by a single CD34-low/negative hematopoietic stem cell. Science. 1996;273 (5272):242–5.
10. Goodell MA, Brose K, Paradis G, Conner AS, Mulligan RC. Isolation and functional properties of murine hematopoietic stem cells that are replicating in vivo. J Exp Med. 1996;183(4):1797–806.
11. Challen GA, Boles N, Lin KK, Goodell MA. Mouse hematopoietic stem cell identification and analysis. Cytometry A. 2009;75(1):14–24.
12. Akashi K, Reya T, Dalma-Weiszhausz D, Weissman IL. Lymphoid precursors. Curr Opin Immunol. 2000;12(2):144–50.
13. Oguro H, Ding L, Morrison SJ. SLAM family markers resolve functionally distinct subpopulations of hematopoietic stem cells and multipotent progenitors. Cell Stem Cell. 2013;13(1):102–16 Epub 2013/07/06.
14. Dykstra B, Kent D, Bowie M, McCaffrey L, Hamilton M, Lyons K, et al. Long-term propagation of distinct hematopoietic differentiation programs in vivo. Cell Stem Cell. 2007;1(2):218–29.
15. Sieburg HB, Cho RH, Dykstra B, Uchida N, Eaves CJ, Muller-Sieburg CE. The hematopoietic stem compartment consists of a limited number of discrete stem cell subsets. Blood. 2006;107(6):2311–6.

16. Yilmaz OH, Kiel MJ, Morrison SJ. SLAM family markers are conserved among hematopoietic stem cells from old and reconstituted mice and markedly increase their purity. Blood. 2006;107(3):924–30.
17. Mikkola HK, Fujiwara Y, Schlaeger TM, Traver D, Orkin SH. Expression of CD41 marks the initiation of definitive hematopoiesis in the mouse embryo. Blood. 2003;101(2):508–16.
18. Goldie LC, Lucitti JL, Dickinson ME, Hirschi KK. Cell signaling directing the formation and function of hemogenic endothelium during murine embryogenesis. Blood. 2008;112 (8):3194–204.
19. North TE, de Bruijn MF, Stacy T, Talebian L, Lind E, Robin C, et al. Runx1 expression marks long-term repopulating hematopoietic stem cells in the midgestation mouse embryo. Immunity. 2002;16(5):661–72.
20. Kim I, Yilmaz OH, Morrison SJ. CD144 (VE-cadherin) is transiently expressed by fetal liver hematopoietic stem cells. Blood. 2005;106(3):903–5.
21. Morrison SJ, Hemmati HD, Wandycz AM, Weissman IL. The purification and characterization of fetal liver hematopoietic stem cells. Proc Natl Acad Sci USA. 1995;92 (22):10302–6.
22. Petrenko O, Beavis A, Klaine M, Kittappa R, Godin I, Lemischka IR. The molecular characterization of the fetal stem cell marker AA4. Immunity. 1999;10(6):691–700.
23. Ivanova NB, Dimos JT, Schaniel C, Hackney JA, Moore KA, Lemischka IR. A stem cell molecular signature. Science. 2002;298(5593):601–4 Epub 2002/09/14.
24. Kiel MJ, Iwashita T, Yilmaz OH, Morrison SJ. Spatial differences in hematopoiesis but not in stem cells indicate a lack of regional patterning in definitive hematopoietic stem cells. Dev Biol. 2005;283(1):29–39.
25. Phillips RL, Ernst RE, Brunk B, Ivanova N, Mahan MA, Deanehan JK, et al. The genetic program of hematopoietic stem cells. Science. 2000;288(5471):1635–40 Epub 2000/06/02.
26. Harrison DE, Zhong RK, Jordan CT, Lemischka IR, Astle CM. Relative to adult marrow, fetal liver repopulates nearly five times more effectively long-term than short-term. Exp Hematol. 1997;25(4):293–7.
27. Rebel VI, Miller CL, Thornbury GR, Dragowska WH, Eaves CJ, Lansdorp PM. A comparison of long-term repopulating hematopoietic stem cells in fetal liver and adult bone marrow from the mouse. Exp Hematol. 1996;24(5):638–48.
28. Tajima F, Deguchi T, Laver JH, Zeng H, Ogawa M. Reciprocal expression of CD38 and CD34 by adult murine hematopoietic stem cells. Blood. 2001;97(9):2618–24.
29. Bowie MB, McKnight KD, Kent DG, McCaffrey L, Hoodless PA, Eaves CJ. Hematopoietic stem cells proliferate until after birth and show a reversible phase-specific engraftment defect. J Clin Invest. 2006;116(10):2808–16.
30. Ogawa M, Tajima F, Ito T, Sato T, Laver JH, Deguchi T. CD34 expression by murine hematopoietic stem cells. Developmental changes and kinetic alterations. Ann NY Acad Sci. 2001;938:139–45.
31. McKinney-Freeman SL, Naveiras O, Yates F, Loewer S, Philitas M, Curran M, et al. Surface antigen phenotypes of hematopoietic stem cells from embryos and murine embryonic stem cells. Blood. 2009;114(2):268–78.
32. Kajiume T, Ninomiya Y, Ishihara H, Kanno R, Kanno M. Polycomb group gene mel-18 modulates the self-renewal activity and cell cycle status of hematopoietic stem cells. Exp Hematol. 2004;32(6):571–8.
33. Ohta H, Sawada A, Kim JY, Tokimasa S, Nishiguchi S, Humphries RK, et al. Polycomb group gene rae28 is required for sustaining activity of hematopoietic stem cells. J Exp Med. 2002;195(6):759–70.
34. Park IK, Qian D, Kiel M, Becker MW, Pihalja M, Weissman IL, et al. Bmi-1 is required for maintenance of adult self-renewing haematopoietic stem cells. Nature. 2003;423(6937):302–5.
35. Palis J, Yoder MC. Yolk-sac hematopoiesis: the first blood cells of mouse and man. Exp Hematol. 2001;29(8):927–36 Epub 2001/08/10.

36. Coskun S, Hirschi KK. Establishment and regulation of the HSC niche: Roles of osteoblastic and vascular compartments. Birth Defects Res C Embryo Today. 2010;90(4):229–42.
37. Ivanovs A, Rybtsov S, Welch L, Anderson RA, Turner ML, Medvinsky A. Highly potent human hematopoietic stem cells first emerge in the intraembryonic aorta-gonad-mesonephros region. J Exp Med. 2011;208(12):2417–27 Epub 2011/11/02.
38. Palis J, McGrath KE, Kingsley PD. Initiation of hematopoiesis and vasculogenesis in murine yolk sac explants. Blood. 1995;86(1):156–63.
39. Liu CP, Auerbach R. In vitro development of murine T cells from prethymic and preliver embryonic yolk sac hematopoietic stem cells. Development. 1991;113(4):1315–23 Epub 1991/12/01.
40. Moore MA, Metcalf D. Ontogeny of the haemopoietic system: yolk sac origin of in vivo and in vitro colony forming cells in the developing mouse embryo. Br J Haematol. 1970;18 (3):279–96.
41. Dieterlen-Lievre F. On the origin of haemopoietic stem cells in the avian embryo: an experimental approach. J Embryol Exp Morphol. 1975;33(3):607–19 Epub 1975/06/01.
42. Chen XD, Turpen JB. Intraembryonic origin of hepatic hematopoiesis in Xenopus laevis. J Immunol. 1995;154(6):2557–67 Epub 1995/03/15.
43. Cumano A, Garcia-Porrero J, Dieterlen-Lievre F, Godin I. Intra-embryonic hematopoiesis in mice. C R Seances Soc Biol Fil. 1995;189(4):617–27.
44. Medvinsky A, Dzierzak E. Definitive hematopoiesis is autonomously initiated by the AGM region. Cell. 1996;86(6):897–906.
45. Yoder MC, Hiatt K, Dutt P, Mukherjee P, Bodine DM, Orlic D. Characterization of definitive lymphohematopoietic stem cells in the day 9 murine yolk sac. Immunity. 1997;7(3):335–44.
46. Yoder MC, Hiatt K. Engraftment of embryonic hematopoietic cells in conditioned newborn recipients. Blood. 1997;89(6):2176–83.
47. Matsuoka S, Tsuji K, Hisakawa H, Xu M, Ebihara Y, Ishii T, et al. Generation of definitive hematopoietic stem cells from murine early yolk sac and paraaortic splanchnopleures by aorta-gonad-mesonephros region-derived stromal cells. Blood. 2001;98(1):6–12.
48. Rossant J, Cross JC. Placental development: lessons from mouse mutants. Nat Rev Genet. 2001;2(7):538–48.
49. Gekas C, Dieterlen-Lievre F, Orkin SH, Mikkola HK. The placenta is a niche for hematopoietic stem cells. Dev Cell. 2005;8(3):365–75.
50. Ottersbach K, Dzierzak E. The murine placenta contains hematopoietic stem cells within the vascular labyrinth region. Dev Cell. 2005;8(3):377–87.
51. Corbel C, Salaun J, Belo-Diabangouaya P, Dieterlen-Lievre F. Hematopoietic potential of the pre-fusion allantois. Dev Biol. 2007;301(2):478–88.
52. Zeigler BM, Sugiyama D, Chen M, Guo Y, Downs KM, Speck NA. The allantois and chorion, when isolated before circulation or chorio-allantoic fusion, have hematopoietic potential. Development. 2006;133(21):4183–92.
53. Rhodes KE, Gekas C, Wang Y, Lux CT, Francis CS, Chan DN, et al. The emergence of hematopoietic stem cells is initiated in the placental vasculature in the absence of circulation. Cell Stem Cell. 2008;2(3):252–63.
54. Tam PP, Behringer RR. Mouse gastrulation: the formation of a mammalian body plan. Mech Dev. 1997;68(1–2):3–25.
55. Kumaravelu P, Hook L, Morrison AM, Ure J, Zhao S, Zuyev S, et al. Quantitative developmental anatomy of definitive haematopoietic stem cells/long-term repopulating units (HSC/RUs): role of the aorta-gonad-mesonephros (AGM) region and the yolk sac in colonisation of the mouse embryonic liver. Development. 2002;129(21):4891–9.
56. Charbord P, Oostendorp R, Pang W, Herault O, Noel F, Tsuji T, et al. Comparative study of stromal cell lines derived from embryonic, fetal, and postnatal mouse blood-forming tissues. Exp Hematol. 2002;30(10):1202–10.
57. Durand C, Robin C, Dzierzak E. Mesenchymal lineage potentials of aorta-gonad-mesonephros stromal clones. Haematologica. 2006;91(9):1172–9.

58. Cumano A, Godin I. Ontogeny of the hematopoietic system. Annu Rev Immunol. 2007;25:745–85.
59. Ema H, Nakauchi H. Expansion of hematopoietic stem cells in the developing liver of a mouse embryo. Blood. 2000;95(7):2284–8.
60. Mikkola HK, Orkin SH. The journey of developing hematopoietic stem cells. Development. 2006;133(19):3733–44.
61. Takeuchi M, Sekiguchi T, Hara T, Kinoshita T, Miyajima A. Cultivation of aorta-gonad-mesonephros-derived hematopoietic stem cells in the fetal liver microenvironment amplifies long-term repopulating activity and enhances engraftment to the bone marrow. Blood. 2002;99(4):1190–6.
62. Klassen LW, Birks J, Allen E, Gurney CW. Experimental medullary aplasia. J Lab Clin Med. 1972;80(1):8–17.
63. Visnjic D, Kalajzic Z, Rowe DW, Katavic V, Lorenzo J, Aguila HL. Hematopoiesis is severely altered in mice with an induced osteoblast deficiency. Blood. 2004;103(9):3258–64.
64. Ohno H, Ogawa M, Nishikawa S, Hayashi S, Kunisada T. Conditions required for myelopoiesis in murine spleen. Immunol Lett. 1993;35(2):197–204 Epub 1993/02/01.
65. Sasaki K, Matsumura G. Hemopoietic cells in the liver and spleen of the embryonic and early postnatal mouse: a karyometrical observation. Anat Rec. 1987;219(4):378–83 Epub 1987/12/01.
66. Yanai N, Satoh T, Obinata M. Endothelial cells create a hematopoietic inductive microenvironment preferential to erythropoiesis in the mouse spleen. Cell Struct Funct. 1991;16(1):87–93 Epub 1991/02/01.
67. Mackie EJ, Ahmed YA, Tatarczuch L, Chen KS, Mirams M. Endochondral ossification: how cartilage is converted into bone in the developing skeleton. Int J Biochem Cell Biol. 2008;40 (1):46–62.
68. Coskun S, Chao H, Vasavada H, Heydari K, Gonzales N, Zhou X, et al. Development of the fetal bone marrow niche and regulation of HSC quiescence and homing ability by emerging osteolineage cells. Cell Rep. 2014;9(2):581–90 Epub 2014/10/15.
69. Lo Celso C, Fleming HE, Wu JW, Zhao CX, Miake-Lye S, Fujisaki J, et al. Live-animal tracking of individual haematopoietic stem/progenitor cells in their niche. Nature. 2009;457 (7225):92–6.
70. Shalaby F, Ho J, Stanford WL, Fischer KD, Schuh AC, Schwartz L, et al. A requirement for Flk1 in primitive and definitive hematopoiesis and vasculogenesis. Cell. 1997;89(6):981–90 Epub 1997/06/13.
71. Shalaby F, Rossant J, Yamaguchi TP, Gertsenstein M, Wu XF, Breitman ML, et al. Failure of blood-island formation and vasculogenesis in Flk-1-deficient mice. Nature. 1995;376 (6535):62–6.
72. Sabin FR. Studies on the origin of blood-vessels and of red blood corpuscles as seen in the living blastoderm of chicks during the second day of incubation. Carnegie Contrib Embryol. 1920;27:214–62.
73. Hirschi KK. Hemogenic endothelium during development and beyond. Blood. 2012;119 (21):4823–7 Epub 2012/03/15.
74. Jaffredo T, Gautier R, Eichmann A, Dieterlen-Lievre F. Intraaortic hemopoietic cells are derived from endothelial cells during ontogeny. Development. 1998;125(22):4575–83.
75. Sugiyama D, Arai K, Tsuji K. Definitive hematopoiesis from acetyl LDL incorporating endothelial cells in the mouse embryo. Stem Cells Dev. 2005;14(6):687–96.
76. Zovein AC, Hofmann JJ, Lynch M, French WJ, Turlo KA, Yang Y, et al. Fate tracing reveals the endothelial origin of hematopoietic stem cells. Cell Stem Cell. 2008;3(6):625–36.
77. Bertrand JY, Chi NC, Santoso B, Teng S, Stainier DY, Traver D. Haematopoietic stem cells derive directly from aortic endothelium during development. Nature. 2010;464(7285):108–11.
78. Boisset JC, van Cappellen W, Andrieu-Soler C, Galjart N, Dzierzak E, Robin C. In vivo imaging of haematopoietic cells emerging from the mouse aortic endothelium. Nature. 2010;464(7285):116–20.

79. Kissa K, Herbomel P. Blood stem cells emerge from aortic endothelium by a novel type of cell transition. Nature. 2010;464(7285):112–5.
80. Taoudi S, Medvinsky A. Functional identification of the hematopoietic stem cell niche in the ventral domain of the embryonic dorsal aorta. Proc Natl Acad Sci USA. 2007;104 (22):9399–403.
81. Boisset JC, Clapes T, Klaus A, Papazian N, Onderwater J, Mommaas-Kienhuis M, et al. Progressive maturation towards hematopoietic stem cells in the mouse embryo aorta. Blood. 2014. Epub 2014/10/11.
82. Swiers G, Baumann C, O'Rourke J, Giannoulatou E, Taylor S, Joshi A, et al. Early dynamic fate changes in haemogenic endothelium characterized at the single-cell level. Nat Commun. 2013;4:2924 Epub 2013/12/12.
83. Chen MJ, Li Y, De Obaldia ME, Yang Q, Yzaguirre AD, Yamada-Inagawa T, et al. Erythroid/myeloid progenitors and hematopoietic stem cells originate from distinct populations of endothelial cells. Cell Stem Cell. 2011;9(6):541–52 Epub 2011/12/06.
84. Nishikawa SI, Nishikawa S, Hirashima M, Matsuyoshi N, Kodama H. Progressive lineage analysis by cell sorting and culture identifies FLK1+VE-cadherin+ cells at a diverging point of endothelial and hemopoietic lineages. Development. 1998;125(9):1747–57 Epub 1998/06/12.
85. Marcelo KL, Sills TM, Coskun S, Vasavada H, Sanglikar S, Goldie LC, et al. Hemogenic endothelial cell specification requires c-Kit, Notch signaling, and p27-mediated cell-cycle control. Dev Cell. 2013;27(5):504–15 Epub 2013/12/18.
86. Nadin BM, Goodell MA, Hirschi KK. Phenotype and hematopoietic potential of side population cells throughout embryonic development. Blood. 2003;102(7):2436–43.
87. Hierlihy AM, Seale P, Lobe CG, Rudnicki MA, Megeney LA. The post-natal heart contains a myocardial stem cell population. FEBS Lett. 2002;530(1–3):239–43 Epub 2002/10/22.
88. Kubota H, Avarbock MR, Brinster RL. Spermatogonial stem cells share some, but not all, phenotypic and functional characteristics with other stem cells. Proc Natl Acad Sci USA. 2003;100(11):6487–92 Epub 2003/05/10.
89. Welm B, Behbod F, Goodell MA, Rosen JM. Isolation and characterization of functional mammary gland stem cells. Cell Prolif. 2003;36(Suppl 1):17–32.
90. Wulf GG, Luo KL, Jackson KA, Brenner MK, Goodell MA. Cells of the hepatic side population contribute to liver regeneration and can be replenished with bone marrow stem cells. Haematologica. 2003;88(4):368–78 Epub 2003/04/12.
91. Mizuochi C, Fraser ST, Biasch K, Horio Y, Kikushige Y, Tani K, et al. Intra-aortic clusters undergo endothelial to hematopoietic phenotypic transition during early embryogenesis. PLoS ONE. 2012;7(4):e35763. Epub 2012/05/05.
92. Murray PDF. The development in vitro of the blood of the Early Chick Embryo1932 1932-10-01 00:00:00. p. 497–521.
93. Choi K, Kennedy M, Kazarov A, Papadimitriou JC, Keller G. A common precursor for hematopoietic and endothelial cells. Development. 1998;125(4):725–32.
94. Kennedy M, Firpo M, Choi K, Wall C, Robertson S, Kabrun N, et al. A common precursor for primitive erythropoiesis and definitive haematopoiesis. Nature. 1997;386(6624):488–93.
95. Fehling HJ, Lacaud G, Kubo A, Kennedy M, Robertson S, Keller G, et al. Tracking mesoderm induction and its specification to the hemangioblast during embryonic stem cell differentiation. Development. 2003;130(17):4217–27 Epub 2003/07/23.
96. Huber TL, Kouskoff V, Fehling HJ, Palis J, Keller G. Haemangioblast commitment is initiated in the primitive streak of the mouse embryo. Nature. 2004;432(7017):625–30.
97. Nishikawa S. Hemangioblast: an in vitro phantom. Wiley interdisciplinary reviews. Dev Biol. 2012;1(4):603–8 Epub 2013/06/27.
98. Lancrin C, Sroczynska P, Stephenson C, Allen T, Kouskoff V, Lacaud G. The haemangioblast generates haematopoietic cells through a haemogenic endothelium stage. Nature. 2009;457(7231):892–5.

99. Padron-Barthe L, Temino S, Villa del Campo C, Carramolino L, Isern J, Torres M. Clonal analysis identifies hemogenic endothelium as the source of the blood-endothelial common lineage in the mouse embryo. Blood. 2014;124(16):2523–32. Epub 2014/08/21.
100. Pereira CF, Chang B, Qiu J, Niu X, Papatsenko D, Hendry CE, et al. Induction of a hemogenic program in mouse fibroblasts. Cell Stem Cell. 2013;13(2):205–18 Epub 2013/06/19.
101. Doulatov S, Vo LT, Chou SS, Kim PG, Arora N, Li H, et al. Induction of multipotential hematopoietic progenitors from human pluripotent stem cells via respecification of lineage-restricted precursors. Cell Stem Cell. 2013;13(4):459–70 Epub 2013/10/08.
102. Myers CT, Krieg PA. BMP-mediated specification of the erythroid lineage suppresses endothelial development in blood island precursors. Blood. 2013;122(24):3929–39 Epub 2013/10/09.
103. Wright DE, Wagers AJ, Gulati AP, Johnson FL, Weissman IL. Physiological migration of hematopoietic stem and progenitor cells. Science. 2001;294(5548):1933–6.
104. Rafii S, Shapiro F, Pettengell R, Ferris B, Nachman RL, Moore MA, et al. Human bone marrow microvascular endothelial cells support long-term proliferation and differentiation of myeloid and megakaryocytic progenitors. Blood. 1995;86(9):3353–63 Epub 1995/11/01.
105. Avecilla ST, Hattori K, Heissig B, Tejada R, Liao F, Shido K, et al. Chemokine-mediated interaction of hematopoietic progenitors with the bone marrow vascular niche is required for thrombopoiesis. Nat Med. 2004;10(1):64–71.
106. Kiel MJ, Radice GL, Morrison SJ. Lack of evidence that hematopoietic stem cells depend on N-cadherin-mediated adhesion to osteoblasts for their maintenance. Cell Stem Cell. 2007;1 (2):204–17.
107. Hooper AT, Butler JM, Nolan DJ, Kranz A, Iida K, Kobayashi M, et al. Engraftment and reconstitution of hematopoiesis is dependent on VEGFR2-mediated regeneration of sinusoidal endothelial cells. Cell Stem Cell. 2009;4(3):263–74.
108. Kopp HG, Hooper AT, Avecilla ST, Rafii S. Functional heterogeneity of the bone marrow vascular niche. Ann NY Acad Sci. 2009;1176:47–54.
109. Yao Y, Song X, Cheng H, Tang G, Hu X, Zhou H, et al. Dysfunction of bone marrow vascular niche in acute graft-versus-host disease after MHC-haploidentical bone marrow transplantation. PLoS ONE. 2014;9(8):e104607. Epub 2014/08/15.
110. Himburg HA, Harris JR, Ito T, Daher P, Russell JL, Quarmyne M, et al. Pleiotrophin regulates the retention and self-renewal of hematopoietic stem cells in the bone marrow vascular niche. Cell reports. 2012;2(4):964–75 Epub 2012/10/23.
111. Nombela-Arrieta C, Pivarnik G, Winkel B, Canty KJ, Harley B, Mahoney JE, et al. Quantitative imaging of haematopoietic stem and progenitor cell localization and hypoxic status in the bone marrow microenvironment. Nat Cell Biol. 2013;15(5):533–43 Epub 2013/04/30.
112. Takaku T, Malide D, Chen J, Calado RT, Kajigaya S, Young NS. Hematopoiesis in 3 dimensions: human and murine bone marrow architecture visualized by confocal microscopy. Blood. 2010;116(15):e41–55 Epub 2010/07/22.
113. Godin I, Garcia-Porrero JA, Dieterlen-Lievre F, Cumano A. Stem cell emergence and hemopoietic activity are incompatible in mouse intraembryonic sites. J Exp Med. 1999;190 (1):43–52 Epub 1999/08/03.
114. Muller AM, Medvinsky A, Strouboulis J, Grosveld F, Dzierzak E. Development of hematopoietic stem cell activity in the mouse embryo. Immunity. 1994;1(4):291–301.
115. de Bruijn MF, Ma X, Robin C, Ottersbach K, Sanchez MJ, Dzierzak E. Hematopoietic stem cells localize to the endothelial cell layer in the midgestation mouse aorta. Immunity. 2002;16(5):673–83.
116. Cumano A, Dieterlen-Lievre F, Godin I. Lymphoid potential, probed before circulation in mouse, is restricted to caudal intraembryonic splanchnopleura. Cell. 1996;86(6):907–16.
117. Li Z, Lan Y, He W, Chen D, Wang J, Zhou F, et al. Mouse embryonic head as a site for hematopoietic stem cell development. Cell Stem Cell. 2012;11(5):663–75 Epub 2012/11/06.

118. Kunisaki Y, Frenette PS. Influences of vascular niches on hematopoietic stem cell fate. Int J Hematol. 2014;99(6):699–705 Epub 2014/04/24.
119. Winkler IG, Barbier V, Nowlan B, Jacobsen RN, Forristal CE, Patton JT, et al. Vascular niche E-selectin regulates hematopoietic stem cell dormancy, self renewal and chemoresistance. Nat Med. 2012;18(11):1651–7 Epub 2012/10/23.
120. Kunisaki Y, Bruns I, Scheiermann C, Ahmed J, Pinho S, Zhang D, et al. Arteriolar niches maintain haematopoietic stem cell quiescence. Nature. 2013;502(7473):637–43 Epub 2013/10/11.
121. Sugiyama T, Kohara H, Noda M, Nagasawa T. Maintenance of the hematopoietic stem cell pool by CXCL12-CXCR4 chemokine signaling in bone marrow stromal cell niches. Immunity. 2006;25(6):977–88.
122. Sacchetti B, Funari A, Michienzi S, Di Cesare S, Piersanti S, Saggio I, et al. Self-renewing osteoprogenitors in bone marrow sinusoids can organize a hematopoietic microenvironment. Cell. 2007;131(2):324–36.
123. Mendez-Ferrer S, Michurina TV, Ferraro F, Mazloom AR, Macarthur BD, Lira SA, et al. Mesenchymal and haematopoietic stem cells form a unique bone marrow niche. Nature. 2010;466(7308):829–34 Epub 2010/08/13.
124. Greenbaum A, Hsu YM, Day RB, Schuettpelz LG, Christopher MJ, Borgerding JN, et al. CXCL12 in early mesenchymal progenitors is required for haematopoietic stem-cell maintenance. Nature. 2013;495(7440):227–30 Epub 2013/02/26.
125. Ding L, Morrison SJ. Haematopoietic stem cells and early lymphoid progenitors occupy distinct bone marrow niches. Nature. 2013;495(7440):231–5 Epub 2013/02/26.
126. Ding L, Saunders TL, Enikolopov G, Morrison SJ. Endothelial and perivascular cells maintain haematopoietic stem cells. Nature. 2012;481(7382):457–62 Epub 2012/01/28.
127. Katayama Y, Battista M, Kao WM, Hidalgo A, Peired AJ, Thomas SA, et al. Signals from the sympathetic nervous system regulate hematopoietic stem cell egress from bone marrow. Cell. 2006;124(2):407–21.
128. Mendez-Ferrer S, Lucas D, Battista M, Frenette PS. Haematopoietic stem cell release is regulated by circadian oscillations. Nature. 2008;452(7186):442–7.
129. Hanoun M, Zhang D, Mizoguchi T, Pinho S, Pierce H, Kunisaki Y, et al. Acute myelogenous leukemia-induced sympathetic neuropathy promotes malignancy in an altered hematopoietic stem cell niche. Cell Stem Cell. 2014;15(3):365–75 Epub 2014/07/16.
130. Arranz L, Sanchez-Aguilera A, Martin-Perez D, Isern J, Langa X, Tzankov A, et al. Neuropathy of haematopoietic stem cell niche is essential for myeloproliferative neoplasms. Nature. 2014;512(7512):78–81 Epub 2014/07/22.
131. Vannucchi AM, Bianchi L, Cellai C, Paoletti F, Rana RA, Lorenzini R, et al. Development of myelofibrosis in mice genetically impaired for GATA-1 expression (GATA-1(low) mice). Blood. 2002;100(4):1123–32 Epub 2002/08/01.
132. Shivdasani RA, Orkin SH. Erythropoiesis and globin gene expression in mice lacking the transcription factor NF-E2. Proc Natl Acad Sci USA. 1995;92(19):8690–4 Epub 1995/09/12.
133. Heazlewood SY, Neaves RJ, Williams B, Haylock DN, Adams TE, Nilsson SK. Megakaryocytes co-localise with hemopoietic stem cells and release cytokines that up-regulate stem cell proliferation. Stem Cell Res. 2013;11(2):782–92 Epub 2013/06/25.
134. Olson TS, Caselli A, Otsuru S, Hofmann TJ, Williams R, Paolucci P, et al. Megakaryocytes promote murine osteoblastic HSC niche expansion and stem cell engraftment after radioablative conditioning. Blood. 2013;121(26):5238–49 Epub 2013/05/15.
135. Bruns I, Lucas D, Pinho S, Ahmed J, Lambert MP, Kunisaki Y, et al. Megakaryocytes regulate hematopoietic stem cell quiescence through CXCL4 secretion. Nat Med. 2014;20 (11):1315–20 Epub 2014/10/20.
136. Storan MJ, Heazlewood SY, Heazlewood CK, Haylock DN, Alexander WS, Neaves RJ, et al. Factors released by megakaryocytes thrombin cleave osteopontin to negatively regulate hemopoietic stem cells. Stem Cells. 2015. Epub 2015/04/14.

137. Joseph C, Quach JM, Walkley CR, Lane SW, Lo Celso C, Purton LE. Deciphering hematopoietic stem cells in their niches: a critical appraisal of genetic models, lineage tracing, and imaging strategies. Cell Stem Cell. 2013;13(5):520–33. Epub 2013/11/12.
138. Wang L, Benedito R, Bixel MG, Zeuschner D, Stehling M, Savendahl L, et al. Identification of a clonally expanding haematopoietic compartment in bone marrow. EMBO J. 2013;32 (2):219–30 Epub 2012/11/29.
139. Brandt JE, Galy AH, Luens KM, Travis M, Young J, Tong J, et al. Bone marrow repopulation by human marrow stem cells after long-term expansion culture on a porcine endothelial cell line. Exp Hematol. 1998;26(10):950–61.
140. Chute JP, Saini AA, Chute DJ, Wells MR, Clark WB, Harlan DM, et al. Ex vivo culture with human brain endothelial cells increases the SCID-repopulating capacity of adult human bone marrow. Blood. 2002;100(13):4433–9.
141. Li W, Johnson SA, Shelley WC, Ferkowicz M, Morrison P, Li Y, et al. Primary endothelial cells isolated from the yolk sac and para-aortic splanchnopleura support the expansion of adult marrow stem cells in vitro. Blood. 2003;102(13):4345–53.
142. Li W, Johnson SA, Shelley WC, Yoder MC. Hematopoietic stem cell repopulating ability can be maintained in vitro by some primary endothelial cells. Exp Hematol. 2004;32 (12):1226–37.
143. Duncan AW, Rattis FM, DiMascio LN, Congdon KL, Pazianos G, Zhao C, et al. Integration of Notch and Wnt signaling in hematopoietic stem cell maintenance. Nat Immunol. 2005;6 (3):314–22.
144. Reya T, Duncan AW, Ailles L, Domen J, Scherer DC, Willert K, et al. A role for Wnt signalling in self-renewal of haematopoietic stem cells. Nature. 2003;423(6938):409–14.
145. Robert-Moreno A, Guiu J, Ruiz-Herguido C, Lopez ME, Ingles-Esteve J, Riera L, et al. Impaired embryonic haematopoiesis yet normal arterial development in the absence of the Notch ligand Jagged1. EMBO J. 2008;27(13):1886–95 Epub 2008/06/06.
146. Robert-Moreno A, Espinosa L, de la Pompa JL, Bigas A. RBPjkappa-dependent Notch function regulates Gata2 and is essential for the formation of intra-embryonic hematopoietic cells. Development. 2005;132(5):1117–26.
147. Kim AD, Melick CH, Clements WK, Stachura DL, Distel M, Panakova D, et al. Discrete Notch signaling requirements in the specification of hematopoietic stem cells. EMBO J. 2014;33(20):2363–73 Epub 2014/09/19.
148. Hadland BK, Huppert SS, Kanungo J, Xue Y, Jiang R, Gridley T, et al. A requirement for Notch1 distinguishes 2 phases of definitive hematopoiesis during development. Blood. 2004;104(10):3097–105 Epub 2004/07/15.
149. Kumano K, Chiba S, Kunisato A, Sata M, Saito T, Nakagami-Yamaguchi E, et al. Notch1 but not Notch2 is essential for generating hematopoietic stem cells from endothelial cells. Immunity. 2003;18(5):699–711.
150. Maillard I, Koch U, Dumortier A, Shestova O, Xu L, Sai H, et al. Canonical notch signaling is dispensable for the maintenance of adult hematopoietic stem cells. Cell Stem Cell. 2008;2 (4):356–66.
151. Mancini SJ, Mantei N, Dumortier A, Suter U, MacDonald HR, Radtke F. Jagged1-dependent Notch signaling is dispensable for hematopoietic stem cell self-renewal and differentiation. Blood. 2005;105(6):2340–2.
152. Delaney C, Heimfeld S, Brashem-Stein C, Voorhies H, Manger RL, Bernstein ID. Notch-mediated expansion of human cord blood progenitor cells capable of rapid myeloid reconstitution. Nat Med. 2010;16(2):232–6 Epub 2010/01/19.
153. Butler JM, Nolan DJ, Vertes EL, Varnum-Finney B, Kobayashi H, Hooper AT, et al. Endothelial cells are essential for the self-renewal and repopulation of Notch-dependent hematopoietic stem cells. Cell Stem Cell. 2010;6(3):251–64.
154. Hadland BK, Varnum-Finney B, Poulos MG, Moon RT, Butler JM, Rafii S, et al. Endothelium and NOTCH specify and amplify aorta-gonad-mesonephros-derived hematopoietic stem cells. J Clin Invest. 2015;125(5):2032–45 Epub 2015/04/14.

155. Poulos MG, Guo P, Kofler NM, Pinho S, Gutkin MC, Tikhonova A, et al. Endothelial Jagged-1 is necessary for homeostatic and regenerative hematopoiesis. Cell Rep. 2013;4 (5):1022–34 Epub 2013/09/10.
156. Gerhardt DM, Pajcini KV, D'Altri T, Tu L, Jain R, Xu L, et al. The Notch1 transcriptional activation domain is required for development and reveals a novel role for Notch1 signaling in fetal hematopoietic stem cells. Genes Dev. 2014;28(6):576–93 Epub 2014/03/19.
157. Chiang MY, Shestova O, Xu L, Aster JC, Pear WS. Divergent effects of supraphysiologic Notch signals on leukemia stem cells and hematopoietic stem cells. Blood. 2013;121(6):905–17 Epub 2012/11/02.
158. Varnum-Finney B, Halasz LM, Sun M, Gridley T, Radtke F, Bernstein ID. Notch2 governs the rate of generation of mouse long- and short-term repopulating stem cells. J Clin Invest. 2011;121(3):1207–16 Epub 2011/02/03.
159. Rattis FM, Voermans C, Reya T. Wnt signaling in the stem cell niche. Curr Opin Hematol. 2004;11(2):88–94.
160. Scheller M, Huelsken J, Rosenbauer F, Taketo MM, Birchmeier W, Tenen DG, et al. Hematopoietic stem cell and multilineage defects generated by constitutive beta-catenin activation. Nat Immunol. 2006;7(10):1037–47.
161. Kirstetter P, Anderson K, Porse BT, Jacobsen SE, Nerlov C. Activation of the canonical Wnt pathway leads to loss of hematopoietic stem cell repopulation and multilineage differentiation block. Nat Immunol. 2006;7(10):1048–56.
162. Mazo IB, Gutierrez-Ramos JC, Frenette PS, Hynes RO, Wagner DD, von Andrian UH. Hematopoietic progenitor cell rolling in bone marrow microvessels: parallel contributions by endothelial selectins and vascular cell adhesion molecule 1. J Exp Med. 1998;188(3):465–74.
163. Wilson A, Trumpp A. Bone-marrow haematopoietic-stem-cell niches. Nat Rev Immunol. 2006;6(2):93–106.
164. Whetton AD, Graham GJ. Homing and mobilization in the stem cell niche. Trends Cell Biol. 1999;9(6):233–8.
165. Dao MA, Nolta JA. Cytokine and integrin stimulation synergize to promote higher levels of GATA-2, c-myb, and CD34 protein in primary human hematopoietic progenitors from bone marrow. Blood. 2007;109(6):2373–9 Epub 2006/11/11.
166. Schweitzer KM, Drager AM, van der Valk P, Thijsen SF, Zevenbergen A, Theijsmeijer AP, et al. Constitutive expression of E-selectin and vascular cell adhesion molecule-1 on endothelial cells of hematopoietic tissues. Am J Pathol. 1996;148(1):165–75 Epub 1996/01/01.
167. Kansas GS. Selectins and their ligands: current concepts and controversies. Blood. 1996;88 (9):3259–87 Epub 1996/11/01.
168. Sipkins DA, Wei X, Wu JW, Runnels JM, Cote D, Means TK, et al. In vivo imaging of specialized bone marrow endothelial microdomains for tumour engraftment. Nature. 2005;435(7044):969–73.
169. Ceradini DJ, Kulkarni AR, Callaghan MJ, Tepper OM, Bastidas N, Kleinman ME, et al. Progenitor cell trafficking is regulated by hypoxic gradients through HIF-1 induction of SDF-1. Nat Med. 2004;10(8):858–64.
170. Ara T, Tokoyoda K, Sugiyama T, Egawa T, Kawabata K, Nagasawa T. Long-term hematopoietic stem cells require stromal cell-derived factor-1 for colonizing bone marrow during ontogeny. Immunity. 2003;19(2):257–67.
171. Nie Y, Han YC, Zou YR. CXCR4 is required for the quiescence of primitive hematopoietic cells. J Exp Med. 2008;205(4):777–83.
172. Omatsu Y, Sugiyama T, Kohara H, Kondoh G, Fujii N, Kohno K, et al. The essential functions of adipo-osteogenic progenitors as the hematopoietic stem and progenitor cell niche. Immunity. 2010;33(3):387–99 Epub 2010/09/21.
173. Yamazaki S, Ema H, Karlsson G, Yamaguchi T, Miyoshi H, Shioda S, et al. Nonmyelinating Schwann cells maintain hematopoietic stem cell hibernation in the bone marrow niche. Cell. 2011;147(5):1146–58 Epub 2011/11/29.

174. Ogawa M, Matsuzaki Y, Nishikawa S, Hayashi S, Kunisada T, Sudo T, et al. Expression and function of c-kit in hemopoietic progenitor cells. J Exp Med. 1991;174(1):63–71 Epub 1991/07/01.
175. Russell ES. Hereditary anemias of the mouse: a review for geneticists. Adv Genet. 1979;20:357–459 Epub 1979/01/01.
176. McCulloch EA, Siminovitch L, Till JE, Russell ES, Bernstein SE. The cellular basis of the genetically determined hemopoietic defect in anemic mice of genotype Sl-Sld. Blood. 1965;26(4):399–410.
177. Barker JE. Sl/Sld hematopoietic progenitors are deficient in situ. Exp Hematol. 1994;22 (2):174–7 Epub 1994/02/01.
178. Kimura Y, Ding B, Imai N, Nolan DJ, Butler JM, Rafii S. c-Kit-mediated functional positioning of stem cells to their niches is essential for maintenance and regeneration of adult hematopoiesis. PLoS ONE. 2011;6(10):e26918. Epub 2011/11/03.
179. Calvi LM, Adams GB, Weibrecht KW, Weber JM, Olson DP, Knight MC, et al. Osteoblastic cells regulate the haematopoietic stem cell niche. Nature. 2003;425(6960):841–6.
180. Ferrara N. Vascular endothelial growth factor: basic science and clinical progress. Endocr Rev. 2004;25(4):581–611 Epub 2004/08/06.
181. Kabrun N, Buhring HJ, Choi K, Ullrich A, Risau W, Keller G. Flk-1 expression defines a population of early embryonic hematopoietic precursors. Development. 1997;124 (10):2039–48.
182. Hattori K, Heissig B, Wu Y, Dias S, Tejada R, Ferris B, et al. Placental growth factor reconstitutes hematopoiesis by recruiting VEGFR1(+) stem cells from bone-marrow microenvironment. Nat Med. 2002;8(8):841–9.
183. Zelzer E, Mamluk R, Ferrara N, Johnson RS, Schipani E, Olsen BR. VEGFA is necessary for chondrocyte survival during bone development. Development. 2004;131(9):2161–71 Epub 2004/04/10.
184. Maes C, Kobayashi T, Selig MK, Torrekens S, Roth SI, Mackem S, et al. Osteoblast precursors, but not mature osteoblasts, move into developing and fractured bones along with invading blood vessels. Dev Cell. 2010;19(2):329–44 Epub 2010/08/17.
185. Gerber HP, Malik AK, Solar GP, Sherman D, Liang XH, Meng G, et al. VEGF regulates haematopoietic stem cell survival by an internal autocrine loop mechanism. Nature. 2002;417 (6892):954–8.
186. Sitnicka E, Ruscetti FW, Priestley GV, Wolf NS, Bartelmez SH. Transforming growth factor beta 1 directly and reversibly inhibits the initial cell divisions of long-term repopulating hematopoietic stem cells. Blood. 1996;88(1):82–8.
187. Garbe A, Spyridonidis A, Mobest D, Schmoor C, Mertelsmann R, Henschler R. Transforming growth factor-beta 1 delays formation of granulocyte-macrophage colony-forming cells, but spares more primitive progenitors during ex vivo expansion of CD34+ haemopoietic progenitor cells. Br J Haematol. 1997;99(4):951–8.
188. Batard P, Monier MN, Fortunel N, Ducos K, Sansilvestri-Morel P, Phan T, et al. TGF-(beta)1 maintains hematopoietic immaturity by a reversible negative control of cell cycle and induces CD34 antigen up-modulation. J Cell Sci. 2000;113(Pt 3):383–90.
189. Soma T, Yu JM, Dunbar CE. Maintenance of murine long-term repopulating stem cells in ex vivo culture is affected by modulation of transforming growth factor-beta but not macrophage inflammatory protein-1 alpha activities. Blood. 1996;87(11):4561–7.
190. Hatzfeld J, Li ML, Brown EL, Sookdeo H, Levesque JP, O'Toole T, et al. Release of early human hematopoietic progenitors from quiescence by antisense transforming growth factor beta 1 or Rb oligonucleotides. J Exp Med. 1991;174(4):925–9.
191. Fortunel N, Batard P, Hatzfeld A, Monier MN, Panterne B, Lebkowski J, et al. High proliferative potential-quiescent cells: a working model to study primitive quiescent hematopoietic cells. J Cell Sci. 1998;111(Pt 13):1867–75.
192. Goey H, Keller JR, Back T, Longo DL, Ruscetti FW, Wiltrout RH. Inhibition of early murine hemopoietic progenitor cell proliferation after in vivo locoregional administration of transforming growth factor-beta 1. J Immunol. 1989;143(3):877–80.

193. Fan X, Valdimarsdottir G, Larsson J, Brun A, Magnusson M, Jacobsen SE, et al. Transient disruption of autocrine TGF-beta signaling leads to enhanced survival and proliferation potential in single primitive human hemopoietic progenitor cells. J Immunol. 2002;168 (2):755–62.
194. Kim SJ, Letterio J. Transforming growth factor-beta signaling in normal and malignant hematopoiesis. Leukemia. 2003;17(9):1731–7.
195. Wierenga AT, Vellenga E, Schuringa JJ. Convergence of hypoxia and TGFbeta pathways on cell cycle regulation in human hematopoietic stem/progenitor cells. PLoS ONE. 2014;9(3): e93494. Epub 2014/04/02.
196. Chute JP, Muramoto GG, Dressman HK, Wolfe G, Chao NJ, Lin S. Molecular profile and partial functional analysis of novel endothelial cell-derived growth factors that regulate hematopoiesis. Stem Cells. 2006;24(5):1315–27.
197. Zhang CC, Kaba M, Iizuka S, Huynh H, Lodish HF. Angiopoietin-like 5 and IGFBP2 stimulate ex vivo expansion of human cord blood hematopoietic stem cells as assayed by NOD/SCID transplantation. Blood. 2008;111(7):3415–23.
198. Jones SA, Scheller J, Rose-John S. Therapeutic strategies for the clinical blockade of IL-6/gp130 signaling. J Clin Invest. 2011;121(9):3375–83 Epub 2011/09/02.
199. Yoshida K, Taga T, Saito M, Suematsu S, Kumanogoh A, Tanaka T, et al. Targeted disruption of gp130, a common signal transducer for the interleukin 6 family of cytokines, leads to myocardial and hematological disorders. Proc Natl Acad Sci USA. 1996;93(1):407–11 Epub 1996/01/09.
200. Yao L, Yokota T, Xia L, Kincade PW, McEver RP. Bone marrow dysfunction in mice lacking the cytokine receptor gp130 in endothelial cells. Blood. 2005;106(13):4093–101.
201. Mohyeldin A, Garzon-Muvdi T, Quinones-Hinojosa A. Oxygen in stem cell biology: a critical component of the stem cell niche. Cell Stem Cell. 2010;7(2):150–61 Epub 2010/08/05.
202. Suda T, Takubo K, Semenza GL. Metabolic regulation of hematopoietic stem cells in the hypoxic niche. Cell Stem Cell. 2011;9(4):298–310 Epub 2011/10/11.
203. Parmar K, Mauch P, Vergilio JA, Sackstein R, Down JD. Distribution of hematopoietic stem cells in the bone marrow according to regional hypoxia. Proc Natl Acad Sci USA. 2007;104 (13):5431–6.
204. Takubo K, Goda N, Yamada W, Iriuchishima H, Ikeda E, Kubota Y, et al. Regulation of the HIF-1alpha level is essential for hematopoietic stem cells. Cell Stem Cell. 2010;7(3):391–402 Epub 2010/09/02.
205. Eliasson P, Jonsson JI. The hematopoietic stem cell niche: low in oxygen but a nice place to be. J Cell Physiol. 2010;222(1):17–22 Epub 2009/09/03.
206. Levesque JP, Winkler IG, Hendy J, Williams B, Helwani F, Barbier V, et al. Hematopoietic progenitor cell mobilization results in hypoxia with increased hypoxia-inducible transcription factor-1 alpha and vascular endothelial growth factor A in bone marrow. Stem Cells. 2007;25 (8):1954–65 Epub 2007/05/05.
207. Kubota Y, Takubo K, Suda T. Bone marrow long label-retaining cells reside in the sinusoidal hypoxic niche. Biochem Biophys Res Commun. 2008;366(2):335–9.
208. Semenza GL. Hypoxia-inducible factors in physiology and medicine. Cell. 2012;148(3):399–408 Epub 2012/02/07.
209. Kocabas F, Zheng J, Thet S, Copeland NG, Jenkins NA, DeBerardinis RJ, et al. Meis1 regulates the metabolic phenotype and oxidant defense of hematopoietic stem cells. Blood. 2012;120(25):4963–72 Epub 2012/09/22.
210. Simsek T, Kocabas F, Zheng J, Deberardinis RJ, Mahmoud AI, Olson EN, et al. The distinct metabolic profile of hematopoietic stem cells reflects their location in a hypoxic niche. Cell Stem Cell. 2010;7(3):380–90 Epub 2010/09/02.
211. Miharada K, Karlsson G, Rehn M, Rorby E, Siva K, Cammenga J, et al. Cripto regulates hematopoietic stem cells as a hypoxic-niche-related factor through cell surface receptor GRP78. Cell Stem Cell. 2011;9(4):330–44 Epub 2011/10/11.

212. Rehn M, Olsson A, Reckzeh K, Diffner E, Carmeliet P, Landberg G, et al. Hypoxic induction of vascular endothelial growth factor regulates murine hematopoietic stem cell function in the low-oxygenic niche. Blood. 2011;118(6):1534–43 Epub 2011/06/15.
213. Rouault-Pierre K, Lopez-Onieva L, Foster K, Anjos-Afonso F, Lamrissi-Garcia I, Serrano-Sanchez M, et al. HIF-2alpha protects human hematopoietic stem/progenitors and acute myeloid leukemic cells from apoptosis induced by endoplasmic reticulum stress. Cell Stem Cell. 2013;13(5):549–63 Epub 2013/10/08.

Tendon Stem Cell Niche

Raquel Costa-Almeida, A.I. Gonçalves, P. Gershovich, M.T. Rodrigues, R.L. Reis and M.E. Gomes

Abbreviations

α-SMA	Alpha smooth muscle actin
BMSCs	Bone marrow-derived mesenchymal stem cells
CD	Cluster of differentiation
COMP	Collagen oligomeric matrix protein
Cx	Connexin
ECM	Extracellular matrix
Egr 1	Early growth response 1 transcription factor
Egr 2	Early growth response 2 transcription factor

The original version of this chapter was revised: The spelling of the first author's name was corrected. The erratum to this chapter is available at DOI 10.1007/978-3-319-21705-5_14

R. Costa-Almeida · A.I. Gonçalves · P. Gershovich · M.T. Rodrigues · R.L. Reis · M.E. Gomes (✉)
3B's Research Group, Biomaterials, Biodegradables and Biomimetics, University of Minho, Headquarters of the European Institute of Excellence on Tissue Engineering and Regenerative Medicine, Avepark - Parque de Ciência e Tecnologia, Zona Industrial da Gandra, Barco GMR, 4805-017 Guimarães, Portugal
e-mail: megomes@dep.uminho.pt

R. Costa-Almeida
e-mail: raquel.almeida@dep.uminho.pt

A.I. Gonçalves
e-mail: ana.goncalves@dep.uminho.pt

P. Gershovich
e-mail: pm.gershovich@gmail.com

M.T. Rodrigues
e-mail: mrodrigues@dep.uminho.pt

R.L. Reis
e-mail: rgreis@dep.uminho.pt

R. Costa-Almeida · A.I. Gonçalves · P. Gershovich · M.T. Rodrigues R.L. Reis · M.E. Gomes
ICVS/3B's Government Associate Laboratory, University of Minho, Braga, Portugal

K. Turksen (ed.), *Tissue-Specific Stem Cell Niche*,
Stem Cell Biology and Regenerative Medicine,
DOI 10.1007/978-3-319-21705-5_10

EGF	Epidermal growth factor
FGF	Fibroblast growth factor
GAGs	Glycosaminoglycans
GDF-5	Growth differentiation factor 5
GFs	Growth factors
IGF	Insulin-like growth factor
MHC-II	Major histocompatibility complex II
Mkx	Mohawk
MSCs	Mesenchymal stem cells
PDGF-BB	Platelet derived growth factor
PLs	Platelet lysates
PGs	Proteoglycans
PRP	Platelet-rich plasma
Scx	Scleraxis
TGF-β	Transforming growth factor-beta
TDSCs	Tendon-derived stem cells
TFs	Transcription factors

1 Introduction: Short Overview on Tendon Biology

Tendons are dense fibrous connective tissues with the main responsibility of transmitting forces between muscles and bones, allowing locomotion and assuring the stability of joints along the musculoskeletal system. Tendon tissue has few cells in its composition and is primarily constituted by an abundant hierarchical extracellular matrix (ECM) network with aligned collagen fiber bundles [1]. Nonetheless, at an embryonic stage, developing tendon appears as a highly cellular structure exhibiting a sparse and disorganized ECM [2]. Despite the hypocellular constitution of mature tissues, tendon matrix is maintained through a continuous process of matrix remodeling by the resident cell populations [3]. The basic cellular unit in tendons is the tenocyte, which appears as a fibroblast-like cell, being frequently called "tendon fibroblast" or "tenoblast" in the literature. Tendons exist in very distinct anatomical locations and, thus, site-specific structural, biochemical and cellular properties are associated. This might also be the result of differences in embryonic development of tendon, as axial tendons have a different somitic origin than that of tendons of the limbs and trunk. The discovery of scleraxis (Scx), a basic helix-loop-helix protein, as a marker of mature tendons and tendon development [4] also enabled the detection of *Scx*-expressing tendon progenitors during chick development [5]. Accordingly, *Scx*-expressing progenitor cells of trunk tendons have been described to appear first between the myotome and sclerotome during somite development, whereas a distinct somitic compartment of *Scx*-expressing progenitor cells of axial tendons has been identified, the syndetome [5]. Further details on tendon developmental origin have been reviewed elsewhere [6].

Nevertheless, this differential origin results in additional and more complex variability among tendons in the human body. This generates a need for integrated knowledge when addressing tendon healing and regeneration strategies.

This chapter reviews the main characteristics of tendons, focusing on the role of tendon stem and progenitor cells in tendon biology, as well as the essential factors that coordinate tenogenic differentiation and maintenance of these cell populations in tendon homeostasis.

2 Tendon-Derived Stem Cells

Tendon cell populations have been traditionally considered to be composed only by tenocytes/tenoblasts. However, a population of stem and progenitor cells was firstly identified in 2007 by Young and colleagues both in human and mouse tendons [7]. These cells exhibited the universal characteristics of mesenchymal stem cells (MSCs), namely clonogenecity, multipotency and self-renewal; however they constitute a heterogeneous population, comprising also progenitor cells and being usually termed tendon stem/progenitor cells [7]. Moreover, tendon-derived stem cells (TDSCs) constitute 1–4 % of the total nucleated cells that reside in tendon tissue [7, 8] and have also been identified in tendon samples from rabbit, rat and horse [8–10]. Generally, TDSCs are isolated through enrichment procedures, which originate cell populations that are heterogeneous in terms of stem cell characteristics and that might also contain other cell types, like tenocytes [11]. There is still great controversy regarding the identity of TDSCs, which is somehow related to the debatable identification of MSCs, as well as fibroblasts, since all these cell types share phenotypic markers. Although TDSCs are mesenchymal cells, they can be distinguished from other cell types, like bone marrow derived mesenchymal stem cells (BMSCs), in terms of gene expression, surface markers and response to diverse growth factors [7, 12]. Indeed, these cell populations are characterized by a combination of markers and characteristics that are presented in Table 1. Additionally, TDSCs from distinct regional locations within tendon tissue have been reported to exhibit differences in terms of colony-forming potential and marker expression [13, 14].

Moreover, TDSCs have been reported to differentiate into tenocytes, chondrocytes, osteocytes and adipocytes upon in vitro induction, as well as to originate tendon-, cartilage-, bone- and tendon-bone junction-like tissues in animal models [7–10, 15]. This ability renders TDSCs of potential to be used as a cell source for tendon regenerative therapies. Therefore, in order to contribute for a broader understanding of the biology of TDSCs, the next section of this chapter will address the key elements that constitute their niche, discussing their relevance in tenogenic differentiation for regenerative medicine applications.

Table 1 Overview on characteristics that distinguish between BMSCs and TDSCs

Characteristics	BMSCs	TDSCs
Phenotypic markers		
Tenomodulin (*Tnmd*)	+	+++
Scleraxis (*Scx*)	+	+++
Collagen I (*Col1A1*)	+	+++
Collagen II (*Col2A1*)	+	+++
Collagen III (*Col3A1*)	+	++
Decorin (*Dcn*)	+	+++
Alkaline phosphatase (*Alpl*)	+	+++
Aggrecan (*Acan*)	+	+
Biglycan (*Bgn*)	+	+++
Stemness markers		
Oct 4	+	+++
Nanog	+	+
Sox 2	+	+
Proliferative potential	High	Higher
Multilineage differentiation potential	High	Higher
Colony-forming ability	Yes	Yes (Higher)

Phenotypic markers presented here refer to mRNA levels, which are labeled as "+", positive expression; "++", positive expression without significant statistical differences; "+++", positive expression, with significant statistical differences. Characteristics of both cell populations are based on biological studies of different species, including mouse, rat and human [7, 12]

3 Tendon Stem Cell Niche

Stem cells reside in an instructive microenvironment—the stem cell niche—that is able to maintain their stemness properties, like self-renewal, simultaneously directing stem cell fate and triggering their differentiation towards a specific cell type [16]. Several cues define this stem cell niche, including physical/mechanical, biochemical and biological signals, generating a bidirectional system where cells are influenced and influence the surrounding microenvironment [17, 18]. Stem cells generally reside in a niche that is mainly composed of other cells; however, tendon stem cell niche is a protein-based niche, consisting largely of extracellular matrix [7]. Herein, the characteristics of tendon stem cell niche will be highlighted, including signaling molecules, cellular crosstalk and the extracellular matrix role in tendon biology.

3.1 Signaling Molecules Associated to Tendon-Derived Stem Cells

TDSCs play a primary role in maintaining tissue homeostasis. Under normal conditions, TDSCs differentiate into tenocytes that are responsible for the maintenance, repair and remodeling of tendons. This process is often induced and regulated by signaling molecules such as transcription and growth factors, TFs and GFs, respectively.

Therefore, understanding the origin of tendon stem and progenitor cells, the influence of signaling molecules in their differentiation process, and the specific markers that identify cells within different stages of the tenogenic lineage are essential for designing novel regenerative strategies.

In the early stage of mesenchymal tissue development, *Sox9* expression is confined within the early skeletal primordia, while *Scx* is broadly expressed in early and later embryonic development of tendon precursor cells [19]. Cells expressing Scx contribute to tendon midsubstance formation, while the co-expression of Scx and Sox-9 is often associated to cells that will originate the tendon fibrocartilage of the enthesis [20].

Moreover, Soeda et al. reported that cells in limb tendons, namely the Achilles and the patellar tendons, are originated from *Sox9*-expressing precursors [21].

Similarly, Mohawk is another transcription factor with particular relevance in developing tendons, being required during embryogenesis. *Mkx* is strongly expressed in tendon cells during the early stages of tendon development and down-regulated upon tenocyte differentiation [22]. Nevertheless, *Mkx* also plays a role in tendon maturation after birth [22], especially in regulating the expression of collagen I and associated molecules in tendon cells [23]. Accordingly, *Mkx* null mice evidence defective hypoplastic tendons [23, 24] throughout their body and abnormal collagen fibrils [25].

Early growth response (Egr)1 and Egr2 transcription factors are also involved in tendon formation. Their expression in tendon cells is correlated with the increase of collagen during differentiation, as demonstrated in embryonic chick and mouse limbs [26]. Egr1 is also involved in inducing the expression of *Scx* in rat MSCs via transforming growth factor (TGF)-β signaling pathway [27].

During growth and after injury in the adult, tendon midsubstance progenitor cells can be identified by the expression of alpha smooth muscle actin (α-SMA), while growth differentiation factor 5 (GDF-5)-expressing progenitor cells were described to contribute to tendon enthesis [28].

Growth factors, such as the members of TGF-β superfamily are considered major regulators of the differentiation and growth of skeletal connective tissues [29, 30]. Moreover, TGF-β has been described to participate in the development, ECM synthesis or healing of tendons [3, 31]. Thus, it is not odd to infer that TDSCs fate is likely to be controlled by the TGF signaling pathway through the induction of *Scx* expression [32, 33], which has been described as a regulator of tendon development. The disruption of TGF-β signaling in TGF-β2 and TGF-β3 double mutant mice, or through the inactivation of the type II TGF-β receptor results in the loss of most tendons and ligaments in the limbs, trunk, tail and head in mouse embryos [30].

Investigating the molecular events associated to the mechanisms of embryonic tendon regeneration, through the influence of GFs and TFs associated to tenogenesis may yield valuable knowledge on scar-free healing approaches, and would contribute to successful regeneration strategies for post-natal injured tendon outcomes.

The tendon milieu is hypoxic and the regulatory mechanisms of TDSCs stemness may be dependent on the oxygen tension available, as hypoxic conditions were described to better maintain TDSC stemness, assessed by stem cell markers as nucleostamin, Oct-4, Nanog and SSEA-4 [34]. Even in normoxic conditions, Holliday et al. [35] described that the addition of insulin-like growth factor (IGF)-1 maintained both TDSCs multipotency and phenotype, while GDF-5 preserved the tenogenic phenotype of TDSCs, but was unable to preserve the adipogenic and chondrogenic potential of these cells. Other factors, such as TGF-β2, were also shown to stimulate the tenogenic potential of TDSCs [36] (Fig. 1).

In spite of the growing number of publications on tendon progenitor cells, these cells are still barely investigated, likely due to the fact that tendon endogenous mechanisms of healing are limited and tendon tissue physiology is not completely understood. However, studies with MSCs from non-tendon sources have succeeded

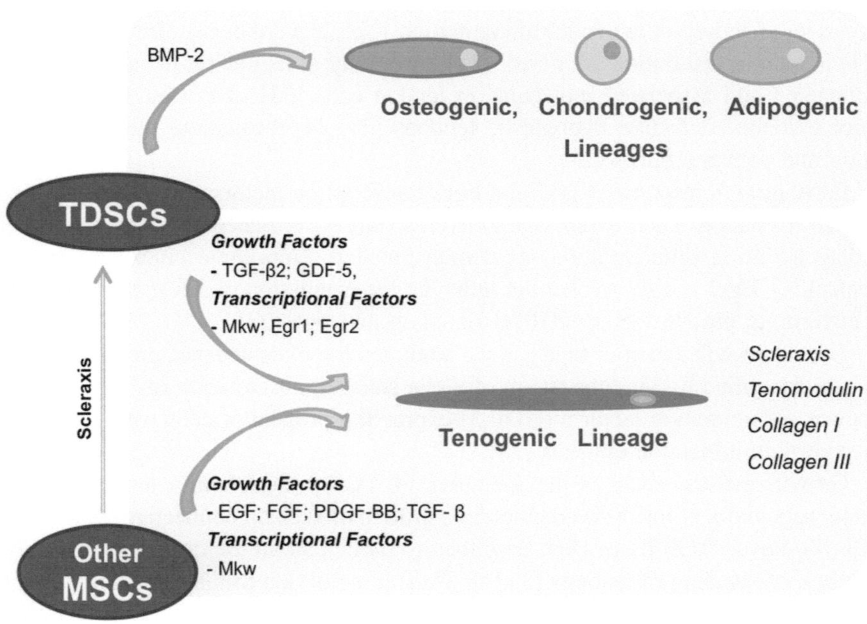

Fig. 1 Schematic representation of biomolecules involved in TDSCs differentiation mechanisms towards the tenogenic phenotype

in bringing new insights on the tenogenic lineage commitment for tendon regenerative applications. Thus, human MSCs from origins as different as adipose tissue, amniotic fluid or bone marrow have been stimulated with exogenous biomolecules towards the tenogenic phenotype [37, 38]. Since tendon specific biomarkers are not established yet for cell and tissue culture proceedings, most tenogenic approaches involve signalling molecules associated to tendon embryogenesis, for instance Scx or Mkx, or GFs that participate in tendon healing and repair, such as epidermal- (EGF), fibroblast- (FGF), GDF-5, platelet derived (PDGF-BB) or TGF-β.

Likewise GFs, transcription factors can also modulate the tenogenic differentiation of non-tendon MSCs. Mkx was recently investigated as a key factor for tenogenic differentiation of rat BMSCs. Mkx expression was enhanced during the tenogenic differentiation of BMSCs both in vitro and in vivo [39]. Mkx also induced the up-regulation of *Scx* and promoted the activation of *Bgn* (biglycan), *Col1a1* (collagen type I), *Col3a1* (collagen type III), *Col5a1* (collagen type V), *Col14a1* (collagen type XIV), *Dcn* (decorin), *Fmod* (fibromodulin), and *Tnc* (tenascin) in MSCs cultured in cell sheets [40]. Furthermore, Mkx enhanced tendon regeneration in a mouse model of Achilles-tendon defect [40]. Mohwak expression was also found to be suppressed in human tendinopathy but activated in tendon repair, and implicated in tendon differentiation [40].

Overall, and despite the fact that different stem cell origins may respond differently to signalling molecules, an ECM rich in tendon related proteins [37, 39, 40] has persistently been produced by non-tendon stem cells indicating that MSCs from different origins could be guided towards the tenogenic lineage.

3.2 Cell-Cell and Cell-Matrix Communication

The anchorage of stem cells is a main function of stem cell niche, which is normally constituted by a group of cells in close synergy with their surroundings. Several types of adhesion molecules are generally involved in these interactions between stem cells and their microenvironment (Fig. 2). Hence, not only cells need to communicate with their substrate, but also among each other, being able to detect and transmit mechanical, as well as chemical signals between them in order to accurately coordinate their biological functions.

3.2.1 Cell-Cell Contacts

Cell adhesive interactions are responsible for the organization of multicellular tissues, having important roles in development, differentiation and tissue homeostasis [41]. Cell-cell junctions are, thus, involved in triggering signaling responses, modulating gene transcription and controlling cell growth [42], constituting a dynamic system in cell communication, rather than simply gluing cells together [41]. Tendon cells are coupled via gap junctions, with connexin (Cx)-32 and Cx-43

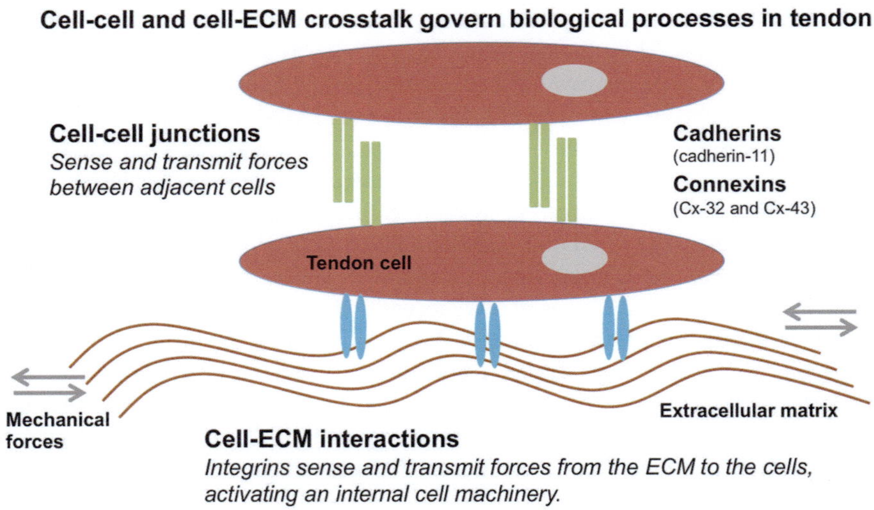

Fig. 2 Cell-cell and cell-ECM interactions in tendon. Adapted from Schiele et al. 2013 [121]

being the most prominent types regulating tenocyte response to mechanical loading [43]. Both gap junction proteins have been found in vivo in rat [44, 45] and horse tendons [46, 47] and, particularly, Cx-43 has been identified in vitro in human tenocytes [48]. Cx-43 was found to be involved in the response to mechanical stimulation through the propagation of a calcium wave between tendon cells [49, 50]. Moreover, Cx-32 and Cx-43 play different roles that are dependent on tissue organization [44]. Tenocytes are arranged in longitudinal rows and exhibit sheet-like extensions surrounding the collagen fiber bundles, which, in turn, separate cells between them, simultaneously bridging cell-cell and cell-ECM interactions [44, 51]. Cx-32 is present between cells of a row, with these junctions being arranged along the major line of loading in tendons; whereas Cx-43 links cells in all directions as it appears both between cells of a row and between cells of adjacent rows due to the cellular sheet-like processes [43]. This suggests the existence of two distinct communication networks in tendons, given that communication via Cx-43 was proven to inhibit collagen secretion by tenocytes exposed to mechanical loading, while communications via Cx-32 junctions had a stimulatory role in ECM synthesis [43].

Additionally to the regulation of cell responses to mechanical stimuli, cell-cell junctions are also involved in the patterning of ECM in tendon. Indeed, a function in forming the parallelism of the ECM has been attributed to an adherens junction molecule, cadherin-11, during embryonic tendon development [52]. The knockdown of cadherin-11 in chicken embryonic tendon has been reported to lead to a loss of cell-cell contacts and a consequent disruption of ECM organization [52]. Here, a community effect might be associated, in which cells need to contact a sufficient number of other neighboring cells in order to direct differentiation within the developing tissue [53–55].

Together, these findings highlight the importance of cell-cell interactions in tendon development, as well as in the maintenance of steady-state tissue physiology, particularly in what concerns to ECM organization.

3.2.2 Cell-ECM Interactions

Besides cell-cell communication, interactions between resident cells and their ECM also govern biological processes in tendon. Cell-ECM crosstalk requires the transduction of signals that can derive from different physical mechanisms based on ECM characteristics, including geometry both at the micro- and nanoscale, as well as matrix elasticity, in addition to external mechanical signals [56].

Indeed, cells need to integrate substrate alterations and transduce them into biological responses. In tendon, matrix stiffness varies during development, with temporal changes occurring in nano- and microscale elastic modulus [2], mechanoregulating the differentiation of tendon stem and progenitor cells.

Cell surface binding to collagens, including collagen I or IV, is mediated by β1 integrins, mainly α1β1 and α2β1 [57]. In fact, four different collagen-binding integrins, α1β1, α2β1, α10β1 and α11β1, have been identified. Either α1 or α2 integrin subunits are able to recognize a specific domain of the collagen triple-helical structure, the peptide motif GFOGER [58, 59]. The same has also been observed in the case of α11β1 [60]. Moreover, β1 and α5, as well as α6 integrin subunits have been reported to appear with an increased expression during early healing in a model of canine intrasynovial flexor tendon repair [61]. α5 subunit increased expression is related to an increase in fibronectin deposition, whereas α6 subunit is associated with capillary-forming cells and consequent angiogenesis augmentation near the wound site [61]. Although poor information exists regarding this topic in tendon biology, this result highlights the role of cell-ECM adhesion molecules in tendon tissue physiology and repair.

Furthermore, these cell-ECM interactions have distinct implications in two-dimensional (2D) culture or in 3D matrices. Indeed, in the first situation, these cell-matrix interactions lead to the formation of focal adhesions; while in 3D matrices, cells develop several adhesive contacts, like filopodia, spikes, lamellae, ruffles and pseudopodia [62]. This allows tendon cells, as well as other cell types, to develop a dynamic crosstalk with the matrix fibrils, and must be taken into consideration when developing strategies to mimic tendon stem cell niche.

3.3 Extracellular Matrix

The extracellular matrix is the ultimate microenvironment of cells within a tissue. The primary function of ECM is to provide tissues with their mechanical and biochemical properties and to form specific niche for resident cells. Cells are responsible for ECM synthesis and maintenance and, in turn, ECM has an impact on cellular functions, by providing tissue structural support, linking tissues of the body together, or regulating intercellular communication.

The importance of ECM for development, normal functioning and regeneration of tendon cannot be refuted, as adult tendon tissues are relatively acellular, avascular and practically non-innervated. The ECM of tendon tissue is a composite material that consists mainly of aligned collagens and residual non-collagenous proteins, including elastic fibers, proteoglycans and water. In addition, tendon ECM not only transmits the developed muscular force to bones, but it also acts as an energy storage device. Elastic energy is stored in tendon through reversible stretching of collagen molecules [63, 64].

3.3.1 Architecture of Extracellular Matrix in Tendon

The extracellular matrix of tendon tissue is organized in hierarchically assembled fibrillar structures arranged in three-dimensional lattice structures (Fig. 3). This type of hierarchical structure aligns fiber bundles with the long axis of tendon and affords the tendon's tensile strength. The division of tendons into individual structural units—fibers with different levels of organization—ensures that extensive damage does not necessarily spreads to the entire tendon. Tendon ECM is comprised mostly of collagen type I, the member of fibril-forming collagens family, as well as of other non-collagenous proteins. The fibril-forming collagens provide the structural framework of tissues and include type I, II, III, V, and XI collagens. All fibril-forming collagen molecules form pentamer cross-striated microfibrils that could be defined by a characteristic banding pattern with a periodicity of about 70 nm. Microfibrils group together to form collagen fibrils with diameters of about 10–500 nm, depending on species, age and anatomical location [65–69]. The ability of collagen molecules to assemble into crosslinked fibrils is an important requirement for the development of tissue strength. The presence of crosslinks between collagen molecules increases the range of elastic fibril deformation. It has been reported that the size of fibrils alter the potential for inter-fibrillar crosslinks and the

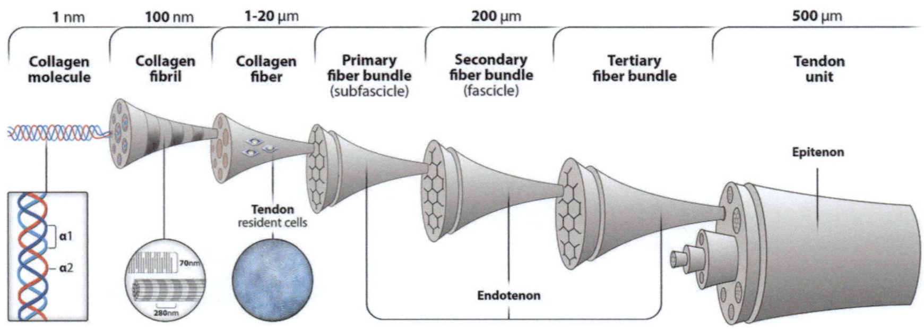

Fig. 3 Structural hierarchy in the tendon. This scheme illustrates the general relationship between collagen molecules, fibrils, fibers, fascicles (sub-fascicles) and tendon units

mechanical properties of tissue [70, 71]. Scanning electron microscopy confirmed that fibrils in tendon form so called crimps. Within the crimps, fibrils are changing their plane of running, and then bend between each other. Fibrillar crimps open when tensional load is applied and bounce when load is removed. The molecular mechanisms of formation of these structures are not well defined; however it is clear that the angle and frequency of such crimps differ in different tendons [72, 73]. The collagen fibrils form fibers with a diameter of approximately 1–20 μm. Fibers have the waveform, most likely originating from the crimps at the level of fibril alignment [74, 75]. The fibers bonded to each other into sub-fascicles and the sub-fasciles into fascicles (20–200 μm). Fascicles are bound by the endotenon, which is a loose connective tissue composed of thin collagen fibrils. A thin connective tissue membrane, known as the epitenon, surrounds the tendon unit formed by fascicles. The combination of mechanical properties at each of these sub-structural levels in this hierarchically organized arrangement results in the overall mechanical behavior of tendon. Thus, the architecture of tendon ECM is extremely important for its ability to efficiently transmit the loads and to store elastic energy during motion activity [76].

3.3.2 Mechanical Properties of Tendon Extracellular Matrix

Tendons are viscoelastic materials, naturally designed to resist tensile forces. They act as biological springs that can stretch, storing and releasing energy during locomotion. Viscoelasticity makes tendons more deformable at low strain rates but less deformable at high strain rates. Therefore, tendons at low strain rates absorb larger amount of mechanical energy but could carry out less mechanical loads [76]. Tendon stress–strain curve has an initial toe region, where the tendon is strained up to 2 % (Fig. 4). It represents the stretching-out of the pattern of crimps formed by collagen fibers. In the linear region of the stress–strain curve, where the tendon is stretched less than 4 %, crimps are lose within the fibers. Tendon force and stress increased curvilinearly as a function of displacement and strain, respectively. The slope of linear region is referred to as the Young's modulus of tendon. Stretching over 4 % causes tearing of collagen fibers; strain beyond 8–14 % causes tendon rupture [77]. Experimental data suggests that different types of tendons differ in their mechanical properties. Moreover, it was found that Young's modulus values vary in an age-dependent manner [78, 79]. Despite all revealed deviations in mechanical properties, molecular composition of ECM in all tendons is very similar.

3.3.3 Collagenous Components of Tendon Extracellular Matrix

Collagen is the most abundant protein in dense fibrous connective tissues and forms essential structural elements in the musculoskeletal system. Tendon is composed predominantly of fibrous collagen that accounts for approximately 75 % of the dry

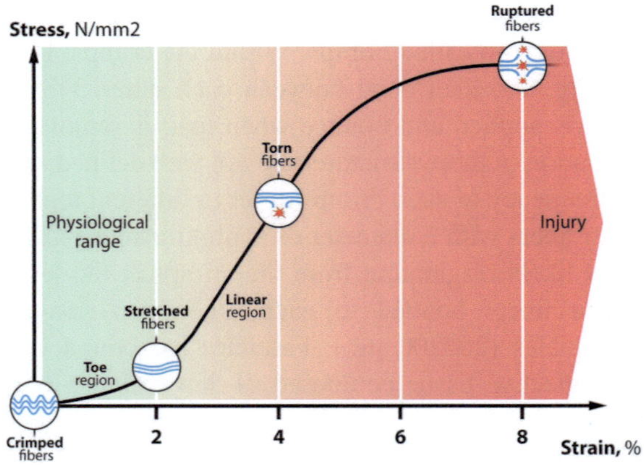

Fig. 4 Tendon stress–strain curve. The mechanical stress versus strain curve which represents an initial toe region, where the tendon is strained up to 2 %, linear region, where the tendon is stretched less than 4 % and slope, where the tendon is stretched over 4 % (microscopic tearing of tendon fibers occurs). Beyond 8–10 % strain, macroscopic failure occurs. In physiologic conditions, most tendons exist in the toe and to some extent in the linear region

weight of the tissue [80]. The predominant collagen type (approximately 95 %) is the fibril-forming collagen type I. The second most abundant is collagen type III (5 % in tendons) [81]. Other members of the collagen family represented in tendon tissue include type II, IV, V, VI, IX, X, XII and XIV collagens. However, these collagens are available in basal amounts within the tendon midsubstance and appear to be associated with fibrocartilage that is located at the tendon-bone junction [76, 82–84].

The rod-like structure of collagen type I provides mechanical stability and determines the mechanical properties of tendon ECM. Initially, collagen type I molecules have been thought to possess little flexibility and high mechanical strength [85]. However, type I collagen molecule was later shown to have numerous bends and not being completely rigid [86]. The flexibility of type I collagen comes from molecular sequences that lack the amino acids proline and hydroxyproline (Fig. 5). Five sites within these sequences where bends can occur in the triple helix have been identified [87]. The results of this modeling study suggest that sequences without proline and hydroxyproline are able to form internal loops that give these regions more flexibility than the other regions of the triple helix [88]. Generally, fibril-forming collagens represented in tendon tissue slightly differ from each other in terms of physical properties. For example, collagens type II and III have slightly higher translational diffusion coefficients and slightly shorter end-to-end distances, being therefore more flexible than type I collagen [86].

Biosynthesis of collagen I begins in resident tendon cells. Collagen is synthesized in the precursor form, pro-collagen, which contains non-triple helical extensions at both ends (Fig. 5). Translation of pro-collagen mRNA occurs on the membrane bound ribosomes of the rough endoplasmic reticulum. Pro-collagen

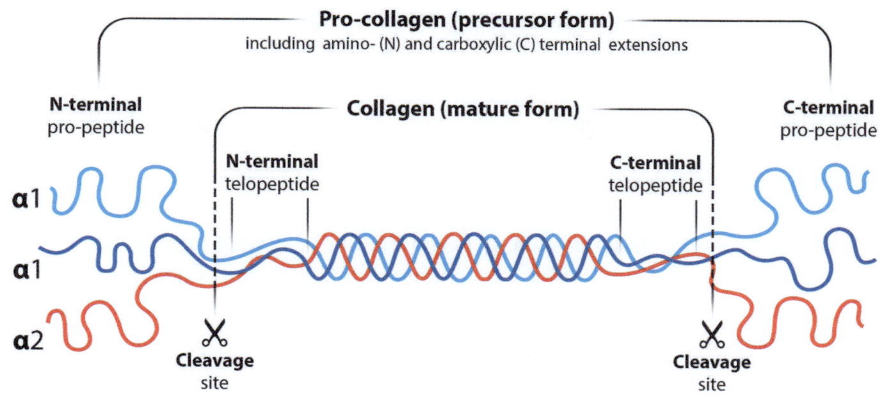

Fig. 5 Structure of collagen molecule. Pro-collagen molecule is a triple helix comprised of two alpha-1 chains and one alpha-2 chain. Pro-peptide domains at the carboxy-terminals and amino-terminals are cleaved in mature collagen (markers of collagen synthesis). When collagen is degraded, during physiological turnover, telopeptides are cleaved and released into the plasma (markers of collagen degradation). Adapted from Fan et al. [122]

assembly also takes place in the endoplasmic reticulum. Collagen precursor form contains amino-(N) and carboxylic (C) terminal extensions, which are essential for initiation of its molecular assembly and assembly of fibrils [76, 89]. Molecular assembly of pro-collagen in native tendon tissue initiates within intracellular vesicles. Synthesized protein undergoes subsequent posttranslational modifications of the polypeptide chains, which contribute to the quality and stability of the molecule [90, 91]. Before the translocation from ribosome occurs, three different hydroxylases convert residues to 4-hydroxyproline or 3-hydroxyproline [92]. The intracellular processing finishes with the glycosylation and synthesis of both intra-chain and inter-chain disulfide bonds [93]. After secretion into the extracellular space, the terminal extensions are cleaved by specific proteinases [94]. Conversion of pro-collagen into collagen is followed by subsequent incorporation into stable crosslinked collagen fibrils.

3.3.4 Non-collagenous Components of Tendon Extracellular Matrix

The organization of collagen within tendon depends on interactions with other non-collagenous proteins. A variety of proteoglycans (PGs) such as decorin, fibromodulin, biglycan, lumican, aggrecan and versican are able to interact with collagen fibers, being also involved in the regulation of collagen fibrils formation process [95, 96]. Structurally, PGs are closely bound with the collagen fibrils being important for the interconnection of the fibrils [97]. For instance, genetic mutations lacking decorin lead to the formation of collagen fibrils with irregular diameters and decreased skin strength. In addition, down-regulation of decorin initiated the development of collagen fibrils with larger diameters and higher ultimate tensile strengths in ligament scar of animals [98, 99]. Decorin also appears to take part in the alignment of collagen molecules in tendon as well, and assists sliding during mechanical deformation [100]. PGs are composed of a protein core and a glycosaminoglycan (GAG) side chain. The most prevalent GAGs of tendon are dermatan sulfate and chondroitin sulfate, which associate with collagen and are involved in the fibril assembly process during tendon development. Dermatan sulfate is thought to be responsible of forming connections between fibrils, while chondroitin sulfate is found in the extracellular volume between the fibrils to keep them separated and help withstand deformation [101]. The dermatan sulfate chains of decorin aggregate in solution and assist with the assembly of the collagen fibrils. When decorin molecules bind to a collagen fibril, their dermatan sulfate chains may extend and associate with other dermatan sulfate chains on another molecule of decorin bounded to separate fibrils, providing interfibrillar bridges and eventually causing parallel alignment of the fibrils [44]. Another type of matrix molecules that may have an important role in tendon and ligament ECM is collagen oligomeric matrix protein (COMP). COMP is a member of the thrombospondin family and was originally described in cartilage. It is a calcium binding glycoprotein present as a pentamer, in which each of the subunits forms a linking coiled-coil domain near N-terminal region. The C-terminal globular domain binds to the molecules of collagen types I, II and IX in a zinc-dependent manner. The function of COMP is not yet clear, but it has been suggested that pentameric molecules provide structural integrity to the ECM, connecting multiple collagen fibrils and expressed as a response to mechanical load [102, 103].

4 Relevance of Signaling Molecules in the Modulation of Tendon Healing Versus Tendon Regeneration

After birth, the limited capability of tendons to regenerate often results in scar-tissue with impaired mechanical properties, lacking the typical tissue organization and functionality of healthy tendons. Interestingly, prenatal tendons have shown an improved response to tendon injury, regenerating with orderly deposition of collagen fibers, or up to a certain degree depending on the extension of the wound [104, 105]. In the process of tendon healing and repair, TDSCs and signalling molecules are likely to be the most important intrinsic and active participants to endogenously regenerate tendons.

Embryonic wounds that heal without a scar have shown low levels of TGF-β1 and TGF-β2, low levels of PDGF and high levels of TGF-β3 [104, 106] (Fig. 6).

In adult tissues, scar healing and regeneration can occur within the same tissue, for instance in human skin or liver, whose main differences are dependent on the type of wound induced. Although adult tendons do not behave accordingly, the differential tissue response suggests that the mechanisms related to scarring and regeneration are similar and likely sharing signalling molecules or a combination of regulatory molecules.

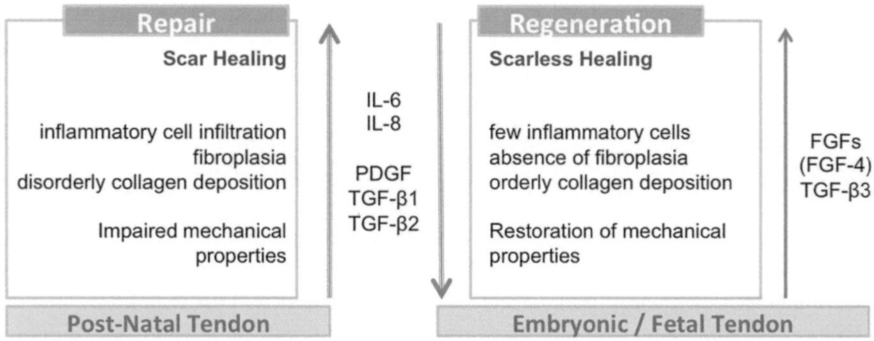

Fig. 6 Schematic representation of the major differences between tendon repair with scarring healing and complete regeneration. These differences are reflected in the increased/decreased levels of biomolecules in the embryonic/fetal tendon ability to regenerate and post-natal tendons to heal

Members of the TGF-β superfamily have been reported to influence tendon lesions, as well as tendon healing mechanisms. TGF-β isoforms (β1, β2 and β3) and the two signalling receptors (TGF-βRI and TGF-βRII) have been investigated in healthy and pathological Achilles tendons [107]. The increased number of tendon cells expressing TGF-β2 in chronic pathology suggests a role for TGF-β2 in mediating cell activity during disease. Fenwick and co-workers also verified that without the presence of both receptors (TGF-βRI and TGF-βRII), exogenous TGF-β is unlikely to be effective [107]. All TGF-β isoforms significantly increased collagen I but only TGF-β3 exhibited the highest potency in stimulating collagen I and collagen III in a rat model of tendon healing [108]. According to histological analysis on healing tendons, aligned collagen fibers were observed to be regenerated 28 days post injury but the maturity of collagen fibers did not return to normal. Twenty-eight days after surgery TGF-β1 expression was diminished and regressed to lateral edges of the wound [108].

Although TGF-β isoforms contribute to the process of collagen synthesis in tendon cells, different TGF-β isoforms may have distinct roles in the regulation of collagen synthesis. It is also possible that TGF-β isoforms interact with one another to control collagen deposition in healing tendons [108].

Despite the relevance of understanding the specific role of each GF involved in tendon healing and regeneration, alternative strategies have been pursued. Platelet-rich plasma (PRP) and platelet lysates (PLs) are endogenous pools of GFs that participate in the inflammatory process, and consequently in healing and repair mechanisms. The presence of platelet-released GFs was described to enhance human tenocyte proliferation and promoted ECM synthesis to levels similar to those found in tendon-bone insertion of normal human rotator cuffs [109].

Moreover, Chen and co-workers found a synergistic effect on tendon healing when TDSCs and PRP were combined in a rat Achilles model [110]. Also in the Achilles model, Fernandez-Sarmiento et al. verified that infiltrations of plasma rich in growth factors onto injured tendons of sheep accelerated the healing response with evidence of collagen bundles organization, lower vascular densities and decreased fibroblast densities, compared to untreated tendons [111].

Despite the fact that the mechanisms involved in healing (scarring) and regeneration (scar-free healing) seem not to be as distinctive as initially perceived, further studies are required to fine tune regulatory molecules and address their role in the precise sequence of events that modulates the regeneration of tendons.

Understanding and mimicking the signaling regulators in this naturally intricated process will provide fundamental tools to bioengineer successful outcomes under different stages of tendon injury or damage. Moreover, it will provide know-how to develop and establish therapies that ultimately will result in restored morphology and regain of complete functionality in a fibrosis-free regeneration mechanism.

The diminished functionality of healed tendons generates a great demand for research focused on tendon tissue engineering and regenerative strategies. Different tendon regeneration approaches can be inspired through the basic understanding of tendon tissue development and biology, as well as of the integrated context of tendon stem cells.

5 Tendon Regeneration Therapies: Using TDSCs and Mimicking Their Niche

Cell-based therapies have the potential of generating a regenerative response, instead of leading to the formation of fibrotic tissue. The use of patient-derived stem cell populations and their conversion into a functional tendon tissue not only obviates immune rejection, but also avoids donor site morbidity related with harvesting tendon grafts [112]. Nonetheless, a relevant aspect that renders TDSCs of potential to be used in cell-based therapies for tendon regeneration consists on the absence of immunoreactions when they are transplanted in vivo [113], similar to the low immunogenicity and to the immunomodulatory effects already attributed to non-tendon MSC populations. Indeed, allogeneic transplantation of TDSCs in a rat model promoted tendon healing from week 1 up to week 16, improving fiber arrangement and cell alignment, simultaneously reducing inflammation [113]. TDSCs lacked the expression of surface markers, like major histocompatibility complex (MHC)-II, CD80 and CD86, that are required for the activation of T-cells and transplant rejection [114]. Thus, allogeneic transplantation of TDSCs did not originate an exacerbated immune response [113, 114].

Additionally, TDSCs can be combined with other cues to develop biomimetic strategies aimed at tendon regeneration. Biochemical factors have been extensively explored herein. Nevertheless, the generation of engineered tendon tissue is also dependent on mechanical/physical signals that replicate ECM organization and, thus, different strategies can be used to produce such biomaterials.

For instance, the combination of a collagen gel with mechanical loading directed tenogenic differentiation of human MSCs [115]. Similarly, the incorporation of tendon-derived ECM in 3D collagen gels combined with uniaxial tension enhanced the proliferation of human adipose stem cells and induced tenogenic differentiation [116].

Furthermore, fiber alignment has proven to be an essential factor in directing tenogenic differentiation of TDSCs, simultaneously preventing their differentiation into osteoblasts when exposed to osteogenic conditions [15]. Indeed, the replication of functional tendon tissue requires cell alignment within a highly organized parallel ECM. For this, decellularized tendon matrices have been similarly explored to better address tendon regeneration [117–119]. Hence, the implantation of TDSCs using a decellularized native ECM has been reported to generate tendon-like tissue in vivo in a rat patellar tendon defect model [120].

In conclusion, and despite the strategies followed, the use of either TDSCs or microenvironmental cues from their niche, as well as a combination of both has beneficial effects on the modulation of tenogenesis.

6 Future Perspectives and Conclusions

The discovery of a stem/progenitor cell population within tendons has brought new insights into the field of tendon regeneration. Despite the inexistence of a unique marker that can specifically identify these cells, TDSCs hold a great potential for being used in regenerative therapies due to their role in tendon formation and healing. The understanding of both their biology and microenvironment constitutes a promise for the development of new therapies overcoming scarring, towards the improvement of functional properties of repaired tendon.

The biomolecules herein discussed play crucial roles in tendon development and in tendon healing and regeneration phenomena, which ultimately may become useful tools for successfully designing tendon regeneration strategies.

Although the expression profile or the optimal spatial/temporal distribution of these bioactive molecules towards tendon progenitor cell processes remain to be assessed, further investigation is required to increase the knowledge on the mechanisms involved in tendon formation and regeneration.

The development of future therapies that more closely recapitulate the intricate tendon microenvironment and its influence in tendon progenitor cells, in which bioactive signaling can be applied as biomolecular guidelines to stimulate local tendon tissue regeneration, may result in significantly better clinical outcomes. Furthermore, the use of biomimetic materials that can replicate the physical microenvironment could complement the application of bioactive molecules. Such materials are now being used in combination with technologies that enable the reproduction at the micro- and nano-scale of the physical cues provided by tendon ECM, having the potential to ameliorate the effects of surgical approaches for tendon healing.

References

1. Gelse K, Poschl E, Aigner T. Collagens—structure, function, and biosynthesis. Adv Drug Deliv Rev. 2003;55:1531–46.
2. Marturano JE, Arena JD, Schiller ZA, Georgakoudi I, Kuo CK. Characterization of mechanical and biochemical properties of developing embryonic tendon. Proc Natl Acad Sci USA. 2013;110:6370–5.
3. Rodrigues MT, Reis RL, Gomes ME. Engineering tendon and ligament tissues: present developments towards successful clinical products. J Tissue Eng Regen Med. 2013;7(9):673–86.
4. Cserjesi P, Brown D, Ligon KL, Lyons GE, Copeland NG, Gilbert DJ, et al. Scleraxis: a basic helix-loop-helix protein that prefigures skeletal formation during mouse embryogenesis. Development. 1995;121:1099–110.
5. Brent AE, Schweitzer R, Tabin CJ. A somitic compartment of tendon progenitors Cell. 2003;113:235–48.
6. Liu C-F, Aschbacher-Smith L, Barthelery NJ, Dyment N, Butler D, Wylie C. What we should know before using tissue engineering techniques to repair injured tendons: a developmental biology perspective. Tissue Eng Part B. 2011;17:165–76.
7. Bi Y, Ehirchiou D, Kilts TM, Inkson CA, Embree MC, Sonoyama W, et al. Identification of tendon stem/progenitor cells and the role of the extracellular matrix in their niche. Nat Med. 2007;13(10):1219–27.
8. Rui Y-F, Lui PPY, Li G, Fu SC, Lee YW, Chan KM. Isolation and characterization of multipotent rat tendon-derived stem cells. Tissue Eng Part A. 2010;16:1549–58.
9. Zhang J. Wang JH-C. Characterization of differential properties of rabbit tendon stem cells and tenocytes. BMC Musculoskelet Disord. 2011;11:10.
10. Lovati A, Corradetti B, Consiglio AL, Recordati C, Bonacina E, Bizzaro D, et al. Characterization and differentiation of equine tendon-derived progenitor cells. J Biol Regul Homeost Agents. 2011;25(2 Suppl):S75–84.
11. Lui PPY, Wong OT. Tendon stem cells: experimental and clinical perspectives in tendon and tendon-bone junction repair. Muscles, Ligaments Tendons J. 2012;2:163–8.
12. Tan Q, Lui PPY, Rui YF, Wong YM. Comparison of potentials of stem cells isolated from tendon and bone marrow for musculoskeletal tissue engineering. Tissue Eng Part A. 2012;18:840–51.
13. Mienaltowski MJ, Adams SM, Brik DE. Regional differences in stem cell/progenitor cell populations from the mouse Achilles tendon. Tissue Eng Part A. 2013;19:199–210.
14. Mienaltowski MJ, Adams SM, Brik DE. Tendon proper- and peritenon-derived progenitor cells have unique tenogenic properties. Stem Cell Res Ther. 2014;5:86.
15. Yin Z, Chen X, Chen JL, Shen WL, Nguyen TMH, Gao L, et al. The regulation of tendon stem cell differentiation by the alignment of nanofibers. Biomaterials. 2010;31:2163–75.
16. Scadden DT. The stem-cell niche as an entity of action. Nature. 2006;441:1075–9.
17. Moore KA, Lemischka IR. Stem cells and their niches. Science. 2006;311:1880–5.
18. Spanoudes K, Gaspar D, Pandit A, Zeugolis DI. The biophysical, biochemical, and biological toolbox for tenogenic phenotype maintenance in vitro. Trends Biotechnol. 2014;32:474–82.
19. Asou Y, Nifuji A, Tsuji K, Shinomiya K, Olson EN, Koopman P, et al. Coordinated expression of scleraxis and Sox9 genes during embryonic development of tendons and cartilage. J Orthop Res: official publication of the Orthopaedic Research Society. 2002;20(4):827–33.
20. Blitz E, Sharir A, Akiyama H, Zelzer E. Tendon-bone attachment unit is formed modularly by a distinct pool of Scx- and Sox9-positive progenitors. Development. 2013;140:2680–90.
21. Soeda T, Deng JM, de Crombrugghe B, Behringer RR, Nakamura T, Akiyama H. Sox9-expressing precursors are the cellular origin of the cruciate ligament of the knee joint and the limb tendons. Genesis. 2010 Nov;48(11):635–44. PubMed PMID: 20806356. Pubmed Central PMCID: 3982414.

22. Liu W, Watson SS, Lan Y, Keene DR, Ovitt CE, Liu H, et al. The atypical homeodomain transcription factor Mohawk controls tendon morphogenesis. Mol Cell Biol. 2010 Oct;30 (20):4797–807. PubMed PMID: 20696843. Pubmed Central PMCID: 2950547.
23. Ito Y, Toriuchi N, Yoshitaka T, Ueno-Kudoh H, Sato T, Yokoyama S, et al. The Mohawk homeobox gene is a critical regulator of tendon differentiation. Proc Natl Acad Sci USA. 2010 Jun 8;107(23):10538–42. PubMed PMID: 20498044. Pubmed Central PMCID: 2890854.
24. Onizuka N, Ito Y, Inagawa M, Nakahara H, Takada S, Lotz M, et al. The Mohawk homeobox transcription factor regulates the differentiation of tendons and volar plates. J Orthop Sci: official journal of the Japanese Orthopaedic Association. 2014 Jan;19(1):172–80. PubMed PMID: 24166359. Pubmed Central PMCID: 3943675.
25. Ito Y, Toriuchi N, Yoshitaka T, Ueno-Kudoh H, Sato T, Yokoyama S, et al. The Mohawk homeobox gene is a critical regulator of tendon differentiation. Proc Natl Acad Sci USA. 2010;107:10538–42.
26. Lejard V, Blais F, Guerquin MJ, Bonnet A, Bonnin MA, Havis E, et al. EGR1 and EGR2 involvement in vertebrate tendon differentiation. J Biol Chem. 2011 Feb 18;286(7):5855–67. PubMed PMID: 21173153. Pubmed Central PMCID: 3037698.
27. Guerquin MJ, Charvet B, Nourissat G, Havis E, Ronsin O, Bonnin MA, et al. Transcription factor EGR1 directs tendon differentiation and promotes tendon repair. J Clin Invest. 2013 Aug 1;123(8):3564–76. PubMed PMID: 23863709. Pubmed Central PMCID: 4011025.
28. Dyment NA, Hagiwara Y, Matthews BG, Li Y, Kalajzic I, Rowe DW. Lineage tracing of resident tendon progenitor cells during growth and natural healing. PLoS ONE. 2014;9: e96113.
29. Lorda-Diez CI, Montero JA, Martinez-Cue C, Garcia-Porrero JA, Hurle JM. Transforming growth factors beta coordinate cartilage and tendon differentiation in the developing limb mesenchyme. J Biol Chem. 2009 Oct 23;284(43):29988–96. PubMed PMID: 19717568. Pubmed Central PMCID: 2785627.
30. Pryce BA, Watson SS, Murchison ND, Staverosky JA, Dunker N, Schweitzer R. Recruitment and maintenance of tendon progenitors by TGFbeta signaling are essential for tendon formation. Development. 2009 Apr;136(8):1351–61. PubMed PMID: 19304887. Pubmed Central PMCID: 2687466.
31. Liu C-F, Aschbacher-Smith L, Barthelery NJ, Dyment N, Butler D, Wylie C. Spatial and temporal expression of molecular markers and cell signals during normal development of the mouse patellar tendon. Tissue Eng Part A. 2012;18:598–608.
32. Brent AE, Braun T, Tabin CJ. Genetic analysis of interactions between the somitic muscle, cartilage and tendon cell lineages during mouse development. Development. 2005;132 (3):515–28.
33. Brent AE, Tabin CJ. FGF acts directly on the somitic tendon progenitors through the Ets transcription factors Pea3 and Erm to regulate scleraxis expression. Development. 2004;131 (16):3885–96.
34. Zhang J, Wang JH-C. Human tendon stem cells better maintain their stemness in hypoxic culture conditions. PLoS ONE. 2013;8:e61424.
35. Holladay C, Abbah S-A, O'Dowd C, Pandit A, Zeugolis DI. Preferential tendon stem cell response to growth factor supplementation. J Tissue Eng Regen Med. 2014;. doi:10.1002/term.1852.
36. Brown JP, Finley VG, Kuo CK. Embryonic mechanical and soluble cues regulate tendon progenitor cell gene expression as a function of developmental stage and anatomical origin. J Biomech. 2014;47:214–22.
37. Goncalves AI, Rodrigues MT, Lee SJ, Atala A, Yoo JJ, Reis RL, et al. Understanding the role of growth factors in modulating stem cell tenogenesis. PLoS ONE. 2013;8(12):e83734. PubMed PMID: 24386267. Pubmed Central PMCID: 3875481.
38. Park A, Hogan MV, Kesturu GS, James R, Balian G, Chhabra AB. Adipose-derived mesenchymal stem cells treated with growth differentiation factor-5 express tendon-specific markers. Tissue Eng Part A. 2010;16:2941–51.

39. Otabe K, Nakahara H, Hasegawa A, Matsukawa T, Ayabe F, Onizuka N, et al. Transcription factor Mohawk controls tenogenic differentiation of bone marrow mesenchymal stem cells in vitro and in vivo. J Orthop Res: official publication of the Orthopaedic Research Society. 2014 Oct 13. PubMed PMID: 25312837.
40. Liu H, Zhang C, Zhu S, Lu P, Zhu T, Gong X, et al. Mohawk promotes the tenogenesis of mesenchymal stem cells through activation of the TGFbeta signaling pathway. Stem Cells. 2014 Oct 21. PubMed PMID: 25332192.
41. Braga VM, Balda MS. Regulation of cell-cell adhesion. Semin Cell Dev Biol. 2004;15:631–2.
42. Giepmans BNG. Gap junctions and connexin-interacting proteins. Cardiovasc Res. 2004;62:233–45.
43. Wagget AD, Benjamin M, Ralphs JR. Connexin 32 and 43 gap junctions differentially modulate tenocyte response to cyclic mechanical load. Eur J Cell Biol. 2006;85:1145–54.
44. McNeilly CM, Banes AJ, Benjamin M, Ralphs JR. Tendon cells in vivo form a three dimensional network of cell processes linked by gap junctions. J Anat. 1996;189:593–600.
45. Ralphs JR, Benjamin M, Wagget AD, Russel DC, Messner K, Gao J. Regional differences in cell shape and gap junction expression in rat Achilles tendon: relation to fibrocartilage differentiation. J Anat. 1998;193:215–22.
46. Stanley RL, Fleck RA, Becker DL, Goodship AE, Ralphs JR. Patterson-Kane JC. Gap junction protein expression and cellularity: comparison of immature and adult equine digital tendons. 2007;211:325–34.
47. Young NJ, Becker DL, Fleck RA, Goodship AE, Patterson-Kane JC. Maturational alterations in gap junction expression and associated collagen synthesis in response to tendon function. Matrix Biol. 2009;28:311–23.
48. Kuzma-Kuzniarska M, Yapp C, Pearson-Jones TW, Jones AK, Hulley PA. Functional assessment of gap junctions in monolayer and three-dimensional cultures of human tendon cells using fluorescence recovery after photobleaching. J Biomed Opt. 2014;19:doi:10.1117/1.JBO.19.1.015001.
49. Banes AJ, Weinhold P, Yang X, Tsuzaki M, Bynum D, Bottlang M, et al. Gap junctions regulate responses of tendon cells ex vivo to mechanical loading. Clin Orthop Relat Res. 1999;367S:S356–70.
50. Wall ME, Banes AJ. Early responses to mechanical load in tendon: Role for calcium signaling, gap junctions and intercellular communication. J Musculoskelet Neuronal Interact. 2005;5:70–84.
51. Ralphs JR, Wagget AD, Benjamin M. Actin stress fibres and cell–cell adhesion molecules in tendons: organi- sation in vivo and response to mechanical loading of tendon cells in vitro. Matrix Biol. 2002;21:67–74.
52. Richardson SH, Starborg T, Lu Y, Humphries SM, Meadows RS, Kadler KE. Tendon development requires regulation of cell condensation and cell shape via cadherin-11-mediated cell-cell junctions. Mol Cell Biol. 2007;27:6218–28.
53. Gurdon JB. A community effect in animal development. Nature. 1988;336:772–4.
54. Gurdon JB, Lemaire P, Kato K. Community effects and related phenomena in development. Cell. 1993;75:831–4.
55. Gurdon JB, Tiller E, Roberts J, Kato K. A community effect in muscle development. Curr Biol. 1993;3:1–11.
56. Guilak F, Cohen DM, Estes BT, Gimble JM, Liedtke W, Chen CS. Control of stem cell fate by physical interactions with the extracellular matrix. Cell Stem Cell. 2009;5:17–26.
57. Adams JC. Cell-matrix contact structures. Cell Mol Life Sci. 2001;58:371–92.
58. Emsley J, Knight CG, Farndale RW, Barnes MJ, Liddington RC. Structural basis of collagen recognition by integrin a2b1. Cell. 2000;101:47–56.
59. Knight CG, Morton LF, Peachey AR, Tuckwell DS, Farndale RW, Barnes MJ. The collagen-binding A-domains of integrins $\alpha1\beta1$ and $\alpha2\beta1$ recognise the same specific amino acid sequence, GFOGER, in native (triple- helical) collagens. J Biol Chem. 2000;275:35–40.

60. Zhang W-M, Kapyla J, Puranen JS, Knight CG, Tiger C-F, Pentikainen OT, et al. α11β1 integrin recognizes the GFOGER sequence in interstitial collagens. J Biol Chem. 2003;278:7270–7.
61. Harwood FL, Monosov AZ, Goomer RS, Gelberman RH, Winters SC, Silva MJ, et al. Integrin expression is upregulated during early healing in a canine intrasynovial flexor tendon repair and controlled passive motion model. Connect Tissue Res. 1998;39:309–16.
62. Tarpila E, Ghassemifar RM, Franzen LE. Fibroblast movements during contraction of collagen lattices—a quantitative study using a new three-dimensional time-lapse technique with phase-contrast laser scanning microscopy. In Vitro Cell Dev Biol—Animal. 1998;34:640–5.
63. Silver FH, Christiansen DL, Snowhill PB, Chen Y. Role of storage on changes in the mechanical properties of tendon and self-assembled collagen fibers. Connect Tissue Res. 2000;42:155–64.
64. Silver FH, Christiansen DL, Snowhill PB, Chen Y. Transition from viscous to elastic dependency of mechanical properties of self-assembled collagen fibers. J Appl Polym Sci. 2001;7:134–42.
65. Ottani V, Martini D, Franchi M, Ruggeri A, Raspanti M. Hierarchical structures in fibrillar collagens. Micron. 2002;33:587–96.
66. Ottani V, Raspanti M, Ruggeri A. Collagen structure and functional implications. Micron. 2001;32:251–60.
67. Hulmes DJ, Miller A. Molecular packing in collagen. Nature. 1981;293:234–9.
68. Gelse K. Collagens—structure, function, and biosynthesis. Adv Drug Deliv Rev. 2003;55:1531–46.
69. Silver FH, Christiansen DL. Introduction to biomaterials science and biocompatibility. In: Silver FH, Christiansen DL, editors. biomaterials science and biocompatibility. New York: Springer; 1999.
70. Buehler MJ. Nanomechanics of collagen fibrils under varying cross-link densities: atomistic and continuum studies. J Mech Behav Biomed Mater. 2008;1:50–67.
71. Parry DA, Barnes GR, Craig AS. A comparison of the size distribution of collagen fibrils in connective tissues as a function of age and a possible relation between fibril size distribution and mechanical properties. Proceedings of the Royal Society B: Biological Sciences. 1978;203:305–21.
72. Shadwick RE. Elastic energy storage in tendons: mechanical differences related to function and age. J Appl Physiol. 1990;68:1033–40.
73. Franchi M, Ottani V, Stagni R, Ruggeri A. Tendon and ligament fibrillar crimps give rise to left-handed helices of collagen fibrils in both planar and helical crimps. J Anat. 2010;216:301–9.
74. Franchi M, Quaranta M, Macciocca M, Pasquale VD, Ottani V, Ruggeri A. Structure relates to elastic recoil and functional role in quadriceps tendon and patellar ligament. Micron. 2009;40:370–7.
75. Zaffagnini S, Muccioli GMM, Franchi M, Bacchelli B, Grassi A, Agati P, et al. Collagen fibre and fibril ultrastructural arrangement of the superficial medial collateral ligament in the human knee. Knee Surgery, Sports Traumatol Arthrosc. 2014;(in press).
76. Silver FH, Freeman JW, Bradica G. Structure and function of ligaments, tendons, and joint capsule. In: Walsh WR, editor. Repair and regeneration of ligaments, tendons, and joint capsule. New Jersey: Humana Press Inc.; 2005.
77. Wang JH. Mechanobiology of tendon. J Biomech. 2006;39:1563–82.
78. Johnson GA, Tramaglini DM, Levine RE, Ohno K, Choi NY, Woo SL. Tensile and viscoelastic properties of human patellar tendon. J Orthop Res. 1994;12:796–803.
79. Maganaris CN, Paul JP. In vivo human tendon mechanical properties. J Physiol. 1999;521:307–13.
80. Birch HL, Bailey AJ, Goodship AE. Macroscopic 'degeneration' of equine superficial digital flexor tendon is accompanied by a change in extracellular matrix composition. Equine Vet J. 1998;30:534–9.

81. Riechert K, Labs K, Lindenhayn K, Sinha P. Semiquantitative analysis of types I and III collagen form tendons and ligaments in a rabbit model. J Orthop Sci. 2001;6:68–74.
82. Graham HK, Holmes DF, Watson RB, Kadler KE. Identification of collagen fibril fusion during vertebrate tendon morphogenesis. The process relies on unipolar fibrils and is regulated by collagen–proteoglycan interaction. J Mol Biol. 2000;295:891–902.
83. Fukuta S, Oyama M, Kavalkovich K, Fu FH, Niyibizi C. Identification of types II, IX and X collagens at the insertion site of the bovine achilles tendon. Matrix Biol. 1998;17:65–73.
84. Kjaer M. Role of extracellular matrix in adaptation of tendon and skeletal muscle to mechanical loading. Physiol Rev. 2004;84:694–8.
85. Engel J. Versatile collagens in invertebrates. Science. 1997;277:1785–6.
86. Silver FH, Birk DE. Molecular structure of collagen in solution: comparison of types I, II, III, and V. Int J Biol Macromol. 1984;6:125–32.
87. Hofmann H, Voss T, Kühn K, Engel J. Localization of flexible sites in thread-like molecules from electron micrographs. Comparison of interstitial, basement membrane and intima collagens. J Mol Biol. 1984;172:325–43.
88. Paterlini MG, Némethy G, Scheraga HA. The energy of formation of internal loops in triple-helical collagen polypeptides. Biopolymers. 1995;35:607–19.
89. Berg RA, Birk DE, Silver FH. Physical characterization of type I procollagen in solution: Evidence that the propeptides limit self-assembly. Int J Biol Macromol. 1986;8:177–82.
90. McAnulty RJ, Laurent GJ. Collagen synthesis and degradation in vivo. Evidence for rapid rates of collagen turnover with extensive degradation of newly synthesized collagen in tissues of the adult rat. Collagen Relat Res. 1987;7:93–104.
91. Vuorio E, de Crombrugghe B. The family of collagen genes. Annu Rev Biochem. 1990;59:837–72.
92. Myllyharju J. Prolyl 4-hydroxylases, the key enzymes of collagen biosynthesis. Matrix Biol. 2003;22:15–24.
93. Kjaer M. Role of extracellular matrix in adaptation of tendon and skeletal muscle to mechanical loading. Physiol Rev. 2004;84:649–98.
94. Prockop DJ, Sieron AL, Li SW. Procollagen N-proteinase and procollagen C-proteinase. Two unusual metalloproteinases that are essential for procollagen processing probably have important roles in development and cell signaling. Matrix Biol. 1998;16:399–408.
95. Scott JE. The nomenclature of glycosaminoglycans and proteoglycans. Glycoconj J. 1993;10:419–21.
96. Iozzo RV, Murdoch AD. Proteoglycans of the extracellular environment: clues from the gene and protein side offer novel perspectives in molecular diversity and function. FASEB J. 1996;10:598–614.
97. Raspanti M, Congiu T, Guizzardi S. Structural aspects of the extracellular matrix of the tendon: an atomic force and scanning electron microscopy study. Arch Histol Cytol. 2002;65:37–43.
98. Danielson KG, Baribault H, Holmes DF, Graham H, Kadler KE, Iozzo RV. Targeted disruption of decorin leads to abnormal collagen fibril morphology and skin fragility. J Cell Biol. 1997;10:729–43.
99. Nakamura N, Hart DA, Boorman RS, Kaneda Y, Shrive NG, Marchuk LL, et al. Decorin antisense gene therapy improves functional healing of early rabbit ligament scar with enhanced collagen fibrillogenesis in vivo. J Orthop Res. 2000;18:517–23.
100. Scott JE. Proteodermatan and proteokeratan sulfate (decorin, lumican/fibromodulin) proteins are horseshoe shaped. Implications for their interactions with collagen. Biochemistry. 1996;35:8795–9.
101. Scott JE, Orford CR, Hughes EW. Proteoglycan-collagen arrangements in developing rat tail tendon. An electron microscopical and biochemical investigation. Biochemistry. 1996;35:8795–9.
102. Hedbom E, Antonsson P, Hjerpe A, Aeschlimann D, Paulsson M, Rosa-Pimentel E, et al. Cartilage matrix proteins. An acidic oligomeric protein (COMP) detected only in cartilage. J Biol Chem. 1992;267:6132–6.

103. Södersten F, Ekman S, Eloranta ML, Heinegård D, Dudhia J, Hultenby K. Ultrastructural immunolocalization of cartilage oligomeric matrix protein (COMP) in relation to collagen fibrils in the equine tendon. Matrix Biol. 2005;24:376–85.
104. Beredjiklian PK, Favata M, Cartmell JS, Flanagan CL, Crombleholme TM, Soslowsky LJ. Regenerative versus reparative healing in tendon: a study of biomechanical and histological properties in fetal sheep. Ann Biomed Eng. 2003;31:1143–52.
105. Herdrich BJ, Danzer E, Davey MG, Bermudez DM, Radu A, Zhang L, et al. Fetal tendon wound size modulates wound gene expression and subsequent wound phenotype. Wound Repair Regen. 2010;18:543–9.
106. Ferguson MW, O'Kane S. Scar-free healing: from embryonic mechanisms to adult therapeutic intervention. Philos Trans Roy Soc B: Biol Sci. 2004;359:839–50.
107. Fenwick SA, Curry V, Harrall RL, Hazleman BL, Hackney R, Riley GP. Expression of transforming growth factor-beta isoforms and their receptors in chronic tendinosis. J Anat. 2001 Sep;199(Pt 3):231–40. PubMed PMID: 11554502. Pubmed Central PMCID: 1468327.
108. Chan KM, Fu SC, Wong YP, Hui WC, Cheuk YC, Wong MW. Expression of transforming growth factor beta isoforms and their roles in tendon healing. Wound Repair Regen: official publication of the Wound Healing Society [and] the European Tissue Repair Society. 2008 May–Jun;16(3):399–407. PubMed PMID: 18471258.
109. Hoppe S, Alini M, Benneker LM, Milz S, Boileau P, Zumstein MA. Tenocytes of chronic rotator cuff tendon tears can be stimulated by platelet-released growth factors. J Should Elbow Surg. 2013;22:340–9.
110. Chen L, Dong S-W, Liu J-P, Tao X, Tang K-L, Xu J-Z. Synergy of tendon stem cells and platelet-rich plasma in tendon healing. J Orthop Res. 2012;30:991–7.
111. Fernandez-Sarmiento JA, Dominguez JM, Granados MM, Morgaz J, Navarrete R, Carrillo JM, et al. Histological study of the influence of plasma rich in growth factors (PRGF) on the healing of divided Achilles tendons in sheep. J Bone Joint Surg. 2013;95 (3):246–55.
112. Liu C-F, Aschbacher-Smith L, Barthelery NJ, Dyment N, Butler D, Wylie C. What we should know before using tissue engineering techniques to repair injured tendons: a developmental biology perspective. Tissue Eng Part B. 2011;17:165–76.
113. Lui PPY, Kong SK, Lau PM, Wong YM, Lee YW, Tan C, et al. Allogeneic tendon-derived stem cells promote tendon healing and suppress immunoreactions in hosts: in vivo model. Tissue Eng Part A. 2014:doi:10.1089/ten.TEA.2013.0713.
114. Lui PPY, Kong SK, Lau PM, Wong YM, Lee YW, Tan C, et al. Immunogenicity and escape mechanisms of allogeneic tendon-derived stem cells. Tissue Eng Part A. 2014:doi:10.1089/ten.TEA.2013.0714.
115. Kuo CK, Tuan RS. Mechanoactive tenogenic differentiation of human mesenchymal stem cells. Tissue Eng Part A. 2008;14:1615–27.
116. Yang G, Rothrauff BB, Lin H, Gottardi R, Alexander PG, Tuan RS. Enhancement of tenogenic differentiation of human adipose stem cells by tendon-derived extracellular matrix. Biomaterials. 2013;34:9295–306.
117. Fox PM, Farnebo S, Lindsey D, Chang J, Schmitt T, Chang J. Decellularized human tendon-bone grafts for composite flexor tendon reconstruction: a cadaveric model of initial mechanical properties. J Hand Surg. 2013;38:2323–8.
118. Farnebo S, Woon CYL, Bronstein JA, Schmitt T, Lindsey DP, Pham H, et al. Decellularized tendon-bone composite grafts for extremity reconstruction: an experimental study. Plast Reconstr Surg. 2014;133:79–89.
119. Farnebo S, Woon CY, Kim M, Pham H, Chang J. Reconstruction of the tendon-bone insertion with decellularized tendon-bone composite grafts: comparison with conventional repair. J Hand Surg. 2014;39:65–74.
120. Zhang J, Li B. Wang JH-C. The role of engineered tendon matrix in the stemness of tendon stem cells in vitro and the promotion of tendon-like tissue formation in vivo. Biomaterials. 2011;32:6972–81.

121. Schiele NR, Marturano JE, Kuo CK. Mechanical factors in embryonic tendon development: potential cues for stem cell tenogenesis. Curr Opin Biotechnol. 2013;24:834–40.
122. Fan D, Takawale A, Lee J, Kassiri Z. Cardiac fibroblasts, fibrosis and extracellular matrix remodeling in heart disease. Fibrogenesis Tissue Repair. 2012;5:15.

Skeletal Stem Cell Niche of the Bone Marrow

Agnieszka A. Janeczek, Edoardo Scarpa, Tracey A. Newman, Richard O.C. Oreffo, Rahul S. Tare and Nicholas D. Evans

Abbreviations

2D	Two-dimensional
3D	Three-dimensional
Ang	Angiopoietin
BMP	Bone morphogenetic protein
BMSC	Bone marrow stromal cell
CAR	CXCL12-abundant reticular cells
Cdh2	N-cadherin (neural)
CFU-Fs	Colony-forming units-fibroblastic
CFU-Ss	Colony-forming units-spleen cells
Col/HA	Collagen/hydroxyapatite
ECM	Extracellular matrix
EPC	Endothelial progenitor cell
ESCs	Embryonic stem cells
G-CSF	Granulocyte colony-stimulating factor
GelMA	Gelatin methacrylate
HSC	Haematopoietic stem cell
HUVECs	Human umbilical vein endothelial cells
ICC	Inverted colloidal crystal
Lepr	Leptin receptor
MK	Megakaryocyte
MSC	Mesenchymal stem cell
NG	Neural/glial antigen
OPN	Osteopontin
OSX	Osterix

A.A. Janeczek · E. Scarpa · T. A. Newman · R.O.C. Oreffo · R. S. Tare · N.D. Evans (✉)
Faculty of Medicine, Bone and Joint Research Group, Centre for Human Development, Stem Cells and Regeneration, Southampton General Hospital, Tremona Road, Southampton SO16 6YD, UK
e-mail: n.d.evans@soton.ac.uk

A.A. Janeczek
e-mail: janeczek.agnieszka@gmail.com

K. Turksen (ed.), *Tissue-Specific Stem Cell Niche*,
Stem Cell Biology and Regenerative Medicine,
DOI 10.1007/978-3-319-21705-5_11

PCL	Polycaprolactone,
PEG	Polyethylene glycol
PEGDA	Polyethylene glycol diacrylate
PLGA	Polylactic-co-glycolic acid
PTH	Parathyroid hormone
PTHr	Parathyroid hormone receptor
RUNX2	Runt-related transcription factor 2
SCF	Stem cell factor
SDF	Stem cell-derived factor
SSC	Skeletal stem cell
VCAM	Vascular cell adhesion molecule

1 The Niche Hypothesis and Development of the Concept

Adult stem cells are self-renewing cells that persist for the lifetime of an animal and contribute to the maintenance and turnover of many different adult tissues. Adult stem cells usually maintain a state of quiescence, are slow-cycling and have a low metabolic rate. This is thought to prevent premature senescence and genetic errors and, to a large extent, these cells rely (in addition to intrinsic) on extrinsic signals from the microenvironment in order to enter the cell cycle and to undergo self-renewal, or to produce daughter progenitor cells [77]. The longevity and multipotency of adult stem cells are tightly regulated within this microenvironment to ensure the lifelong maintenance of the stem cell pool and preservation of the long-term homeostasis and regeneration of tissues.

As tissues have various developmental needs and complexities of cellular hierarchies, there are fundamental tissue-specific differences in how quickly adult stem cells self-renew in vivo. For instance, hair follicle stem cells only fuel the growth phase, remaining in the quiescent state for the most of the hair follicle cycle (which can last months), whereas intestinal stem cells must divide daily to maintain the rapid turnover of the small intestine. The bone marrow is another of most actively proliferating tissues in the body, generating $\sim 10^{11}$–10^{12} mature blood cells per day [53]. These arise from short-lived highly proliferative hematopoietic progenitors which in turn descend from a rare population of haematopoietic stem cells (HSCs). Currently, two populations of HSCs are thought to exist, with different division rates, where a fast dividing population contributes to blood cell homeostasis, and where another separate, dormant population is responsible for repopulation of the bone marrow stem cell niche after a challenge such as ablation [162]. The underlying principle for all these tissue types is the same: all stem cell microenvironments, the so-called niches, are implicated in providing relevant cues for stem cells to self-renew and sustain long-term tissue specific regeneration during homeostasis as well as upon injury [50].

Because of the importance of the niches, research has been directed at locating stem cells in vivo and at identifying more accurately the extracellular ‘milieux’ in which they reside. The “stem cell niche” concept was first coined by Schofield in 1978. In a seminar publication, this author showed that the bone marrow as a site of residence for HSCs was key to their function. When isolated from the spleen rather than from the bone marrow, the colony forming units that these cells produced, called “colony-forming units-spleen cells” (CFU-Ss), were not capable of reconstituting all marrow populations. In contrast, their bone marrow-derived cousins *were* able to do so, leading to Schofield proposing a fundamental requirement for hematopoietic stem cells to be resident in the bone marrow microenvironment for their clonal behaviour—if they were removed from their niche, they would lose their “immortality”, becoming less potent cells, like CFU-Ss [129]. This concept, that the niche can effectively drive and maintain cell state, agreed with previous research in the field, showing that primitive cells from hematopoietic cultures could only be sustained when co-cultured in the presence of bone marrow stromal cells as a ‘feeder’ layer [34].

Since then, stem cell niches have been identified in many other tissues, including but not limited to the nervous system [38], the skin [149] and the intestine [9], validating the general niche concept, which states that stem cell maintenance and function is reliable on cues from the extracellular microenvironment. The niche provides a dynamic and diverse microenvironment that has the ability to control stem cell behaviour in a well-orchestrated manner, often by regulating conflicting mechanisms. For example, while the primary role of the niche is to support homeostatic self-renewal and multipotency of resident stem cells, it must also promote the continuous formation of differentiated progenitors in response to appropriate signals, such as growth or injury, maintaining a balance between quiescence, self-renewal and differentiation [49, 97]. Furthermore, under physiological conditions, the niche might provide protection from the accumulation of genetic mutations, whereas in pathology it may lead to malignant transformation of stem cells and diseases associated with tissue degeneration and ageing [60]. Many of the emerging concepts of the stem cell niche in homeostasis and disease have been comprehensively reviewed elsewhere [127].

Today, the niche is defined as a specialized microenvironment with a precise anatomical localisation, composed of stem cells, their progeny, supportive stromal cells and their extracellular matrix (ECM) [75]. Cell communication within the niche may be through junctions, through secreted soluble factors or/and via cell surface receptors, while the ECM may provide biochemical, physical, structural, and mechanical information. In addition, sympathetic nerves are important in transmitting distant physiological cues, and the systemic circulation may carry signals though blood vessels, which influence migration and homing of cells both outside and inside the niche. The communication between stem cell populations and their niche works in both directions, and the interactions are reciprocal. Niche components influence stem cell behaviour, but stem cells are also able to actively remodel their niche by degrading or secreting ECM components [158]. Therefore, to discover more about stem cell biology, it is important that both interactions should be studied.

Over the past three decades, our understanding of the relationship between stem cells and their niche has expanded dramatically. Consequently, one needs to keep in mind that the list of niche elements that participate in their interactions can be vastly more complex than just a single type of stem cell, its supporting cells and the ECM. For example, niches may range in variety from the well understood model of the distal tip cell in *Caenorhabditis elegans,* which forms a unicellular niche controlling germline stem cells, to the arguably more sophisticated network of the mammalian hair follicle, where the cyclic production of new hairs is controlled by the cooperation of many types of stem cells and their separate niches [67, 114]. In the case of the bone marrow niche, two stem cell populations (HSCs and Skeletal stem cells, SSCs; also referred to as mesenchymal stem cells) interact to achieve a steady-state reconstitution of the blood cell subsets, while at the same time sustaining homeostasis of the connective tissues of the bone (bone, cartilage and fat) by regulating bone turnover, in cooperation with various committed stromal cells [89].

How these often contradictory and complex dynamic processes are driven by the bone marrow niche continues to be an intriguing question in the field of stem cell research. A solution could help us understand how better to use the stem cells therapeutically and how to intervene in bone marrow dysfunction. In this chapter, we review recent research on the importance of spatial localisation of skeletal stem cells within their bone marrow niche, cell-cell interactions present in the niche and key growth factors and cytokines. The Wnt family of growth factors, and the role of both canonical and non-canonical Wnt signalling pathways in regulating the bone marrow stem cell niche will be emphasised. Pathologies arising from dysregulation of this niche will also be highlighted, along with therapeutic targets. Finally, attempts to recreate the bone marrow microenvironment in vitro will be discussed.

2 The Bone Marrow Stem Cell Niche

The identification and characterization of niches within various tissues has revealed that many fundamental components are conserved between niches. However, mechanisms that regulate how niches are established, maintained and modified to support specific tissue stem cell functions are still not fully understood, and remain an exciting topic of current research. The bone marrow stem cell niche has been a subject of studies for many years now, and indeed is the microenvironment the niche hypothesis was first based upon.

Bone marrow is the soft tissue residing in the cavities of the bones and within the cortical shell of long bone. More than 50 % of the volume of non-cortical bone is taken up by bone marrow, with trabecular bone occupying the rest of the space [58]. Bone marrow is divided into two separate types; red or "haematopoietic" marrow, which is highly vascularised and is the site of blood cell differentiation, and yellow or "fatty" marrow, so-called because of its high number of fat-storing adipocytes. During childhood, almost all bones in the human body retain a "hematopoietically" active red bone marrow. After reaching maturity, active bone marrow is restricted to the sternum, ribs, vertebrae, ilium, and femoral heads; the rest of the bones are filled with "inactive" yellow bone marrow. Both types of bone marrow contain numerous blood vessels and capillaries and the organization of the marrow reflects the vascular supply. In a tubular bone, the nutrient artery (*arteria nutricia*) enters the marrow cavity, runs longitudinally in the centre of the marrow cavity before branching out toward the endosteum of the surrounding bone, finally leading to specialized vascular structures known as sinuses or sinusoids. Several of these sinuses may then combine to form collecting sinuses which lead to the central sinus or vein. This vein runs longitudinally next to the nutrient artery. Blood in marrow flows from the centre toward the bone cortices and then returns back to the centre. This structural configuration yields high numbers of vessels and sinuses in the periphery, where haematopoietic activity is highest. Within the bone marrow, signals between different cellular elements are transmitted over relatively short distances (up to hundreds of microns), given the well-defined, compact architecture of the bone [148].

The structure of this specialized vascular anatomy provides a suitable habitat for both HSCs and SSCs, supporting their maintenance and differentiation (Fig. 1). The complexity of this biological *milieu*, with its intricate network of interactions between cellular components, ECM and signalling factors, as well as the dynamic mechanical and physical cues it experiences, enables the marrow to perform its physiological role, but in disease processes may become perturbed. Therefore a thorough understanding of such complexity is critical to developing therapies to address disease mechanisms.

2.1 Organisation of the Bone Marrow Niche

Our knowledge of the constituents of the bone marrow niche has improved significantly over the last few years and increasing levels of complexity between all the elements have been recognised, adding insight into the regulation of this micro-domain. The bone marrow microenvironment is composed of different cell types which communicate through the use of physical contact and soluble factors. Decrypting the role of each independent component within this intricate, interactive system is challenging [98]. It is home for two distinct stem cell populations, HSCs and SSCs, which like all stem cells have the ability to self-renew or give rise to many specialised cell types.

Haematopoietic stem cells are multipotent progenitors that give rise to all types of mature blood cells. They are most stringently defined by their ability to self-renew and support long-term multi-lineage haematopoietic engraftment in lethally irradiated mice [162], but can also be identified and isolated prospectively by multi-parameter flow cytometry, such as by complex cell surface marker expression (for example $Lin^{-}Sca1^{hi}c\text{-}Kit^{+}CD34^{-}CD48^{-}CD150^{hi}$) [57]. In initial efforts to locate more primitive, stem cell-like cell populations within the bone marrow, Lord and Hendry found that more mature, granulocytic cells were

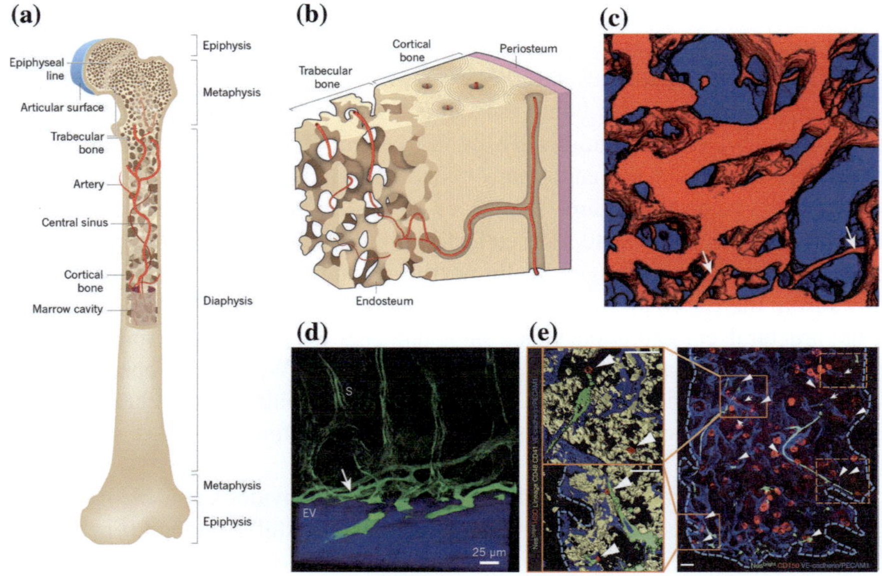

Fig. 1 Bone marrow anatomy. Bone marrow is a complex organ, containing many different haematopoietic and non-haematopoietic cell types, and which is surrounded by a shell of vascularized and innervated bone. **a** Projections of bone (*trabeculae*) are found throughout the metaphysis and epiphyses, increasing the internal endosteal bone surface area, and providing an environment where many cells in the marrow are close to a bone surface. **b** The interface of bone and bone marrow is known as the endosteum, which is covered by bone-lining cells that include bone-forming osteoblasts and bone-resorbing osteoclasts. Arteries carry oxygen, nutrients and growth factors into the bone marrow, before feeding into sinusoids, which coalesce as a central sinus to form the venous circulation. Sinusoids are specialized venules that form a reticular network of fenestrated vessels that allow cells to pass in and out of circulation. There is a particularly rich supply of arterioles, as well as sinusoids, near the endosteum. **c** A three-dimensional reconstructed photomicrograph from the bone marrow in the vicinity of the endosteal surface (50 μm below the surface, marked in *blue*), revealing the rich network of vessels (*red*). Smaller arteriolar vessels (*white arrows*) lead into larger sinusoidal vessels. The field of view is 350 × 350 μm. **d** A cross-sectional view of blood vessels that run along the endosteal surface (EV) and that transition (*white arrow*) into sinusoids (S) that then course towards the central sinus. **e** Illustrative example of whole-mount sternal bone marrow with three-dimensionally reconstructed images. Arrowheads denote HSCs, *arrows* show CD150+Lin+CD48+CD41+cells. HSCs reside in close contact with vascular and perivascular cells as well as other haematopoietic cells. Scale bars, 50 μm. HSCs, haematopoietic stem cells. Adapted by permission from Macmillan Publishers Ltd: Nature [98], Macmillan Publishers Limited (2014) and Nature [73], Macmillan Publishers Limited (2013) and Nature Cell Biology [105], Macmillan Publishers Limited (2013)

increased in frequency in the vicinity of the central arteriole, whereas more immature cells, which had stem cell-like properties based on colony forming assays (CFU-Ss), were increased towards the periphery [81]. These experiments showed for the first time that stem cells, like HSCs, might be confined to specialised regions of the marrow, and has led to research on defining these sub-niches and the supporting cell populations that they contain.

2.2 The Endosteal Niche Hypothesis

Early studies demonstrated an active role for osteoblasts and the endosteal region in particular for influencing the resident pool of HSCs in the bone marrow. To investigate a possible role for osteoblasts in maintaining HSCs, Zhang et al. perturbed the osteoblast population by removing bone morphogenetic protein (BMP) receptor in transgenic mice. Surprisingly, this was found to induce uncontrolled formation of ectopic trabecular bone-like tissue, together with a significant increase in osteoblasts, identified by the phenotype N-cadherin$^+$ CD45$^-$. But in parallel, these researchers found a marked increase in the frequency of HSCs as well. Furthermore, HSCs were found in close proximity to the N-cadherin$^+$ osteoblasts, perhaps suggesting a possible direct control of the latter on HSC proliferation [165]. Similar conclusions were obtained inducing the constitutive activation of the parathyroid hormone receptor (PTHr) exclusively in osteoblasts. PTHr-activated osteoblasts increased in number and augmented their expression of the Notch ligand Jagged-1, which in turn induced an increase in the number of HSCs [15]. These two studies represented the first indications of the bone marrow niche hypothesis in which the osteoblasts played a fundamental role in controlling HSCs.

Further evidence for the interactive roles that osteoblasts might play in the bone marrow niche came from studies showing close communication between osteoblast lining cells and HSCs via Tie2$^+$/angiopoietin (Ang)-1 signalling [4] coupled with secretion of thrombopoietin from osteoblasts [164]. This, combined with the data implicating the importance of osteoblasts located at the endosteal site of the niche, identified key molecular interactions that osteoblasts might have with HSCs, responsible for regulating their quiescence by inhibiting their apoptosis. Another study highlighted a potential role for osteopontin (OPN) in HSC niches. OPN is a matrix protein important for bone remodelling [33], which is produced by osteoblasts at the endosteal surface. Interestingly, Nilsson and colleagues showed that HSCs were able to bind Opn via β1 integrins in vitro, and this interaction promoted HSC quiescence, here demonstrated through an inhibition of proliferation. Moreover, HSCs engrafted into osteopontin knock-out mice failed to localise at the endosteum front, suggesting a chemotactic function of Opn [104]. Another chemoattractant that has been found to facilitate localization of HSCs adjacent to osteoblasts at the endosteum is a high local concentration of Ca^{2+} [1]. HSCs in which the calcium ion-sensing receptor was deleted could not lodge efficiently in the bone marrow, indicating a requirement for this cation in mediating HSC niche residence. These data collectively suggest that the endosteal region, through the presence of osteoblasts and extracellular matrix, has a functional role in regulating the fate and localisation of HSCs (see Table 1 for a list of factors that have been proven to be genetically necessary for the maintenance of normal numbers of HSCs in the bone marrow).

Despite these data, further studies into the HSC-supportive cell type within the marrow niche have undermined the potential importance of osteoblasts. Investigating the direct role of the osteoblast in controlling HSCs, Greenbaum and

Table 1 Adhesion molecules and cell extrinsic factors affecting HSCs in the bone marrow niche and dependent on stromal cell populations

Factor[a]	Evidence for action in the bone marrow *milieu*	Reference
Angiopoietin	Expressed by osteoblasts. Combined loss of Tie2 (the receptor for angiopoietin) as well as Tie1, leads to defects in postnatal HSCs; angiopoietin appears to promote the maintenance of quiescent HSCs	[116] [4]
Ca^{2+} ions	Deletion of the Ca^{2+}-sensing receptor leads to reduced bone-marrow cellularity and HSC content with increased progenitor-cell mobilization into the circulation and spleen	[1]
CXCL12	Highly expressed by putative SSCs, poorly by osteoblasts. Mice deficient in the chemokine CXCL12 or its receptor CXCR4 show disrupted colonization of the bone marrow, whereas conditional deletion of CXCR4 in adult mice leads to reductions in HSC numbers in the bone marrow and reduced HSC activity upon transplantation	[140] [168] [99]
Osteopontin	The matrix glycoprotein osteopontin is expressed at the endosteum by bone-lining cells (osteoblasts) and negatively regulates HSC numbers; osteopontin-deficient mice have moderately increased HSC numbers in the marrow	[137] [104]
SCF	Mice with mutations in SCF (Sl/Sld, steel dickie mutants) or in its receptor KIT (W/Wv, dominant spotting mutants) have fewer HSCs and exhibit less HSC function	[64] [8]
Tenascin-C	Expressed in stromal and endothelial cells. Knockout mice lacking tenascin had normal steady-state haematopoiesis but failed to reconstitute haematopoiesis after bone marrow ablation and showed high lethality	[100]
Thrombopoietin	Thrombopoietin is synthesized in the liver, kidney, bone-marrow stroma and by osteoblasts and may be transported into the bone marrow through the blood; mice deficient in thrombopoietin or the thrombopoietin receptor c–Mpl have profound reductions in HSC numbers	[164] [56] [141] [117] [68]
Wnt	HSCs from transgenic mice with increased expression of Dkk-1 (Wnt inhibitor) in osteoblasts lost the serial ability to reconstitute irradiated marrow, indicating loss of subfraction of quiescent long-term HSCs	[46]

[a]In addition to these factors that are genetically necessary for the maintenance of normal numbers of HSCs in the bone marrow, there are additional factors including VCAM, collagen IV, collagen VI and fibronectin, that are likely to regulate HSC maintenance based on over-expression experiments or experiments performed in culture, but which have not yet been fully tested for necessity in vivo. CXCL12, CXC-chemokine ligand 12; CXCR4, CXC-chemokine receptor 4; HSC, haematopoietic stem cell; SSCs, skeletal stem cell; OPN, osteopontin; SCF, stem-cell factor. Adapted by permission from Macmillan Publishers Ltd: Nature Reviews Immunology [65], Macmillan Publishers Limited (2008)

colleagues created a transgenic mouse expressing Cre recombinase (Cre) under the control of the osteogenic marker osterix (Osx) to deplete *Cdh2*, the gene encoding for N-cadherin. They demonstrated that haematopoiesis, long-term repopulating activity, haematopoietic engraftment, as well as HSC quiescence were not affected by the lack of the adhesion molecules on osteoblastic cells [55]. Furthermore, Kiel and colleagues obtained similar results with the conditional deletion of *Cdh2* from HSC, which did not impair stem cell function [66]. Also, through the use of sophisticated imaging techniques, second harmonic generation microscopy, it has been possible to demonstrate that HSCs are located in proximity of osteoblastic cells, but the direct contact between the two is not required for the HSC proliferation or homing following transplantation in mice [80].

2.3 The Vascular Niche Hypothesis

Given the undeniably important role of the endosteal area in the bone marrow niche, combined with the observation that the effect of osteoblasts on HSCs is indirect, what then might be the *direct* regulator of HSCs in vivo? It is plausible to speculate that the discrepancies observed in the aforementioned studies may be ascribable to the presence of a more immature phenotype of bone progenitor in the niche—potentially SSCs themselves.

Although four decades have passed since Friedenstein's pioneering work on colony-forming cells isolated from bone marrow (CFU-Fs), a thorough understanding and characterisation of skeletal stem cells, as compared to haematopoietic stem cells, still remains elusive [48, 107]. Like HSCs, the putative SSC or the progenitors they give rise to, may be inhomogeneous, with several different lineages and stages of commitment. Although stromal cells from the bone marrow, defined by us as bone marrow stromal cells (BMSCs), but also referred to as mesenchymal stem cells (MSCs), have been described by a variety of non-specific cell surface markers and their ability to differentiate into osteoblasts, chondrocytes and adipocytes, more stringent criteria suggest that any stem cells within these populations must be defined by their ability to form ossicles composing cartilage, bone and fat in host animals after single cell implantation [12, 39]. Despite arguments over definition, it is likely that the putative SSC may serve as a support for haematopoiesis and osteoclastogenesis in vivo, and they are therefore critical for bone marrow function, bone development and bone remodelling throughout life [113]. More recent efforts have focused on identifying and isolating bone, cartilage, and stromal progenitors by rigorous functional characterization, not only in vitro but also in vivo [19, 91, 111].

Supporting a role for SSCs in the regulation of HSC niche, Mendez-Ferrer and colleagues showed that a population of cells within the bone marrow, marked by the intermediate filament protein Nestin (Nestin$^+$ cells) behaved in a way that might be expected for SSCs, with the ability to self-renew and differentiate into multiple lineages. These cells were found to be directly adjacent to HSCs in bone marrow sections. Additionally, the expression of genes associated with the HSC preservation, like Ang-1 and Opn, was higher in these Nestin$^+$ cells than any other cell type in the bone marrow, including osteoblasts [91]. Moreover, Nestin$^+$ cells were responsive to parathyroid hormone (PTH) which induced their proliferation and differentiation into osteoblasts, as in the study by Calvi et al. described earlier [15]. Previous studies had shown that putative SSCs express elevated levels of the stem cell factor (SCF) and CXC chemokine ligand CXCL12, also known as stromal cell-derived factor 1 (SDF-1). Both of these molecules have a role in chemotaxis, homing, maintenance and retention of HSCs in the bone marrow [3, 140], and are especially highly expressed in SSCs found in perivascular regions [37, 91, 125], but are poorly secreted by osteoblasts [37].

In fact, a great deal of data now suggests that SSCs may be components of the vascular wall, with some reports suggesting close similarity of SSCs to pericytes, a cell type found in close apposition to the vascular walls [16, 26, 28, 131]. In addition Nestin^{+} cells, also form part of the endoglin (CD105^{+}) population in mice (a marker of endothelial cells), and contain all of the fibroblastic colony-forming unit (CFU-F) and multipotent mesenchymal sphere (mesensphere) formation capacity. The situation is similar in human bone marrow, where a population positive for CD105^{+}, CD146^{+} and Nestin^{+} (but negative for CD45^{-}CD71^{-}CD31^{-}) contains all the SSC-like CFU-F- and mesensphere-forming capacity [59].

In mice, high levels of CXCL12 expression have been found in a second population of cells, described as CXCL12-abundant reticular cells (CAR), which again are found in close proximity to vessels, surrounding endothelial cells of the sinusoids in proximity to the endosteum [140]. This population supposedly contains adventitial reticular cells and mesenchymal progenitors. Despite similar levels of CXCL12 production, it is plausible that Nestin^{+} SSC, which are approximately four times less represented in the bone marrow, are a distinct and more undifferentiated subpopulation of the marrow niche compared to CAR cells [40].

2.4 *Interaction of Endosteal and Vascular Domains and Further Characterisation of the Niche*

A potential hypothesis that has emerged from these studies is that there may be two distinct stromal niches coexisting in the bone marrow, both of which interact in synergy to regulate HSCs. One being the endosteal (osteoblastic) niche comprising relatively few HSCs ($\sim$14 %) and one being the perivascular niche, with a greater abundance of HSCs at sites surrounding blood vessels [63]. It has been proposed that the two niches are associated with the dormant and activated HSCs, respectively [161]. In support of this theory, it has been shown that the production of CXCL12 in CAR cells and osteoblast precursors at the endosteal site is necessary for the support of B-lymphoid progenitors (a differentiated phenotype of HSC) rather than HSCs per se, an idea that is supported by the fact that the conditional ablation of CXCL12 from osteoblasts does not affect HSC [54]. Similarly, conditional depletion of *Scf* in osteoblasts did not affect HSC cell populations, while in direct contrast controlled depletion of cells producing SCF (SSC and endothelial cells) at the perivascular site completely eliminated the HSC population. This data strongly suggests a role for the perivascular niche in sustaining HSCs [36]. Figure 2

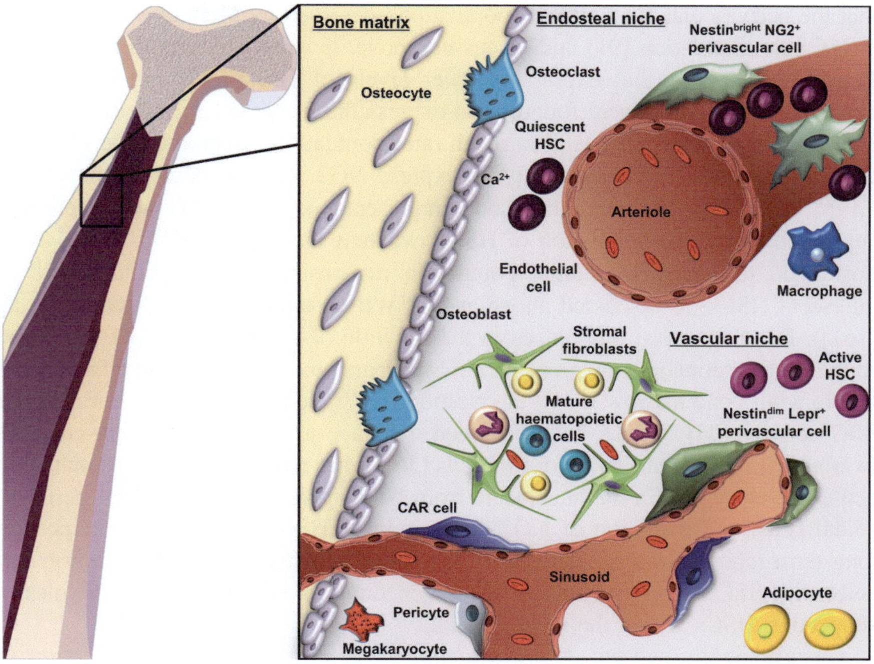

Fig. 2 Key components of the bone marrow niche. In murine models, various cell types have been implicated for their roles in promoting HSC maintenance, including Nestin$^+$ perivascular cells, CAR cells, which have been reported to be more differentiated (containing adventitial reticular cells and mesenchymal progenitors), endothelial cells and macrophages. Adipocytes, which have been shown to negatively affect HSC maintenance, are increasingly present after chemotherapy or radiation, as well as due to ageing. Quiescent HSCs associate with arterioles ensheated with Nestin^{bright} NG2$^+$ pericytes (putative SSCs). After activation HSCs relocate near the Nestin^{dim} Lepr-expressing perisinusoidal area. Osteoblast precursors and CAR cells in the vicinity of the endosteal niche support the differentiated phenotypes of HSCs. Identification of these various bone marrow niche components have advanced the field considerably, but understanding such interactions in the context of human physiology, with cell surface markers present in human cells may help understand other cell populations which might be involved in functioning of the bone marrow niche. CAR cells, CXCL12-abundant reticular cells; HSC, haematopoietic stem cell; SSCs, mesenchymal stem cells; NG, neural/glial antigen

depicts an overview of interactions between the various cellular components of the bone marrow.

Recent developments in murine genetics have enabled us to have a closer look at the interaction between HSCs and SSCs and other supporting cells. These studies have used technology to enable the conditional deletion of various maintenance factors in restricted cell populations by the use of Cre recombinase expressed under the control of cell type-specific promoters. The most commonly used marker genes for cells with SSC-like properties, reviewed by Mendez-Ferrer and colleagues, are Osx, Nestin, neural/glial antigen 2 (NG2) and Leptin receptor (Lepr) [92]. With the use of different recombination techniques in different laboratories, it has been noted that the perivascular Nestin-GFP$^+$ population, identified by Mendez-Ferrer et al. as a SSC-like population [91] and discussed above, is heterogeneous in nature [73]. Perisinusoidal cells that expressed lower levels of GFP, reflecting low Nestin expression, (GFP^{dim}), were found to co-express the receptor Lepr, while cells that expressed higher levels of GFP (a much rarer population of periarteriolar GFP^{bright} cells) were found to preferentially co-express NG2. In these studies, the former GFP^{dim}/Lepr$^+$ cells were hypothesised to represent a population of stromal cells that supported the active population of HSCs, whereas the latter GFP^{bright}/NG2$^+$ were postulated to represent stromal population responsible for the maintenance of quiescent HSCs (as discussed above and by Wilson et al. [73] and Kunisaki et al. [161]). Developmental differences in the function of stromal cells have also been found. In the neonatal bone marrow, SSCs appeared to be perisinusoidal Lepr$^+$ cells and in adult bone marrow it was the Nestin-GFP^{bright} cells which were enriched for SSC activity [94]. Others have provided evidence to suggest that Lepr-cre cells overlap with sinusoidal Nestin-GFP$^+$ cells [36]. It has also been suggested that upon cell-cycle entry, HSCs might relocate from NG2$^+$ periarteriolar to Lepr$^+$ perisinusoidal cells, and depletion of NG2$^+$ cells induces HSC cycling and impairs their long-term repopulation capacity [73]. Together, such study support the notion that cells with SSC-like properties are heterogeneous in nature and have key roles in supporting HSCs within the niche (Fig. 2).

Unsurprisingly, subsets of SSCs identified by different means (expression of genes identified by conditional reporters) seem to differ in their functionality in different studies, possibly because conditional genetic lineage tracing models for the bone marrow populations may be subject to error. For example, the degree of recombination might be unsatisfactory and there might be off-target effects of the recombination-inducing drug. Tamoxifen is known to cause apoptosis in certain haematopoietic progenitor subpopulations and can have an anabolic effect on bones. Complete elimination of a cell population also removes any progeny, may affect other interacting cells, and also could potentially induce local inflammatory responses, leading to further non-specific perturbations. Finally, compensatory feedback mechanisms have to be taken into consideration as well, when analysing the data from reporter strains [61, 92].

In addition to the above, another difficulty arises in examining the relationship between SSCs and HSCs in vivo due to the very dynamic, migratory behaviour of HSCs. This has been illustrated in a parabiotic murine model, where the bloodstream of two animals is physically linked [163]. Under normal physiological conditions adult HSCs constitutively recirculate, passing from the bone marrow into the bloodstream, and returning to the marrow to re-engraft and seed ongoing haematopoiesis. The dynamic aspect of the sub-niches has also been proposed, stating that HSCs can move among the endosteal and vascular niches within the bone marrow quite readily and receive inputs from both of them simultaneously [157].

2.5 Additional Players in the Niche Environment

Another key player in regulating HSCs in the perivascular domain is the nervous system. For example, it has been shown that the release of HSCs from the marrow niche is regulated rhythmically by the circadian clock. In this study, noradrenaline released by the sympathetic nervous system was found to activate the β_3-adrenergic receptor on perivascular cells, inducing a down-regulation of *Cxcl12* and mobilisation of HSCs [90].

Other cells also contribute to the niche microenvironment. For example osteoclasts with their resorption activity reduce the presence of Opn on endosteum inducing a mobilization of haematopoietic progenitors [71]. Similar mobilization has been proposed following CXCL12 down-regulation induced by administration of granulocyte colony-stimulating factor (G-CSF) or through a loss/depletion of resident macrophages, called osteomacs [62]. Chow and colleagues showed that the lack of mononuclear phagocytes drastically affects the production of CXCL12, SCF and Ang-1 by perivascular SSCs (Nestin^{+}), but not by osteoblasts suggesting that macrophages indirectly regulate the marrow niche through the SSCs [23]. A role for megakaryocytes in promoting osteoblastic HSC niche expansion and stem cell engraftment after radioablative conditioning has also been proposed [106]. Recently it has become clear as well, that adipocytes (the number of which increases with age, obesity or after chemotherapy or irradiation) apart from filling the space in the bone marrow, also actively regulate HSC function. This regulation is of an inhibitory nature, as HSCs isolated from fat-rich vertebrae showed reduced activity in transplantation assays, and obese leptin-deficient mice show impaired haematopoiesis [24, 101].

Compared to the cellular and growth factor aspects of the bone marrow niche, the effect of ECM components in the bone marrow niche has been studied rather less thoroughly. Where it has been investigated, it has been mostly by in vitro functional assays. Extracellular matrix molecules including tenascin-C, vascular cell adhesion molecule (VCAM), collagen VI, collagen IV and fibronectin (all secreted by the stromal cells) have been shown to contribute to the maintenance of HSCs [51, 70, 91, 100]. Bone marrow ECM also can dictate the fate of SSCs; bone marrow cells isolated from knockout mice lacking biglycan presented with defects in the ability to differentiate into osteoblasts [20]. Therefore further studies into the role of bone marrow ECM are needed to complement the knowledge of interactions present in this microenvironment.

2.6 Wnt Signalling in the Bone Marrow Niche

Wnt proteins regulate development, cell polarity, proliferation, motility and cell fate determination [79]. The Wnt signalling pathway is one of the most important in the self-renewal of stem and progenitor cells and deregulation of Wnt signalling has been linked to tumorigenesis in different tissues, potentially by aberrant division of stem cells [122]. The Wnt signalling cascade is involved in HSCs self-renewal, as demonstrated in lethally irradiated mice, where Wnt proteins supported the ex vivo culture of HSCs and subsequently promoted the reconstitution of the haematopoietic system [160]. Wnt signalling in HSC has been shown to either maintain the stem cell phenotype of HSCs or induce their differentiation, depending on whether the non-canonical or canonical pathway was involved, respectively [83, 139]. The switch from canonical or non-canonical Wnt signalling is also thought to be involved in HSC ageing [47]. Wnt signalling has also been considered to influence the HSC function indirectly, through the niche [120]. There is evidence that stem and progenitor cells of the haematopoietic system may be responsible for regulating mesenchymal stem and progenitor cells within the niche, based on gene-expression analysis. HSCs and progenitors have been shown to express various factors associated with skeletogenesis, including BMP2, BMP7 and Wnt3a, receptors of which are highly expressed by BMSCs and their progeny [19]. Additionally, several Wnt ligands have been identified as being expressed by BMSCs, such as Wnt2, Wnt4, Wnt5a, Wnt11, Wnt16 and various Wnt inhibitors [43]. A comprehensive investigation into Wnt expression patterns in the developing bone and its contribution to the osteolineage identified that Wnts produced by Osx-expressing cells regulated their proliferation and differentiation [145]. For detailed information on Wnt signalling in bone homeostasis, reader is directed to a review by Baron and Kneissel [10].

2.7 Acellular Factors—Mechanical Effects

An increasing body of evidence suggests that mechanical environment of the bone marrow niche can directly modulate the function of the cells within it [18, 115]. As reviewed comprehensively by Gurkan and Akkus [58], BMSCs have been proven to be responsive to mechanical signals, such as hydrostatic pressure, fluid flow-induced shear stress, and the viscosity of their environment. The hydrostatic pressure values of bone marrow reported in the literature vary in the range of 10.7–120 mmHg (1.4–16 kPa) in mammals, which is generally accepted to be around one quarter of the systemic blood pressure. The viscosity of bone marrow has been reported to range between 37.5 and 400 cP (0.038–0.40 Pa s) in mammals, which is further dependent on marrow composition and temperature. Such mechanical and compositional properties of the bone marrow are thought to be changed during bone diseases or with ageing. And importantly the viscosity of bone marrow decreases with increasing fat content, as may be seen in the onset of osteoporosis, and in ageing [25, 167].

External mechanical influences such as physiological activity, leading to loading or bending of bones, may cause strain and local pressure gradients that drive interstitial fluid flow within the bone marrow. This subsequently results in a shear stress exerted on the endosteal surface. Mechanical loading and hydrodynamic forces are known to regulate BMSC differentiation in vitro [5, 17], and so such forces may act to alter bone remodelling, bone formation and bone adaptation via an action on stem cells. Such mechanical signals within the bone marrow niche have the ability to activate osteogenic signalling pathways in BMSCs, for example through the Wnt signalling pathway, and through kinases such as receptor tyrosine kinase-like orphan receptor 2 (Ror2), and downstream transcription factors such as Runt-related transcription factor 2 (Runx2) [18]. Finally, the material properties of the ECM in the bone marrow may have profound effects on the differentiation and regulation of stromal cells. BMSCs have been shown to differentiate into tissues as distinct as neural, fat and bone lineages solely based on the stiffness of the ECM to which they attach [42]. Changes in bone marrow adiposity in relation to age or disease status may lead to changes in marrow stiffness that leads to changes in the cellular make-up of the SSC-HSC niche. In all these circumstances, delivery of mechanical stimuli to cells has important implications for bone tissue engineering and regeneration applications and targeted mechanical stimulation of SSCs in the bone marrow could have potential in the clinical treatment of bone diseases, such as osteoporosis [84].

2.8 Acellular Factors—Oxygen Tension

Low oxygen tension (hypoxia) is thought to be a crucial niche characteristic, responsible for maintaining multiple stem cell types in a quiescent state [96]. Indirect evidence, including proteomic analysis, expression of hypoxia inducible factors or staining for hypoxic markers, supports this view [112, 143, 150]. Recent research providing direct in vivo measurements of local oxygen tension in the bone marrow of mice confirmed the low pO_2 of the bone marrow despite the very high vascular density in this tissue [134]. Further heterogeneities based on location within the marrow were also discovered, indicating that deeper peri-sinusoidal regions (pO_2 = 9.9 mm Hg; 1.32 kPa; 1.3 %) are more hypoxic, possibly due to these regions being rich in proliferating cells which consume oxygen more avidly than the endosteal region rich in Nestin$^+$ arteries (pO2 = 13.5 mm Hg; 1.80 kPa; 1.8 %), which are home to quiescent HSCs. These values can change drastically after radiotherapy and chemotherapy, resulting in the disappearance of the pO_2 gradient between the sinusoids and the arteries, indicating a role for oxidative stress in altering the stem cell metabolic environment.

In summary, the bone marrow niche is composed of several stem/progenitor cells comprising a complex network of cellular communication mediated by ECM, soluble molecules and mechano/physical signals. In physiological conditions, this niche provides the stimuli for the maintenance or commitment of both HSCs and SSCs. The continuous investigation of this fascinating microenvironment may result in novel theories concerning its *modus operandi*, which will become amenable to further testing in the years to come as our methods of analysis of such complex systems improve.

3 Pathologies in the SSC Bone Marrow Niche

3.1 Ageing

The bone marrow microenvironment, like all tissues, changes substantially with age. Differences in the cellular composition of young *vs.* aged marrow may imply a role for the niche in ageing [154]. From a SSC perspective, previous studies in rats noted an age-related decline in the fraction of $CD90^{+}CD105^{+}$ BMSCs, but the ability of this purified fraction to alter niche function with age in mice and humans still remains to be elucidated [11, 147]. Human BMSCs proliferative capacity has been shown to decline as we grow older [136], which may contribute to the poorer healing rate of bone in older people [118]. Another characteristic of bone marrow in the elderly is an increase in adipocytes. This is further associated with a diminished potential of HSC in generating the blood lineages, and may also cause a disturbance of BMSC differentiation [101]. Moreover, BMSCs obtained from aged human bone marrow show elevated reactive oxygen specie (ROS) levels and markers of cellular senescence [138]. Further studies are necessary to link age-related metabolic changes in BMSCs to their bone marrow niche as well as their function as a bone marrow niche component.

3.2 Malignancies

A key issue that remains to be addressed is whether the niche for malignant or dysplastic cells is the same as or similar to that of normal stem cells. Evidence shows that transformed cells do not exhibit the same degree of niche dependence; however some interactions are still present, leukaemia is not an extra-medullary disease and so some characteristics of the bone marrow niche must be necessary for its persistence [52]. In addition to this, most neoplastic cells develop in their primary sites likely because of favourable interactions with supporting microenvironmental cells or matrix. Such interactions in the context of other tissues can lead to the spread of cancer. When bone marrow host site interactions are weakened in leukaemia, tumour cells tend to migrate to other tissues, and this metastasis is greatly dependent on adhesion molecule expression on the stromal cells, i.e. $\alpha4\beta1$, $\alpha5\beta1$, $\alpha L\beta2$, CD44, or CXCR4/CXCL12, in other environments influencing survival and attenuating the effectiveness of chemotherapy [110]. BMSCs have a strong tropism for tumours and can engraft and either promote or supress tumour progression by differentiating into supportive stromal cell types [7]. The idea of a putative use of BMSCs as "magic bullets" against tumours needs to be validated as we gain more insights into the metastatic niche formation by these cells.

3.3 Therapeutic Targets for Regenerative Medicine Purposes

Emerging targeted therapies for cancer treatment within the bone marrow niche mainly focus on inducing cytotoxicity. Another approach would rely on the assumption that the niche properties could be modulated with drugs to preferentially support normal over malignant or dysplastic cells. Such studies would require an in-depth analysis of the interactions in an in vitro bone marrow niche mimicry model (discussed later on in this chapter).

For regenerative medicine purposes on the other hand, a potential aim would be to deliver factors influencing the stem cells or their bone marrow niche in a spatial-temporal manner. For this purpose, intelligent hydrogel matrices, such as multiphase composites and affinity hydrogels, could be employed [78]. Alternatively, the niche could be targeted by the use of drug delivery vehicles such as nanoparticles. Nanoparticles, such as liposomes and polymersomes, have been used for many years in the treatment of cancer. Due to ability of nanoparticles to facilitate the transport of often insoluble lipophilic drugs, combined with their propensity to accumulate at tumour sites (rather than healthy tissue) due to a tumour's often leaky vasculature, nanoparticles can be an effective chemotherapeutic weapon [14]. However, nanoparticles may also be a useful way of delivering drugs to the bone marrow niche. For example, nanoparticles have been shown to home to the bone marrow niche via both passive and active mechanisms. In the first case, nano-sized lipidic particles (liposomes) administered systemically in dogs were observed to passively accumulate in the bone marrow [128]. Liposomes travelling in the blood flow together with plasma components are free to extravasate into the bone marrow parenchyma taking advantage of the highly branched nature of the vasculature [95]. Once in the bone marrow, nanoparticles can be taken up by resident macrophages as a direct consequence of their phagocytic activity [133]. However, the amount of nanoparticles passively homing into the bone marrow is less than 2 % [128], hence alternative "active" strategies have been implemented in order to augment the specific targeting of these nanocarriers. This is achieved by functionalising the surface of the nanoparticle with specific ligands that induce selectivity for specific organism locations. For example, polymeric nanoparticles conjugated with bisphosphonate molecules were demonstrated to actively target and accumulate in murine bone marrow, where they mediated the delivery of chemotherapeutic agents [142]. Similar fates was also ascribed to porous silicon nanoparticles conjugated with a ligand of E-selectin, which is constitutively expressed on bone marrow endothelium or to liposomes functionalised with succinic acid [86, 132]. Overall, the use of nanoparticles allows for a targeted and spatio-temporally controlled delivery of stimulatory factors, therefore they may represent a novel therapeutic perspective in the modulation of the bone marrow niche for regenerative purposes.

4 In vitro Recreation of the Bone Marrow Stem Cell Niche

Ideally, the most accurate information about the biology of the stem cell niche should come from in vivo studies. However, the complexity of the niche within an organ in vivo on a background of an additionally complex organismal physiology makes these studies very challenging indeed; it is difficult to dissect the factors responsible for functions of interest due to both indirect effects and the presence of compensatory mechanisms. To overcome these challenges, in vitro culture systems that recreate the specialised bone marrow niche are of particular attraction. This has been attempted for several different applications using a variety of technologies.

4.1 The Need for Bone Marrow Mimicry

One reason for recreating the bone marrow stem cell microenvironment in vitro, is to study and culture-expand stem or progenitor cells which are otherwise hard to maintain or are prone to senescence or spontaneous commitment to a particular cell lineage in a standard culture system. For example, for HSCs it has been shown that ex vivo cultivation in suspension culture with a cytokine cocktail only promoted the expansion of cells with short-term, rather than long-term, bone marrow repopulation activity. This resulted in the absence of durable in vivo engraftment in subsequent in vivo experiments [152]. For BMSCs, culture on conventional two-dimensional (2D) substrates (e.g. tissue culture plastic flasks) causes spontaneous differentiation into more committed cell types and gradual loss of "stemness" [6]. And prolonged expansion of these cells to generate sufficient numbers for clinical applications is also an important issue, as it often results in their senescence and lack of functional capacity [155]. Moreover, 8 % of prolonged BMSC cultures have been reported to undergo spontaneous malignant transformation [124]. So different approaches for expanding stem cells in vitro are urgently required.

Another reason for creating ex vivo niches is as a model to extend basic biological knowledge about how stem cells interact with their microenvironment at a molecular level, both in tissue development and in disease processes. ECM, as mentioned before, is a key component of stem cell niches and is involved in various aspects of cell behaviour [51]. Interactions between ECM and cells occur predominately through a class of receptors known as integrins. Integrin signalling can stimulate intracellular signalling pathways for migration, differentiation, and survival and therefore ECM has a major influence on tissue homeostasis and regeneration, both in physiological and pathological conditions [42]. In vitro engineered stem cell niches are invaluable for examining stem cell-ECM and secreted growth factor interactions, since it is possible to deconstruct the effects of specific molecules on a single-cell level by using such reductionist approaches [93]. Recent advances in biomaterials for stem cell culture, fabrication of three-dimensional (3D) scaffolds with micro- or nanoscale topography or "bio-click and bio-clip" reactions for selective addition of biomolecules, and microfluidic bioreactors capable of quantitative studies, all serve to increase our knowledge on stem cell physiology [87, 146].

Finally stem cell niche construction in vitro allows for closer study of therapeutic targets, potential drug-cell interactions and regeneration mechanisms, under conditions that predict the in vivo context. These biomimetic systems could offer an in vitro platform for therapeutic screening [153]. Recent studies have clarified the role of the marrow microenvironment in the pathogenesis of haematologic tumours, underscoring the need for therapeutic targeting of the niche to achieve a complete or at least long-term remission. Establishment of a culture system that closely resembles marrow physiology may speed up the development of drugs which specifically target molecules to leukaemic stem cells, without adversely affecting normal stem cell self-renewal [72]. Such an approach can also be used to study the events at the site of bone injury, and to design therapies to specifically target typically pleiotropic molecules to increase the regenerative potential of mesenchymal stem cells, without leading to side-effects on other cell types. Figure 3 is an overview of the different methods of mimicking the bone marrow niche.

4.2 Efforts Towards Bone Marrow Niche Recapitulation

4.2.1 2D Cultures

Co-culture has been used for many years for facilitating and studying cell-cell interactions in a simplified manner. For example, cultivation of embryonic stem cells (ESCs) on feeder fibroblast cell layers remains a gold standard for keeping ESCs cells in an undifferentiated state, although synthetic matrices and medium components can now substitute, albeit at greater expense [151]. In addition to growing cells in direct contact, however, cells can be separated by using filter membranes in culture dishes. With such innovations, cell contact vs. paracrine dependence of cellular interaction can be assessed quantitatively, which has been helpful in modelling and understanding cell interactions between HSCs and SSCs in the bone marrow niche.

In order for HSCs to be provided with the relevant niche signals to ensure their correct function, HSCs can be cultured either on BMSC populations, or alternatively in media supplemented with high concentrations of growth factors. MSCs provide a more suitable cellular environment for in vitro expansion of HSCs, supporting the maintenance of a more primitive, self-renewing phenotype, in comparison to suspension cultures fed with a cocktail of cytokines [156]. Co-cultivation of human umbilical vein endothelial cells (HUVECs) with BMSCs and human primary osteoblasts has shown an increased proliferation of osteoblasts and BMSCs and their reduced apoptosis under low serum conditions, when cultured in direct contact [135].

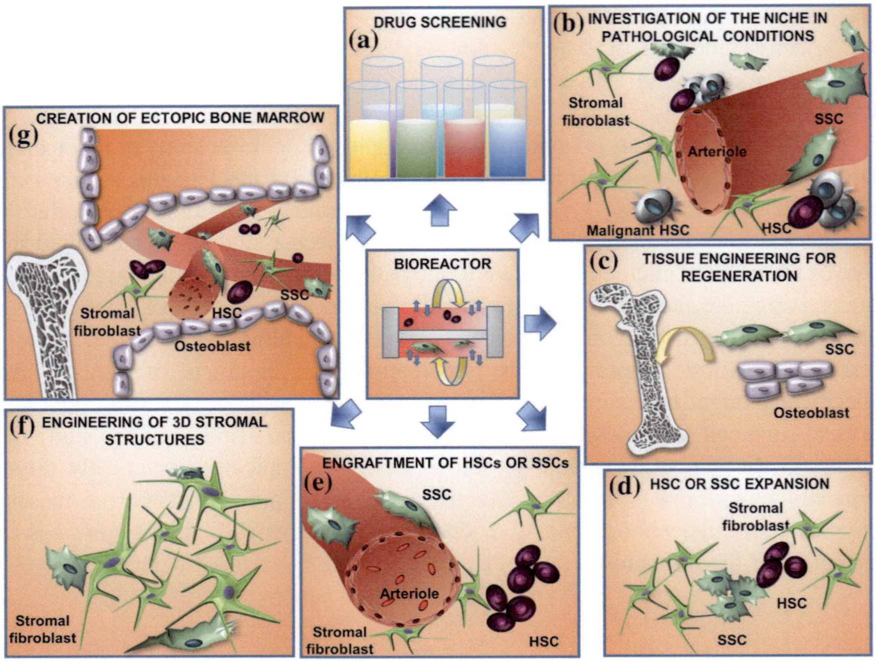

Fig. 3 Ex vivo reconstruction of the bone marrow niche. 3D bioreactors can be used to reconstitute the bone marrow niche in vitro. Such cultivation systems may have applications in different fields, including: **a** Drug screening, to investigate the impact of different molecules on the bone marrow niche; **b** Investigation of the niche in pathological conditions, in order to elucidate the mechanism behind diseases; **c** Tissue engineering for regeneration, to aid the creation of advanced grafts; **d** HSC and MSC expansion for bone marrow transplantation for tissue regeneration; **e** Engraftment of HSCs or SSCs to investigate the regulatory role of all the cells comprising the bone marrow niche; **f** Engineering of 3D stromal structures in order to dissect the function of the extracellular matrix in the niche; **g** Creation of ectopic bone marrow as an alternative environment for the cultivation of stem cells in case of disease. HSC, haematopoietic stem cells; SSCs, skeletal stem cells. Figure adapted from [35]

A very recent study incorporated the simple idea of mimicking the bone marrow microenvironment by enhancing cell-cell interactions. This involved simply seeding endothelial progenitor cells (EPCs) from the rat bone marrow at high densities as "dot cultures", which abolished the need for the use of expensive additional growth factors in the media. This provided an advantageous culture method, as the EPCs could be expanded and enriched efficiently. They further exhibited superior angiogenic potential by forming better tubular networks in vitro and by demonstrating their potent ability to rescue ischemic limbs in vivo [82].

Other attempts sought to solve the issue of sufficient matrix for BMSC cell growth by culturing cells on a cell-free ECM. ECM made by murine bone marrow cells comprised of collagen types I, III and V, syndecan-1, perlecan, fibronectin, laminin, biglycan, and decorin, facilitated the replication and maintained the multipotentiality of murine MSCs, while still preserving their ability to differentiate into more specialised cell types, such as osteoblasts and adipocytes [21]. Although this culture technique did not prevent a decrease in the proportion of progenitors compared to freshly isolated bone marrow cells, ECM-expanded BMSCs transplanted in vivo showed superior bone formation and bone marrow volume. A study conducted on human BMSC-derived ECM produced very similar results [74]. More recent research on ECM derived from fetal BMSCs showed the superiority of this cell type for culture expansion approaches, due to enhanced ECM deposition capabilities [102].

Finally, some seminal studies from Dalby and Oreffo have shown the ability of non-biomimetic disordered nanotopography to direct mesenchymal stem cells towards the osteogenic lineage [29, 88]. In this way, a disordered arrangement of nanopits could be harnessed to drive and switch immunoselected skeletal osteoprogenitors to a directed osteogenic phenotype in the absence of biochemical cues. In contrast, an alternative, highly ordered nanotopography was observed to be capable of promoting skeletal stem cell self-renewal, by enhancing symmetric cell division. Further insight into the importance of the niche in stem cell function is presented by recent findings from Kingham et al. demonstrating the differentiation of human embryonic stem cells also on disordered nanotopographical substrates along the mesoderm lineage and enhanced expression of stromal markers in the absence of soluble, chemical differentiation-inducing factors [69].

As the SSCs have been proposed to reside within the perivascular niche on fenestrated sinusoidal capillaries, it is striking that the endothelial fenestrations of the cells of the sinusoid walls are nanoholes typically 100 nm in diameter comparable to the nanopits detailed above and implicating a key role for such a topography in stem cell niche interaction [30].

4.3 3D ECM Models

Although 2D models allow a certain level of maintenance and expansion of HSCs and BMSCs, by definition they lack the three dimensional architecture of the in situ bone marrow niche. To try and recapitulate three dimensions in a culture dish, one option has been to seed cells at high density as ‘hanging drop’ cultures. Such techniques have been used for decades to form spheroids, embryoid bodies and other stem cell aggregates for cell differentiation studies.

In the context of bone marrow, another 3D in vitro model has been developed to mimic the subendosteal region of the bone. This involved culturing mixed self-reassembled spheroids of active osteoblasts and BMSCs in order to study the proliferation, migration and anchoring of HSCs. They were found to move dynamically in and out of these complex spheroids, and after 48 h lodged within discrete regions [32]. These studies showed that both BMSCs and osteoblasts produce different ECM components and distributed them to resemble the in vivo structure of the endosteal niche (trabecular area close to the bone surface), but with a clear boundary between these two populations. BMSCs cultured by this method became quiescent and showed cytoskeletal changes, with less pronounced stress fibres. Overall, these spheroid cultures formed submicroenvironments that were recognised differentially by HSCs, thus influencing their positioning near the hypoxic insides and inducing quiescence. Another study used a 3D hematopoietic progenitor cell culture system utilizing natural cancellous bone as a scaffold, seeded with osteoblasts and other bone marrow stromal cells [144]. In this system, certain parts of the ECM of the scaffold were still preserved, and the system demonstrated that it could maintain and expand haematopoietic progenitors efficiently.

4.4 Hydrogel 3D Matrices

The 3D culture of bone marrow cells calls for a biomaterial that would be able to restore cell-matrix interactions, direct cell alignment and migration, and apply physical signals, such as flow-induced shear or mechanical stress. It has previously been demonstrated that stem cells are extremely sensitive to the elasticity of the substrate, and stiffer matrices which mimic bone are osteogenic [42]. Hydrogels are a network of interacting polymer chains with a high water content, whose elasticity is easily chemically tuneable to resemble that of natural or pathological tissues, thus enabling control over differentiation and self-renewal of stem cells. In addition, hybrid hydrogels are available, which offer an advantage due to their ability to recreate a stiffness gradient.

Self-assembled spheroid cultures of BMSCs and HUVECs in methylcellulose are one of the simplest 3D co-culture models enabling the study of cell-cell interactions and influence on MSC differentiation [126]. Another 3D method of co-culturing BMSCs and HSCs within collagen scaffolds provided data to suggest the presence of two subpopulations of haematopoietic cells: highly proliferative with a tendency towards lineage commitment; or self-renewing and of immature phenotype [76]. A study by Sharma *et al.* using a hydrogel approach, dissected the interactions responsible for maintaining these two distinct phenotypes. Co-culture of placenta- or marrow-derived stroma cells with HSCs on a 3D pre-set synthetic peptide-based nanofibrous hydrogel (Puramatrix) scaffold resulted in a higher expression of ECM molecules, such as fibronectin, collagen IV, vitronectin, laminin, and integrin α4β1. The 3D MSCs provided a favourable support for HSC growth, resulting in a more robust multi-lineage haematopoiesis. They also supported maintenance of the stem cell pool of primitive HSCs with superior subsequent *in vivo* engraftment potential, due to increased migration [130]. Therefore, this system could perhaps be used for drug-discovery for HSC mobilization agents.

The use of 3D fibrin scaffolds in combination with BMSCs and cytokine supplementation has been shown to be superior in the maintenance of primitive HSC phenotype and long-term engraftment in comparison with polymeric:

polycaprolactone (PCL), poly(lactic-co-glycolic acid) (PLGA) and collagen scaffolds [45]. Cuddihy et al. provide a comparative study of three different hydrogel 3D systems for replication of the bone marrow niche in maintaining the quiescent state of HSCs, including Matrigel, Puramatrix and inverted colloidal crystals (ICC). The latter provided the best microenvironment for this purpose, underscoring the advantage of non-encapsulating pre-set hydrogels [27]. ICCs are organized structures similar to hexagonally packed lattices of spheres, but with the spheres replaced by cavities, while the interstitial spaces are filled. This open geometry and high porosity, as well as full interconnectivity, is similar to the 3D morphology of supporting bone marrow tissue in a trabecular bone. However, the production of inverted colloidal crystals can be technically challenging, especially in a layer-by-layer surface modification approach [103]. In a different study Eng et al. used microsized 3D gelatin methacrylate (GelMA) hydrogels, shape-coded for their biological and physical properties, encapsulating HUVECs and mesenchymal stem cells and molecules, docked onto shape-matching hydrogel templates. This technique was designed to resemble the biology of bone marrow tissue, composed of cellular and molecular "building blocks" that cooperate to provide tissue-specific functions in forming unique spatial gradients to test cellular migration in stem cell differentiation and tumour metastasis [41]. More recent efforts have sought to incorporate microfluidic mixing of collagen hydrogels to create opposing gradients of multiple cell populations (for example osteoblasts and HSCs) in an easy-to-analyse 3D platform [85]. Also macroporous poly(ethylene glycol) diacrylate (PEGDA) hydrogels biofunctionalised with Arg-Gly-Asp (RGD) peptide have been used for MSC and HSC co-culture, enabling effective preservation of the hematopoietic progenitors' stemness [119]. Another approach by Metzger et al. used 3D poly(ethylene glycol) (PEG) matrices for localised predefined docking of BMP-2 and subsequent seeding with BMSCs, to promote spatially specific differentiation areas. Interestingly, this approach can be tuneable, as the bioactive molecule linking system allows strong interactions with the matrix, thus providing a tightly controlled effect on differentiation of cells in predefined areas, while preventing the influence on other tissue milieux [93]. This tool could be further adapted to differentially localise various cell-instructive growth factors to better recapitulate the architectures of native tissues. Also, silk scaffolds were used to culture BMSCs or HUVECs with a multiple myeloma cell line to model cancer growth within the bone [121]. This platform demonstrated myeloma support of capillary-like assembly of endothelial cells and cell adhesion-mediated drug resistance, and can be further advanced to screen anti-cancer compounds. In another approach, an enzymatically-degradable hydrogel-based 3D culture platform engineered to allow live-cell retrieval was used in order to investigate the interactions between BMSCs, osteoblasts, and adipocytes under hyperglycaemic conditions in a model of diabetes-related osteoporosis [123].

4.5 Combined 3D Culture Systems

Stromal stem cell-derived ECM deposited inside a collagen/hydroxyapatite (Col/HA) scaffold, which exhibits properties similar to those of trabecular bone (particularly with regards to tissue architecture and composition) have been employed to study the importance of stromal ECM-cell interactions [2]. This approach promoted proliferation of BMSCs and preserved their in vitro differentiation capacity to a higher degree than a 3D culture in a Col/HA scaffold without the inclusion of stromal ECM. However, after in vivo implantation of these scaffolds, it became clear that the preparations with ECM resulted in significantly less mineralized tissue formation than Col/HA without ECM. Conversely, culture of BMSCs on highly porous Col/HA scaffolds alone has been shown to induce sufficient osteogenic differentiation, while a "softer" collagen type I-incorporating scaffold with smaller porosity, and therefore lower oxygen tension, promoted chondrogenic differentiation [31]. HA ceramic scaffolds, suspended in a microbioreactor, have also been shown to be highly osteoinductive [13]. Together, this indicates that culture systems involving ECM preserve the quiescence of mesenchymal stem cells, much like a true stem cell niche, whereas scaffolds relying on higher stiffness and nutrient and oxygen diffusion promote BMSC differentiation. The latter could be of more use in regenerative medicine strategies for streamlining manufacturing of osteoinductive grafts, whereas the former recapitulate better the niche in vitro, and hence represent a better tool for studying biological interactions.

4.6 Microfluidics

In addition to considerations of the 3D structural composition of the bone marrow niche, it is also crucial to consider the fluidic component of this tissue. In this context, microfluidic devices provide the advantage of precise control in space and time of very small amounts of fluids, down to the attolitre level (10^{-18} litres) [159]. This results in the fine regulation of the culture conditions, enabling the investigation of tissue dynamics and development in a more cost-effective way [22]. For example, a 3D microfluidic device incorporating HUVECs embedded in fibrin gels and BMSCs was used to study the formation of a stable vascular network with successful perivascular recapitulation of MSC localisation, such as that observed in the bone marrow niche [16]. A different group successfully built a bioreactor which could reproduce the bone marrow microenvironment for platelet production from $CD34^+$ cell-derived megakaryocytes (MKs). This was done by employing a collagen type I hydrogel to represent the "osteoblastic" niche, silk microtubes coated with von Willebrand Factor and fibrinogen to represent the "vascular" niche, and medium flow applied within the silk microtubes to mimic blood flow [108]. Migration of the MKs from the osteoblastic to vascular niche in this 3D system was observed, and functional platelets were produced in the "blood flow" mimicking niche. This 3D niche-recapitulating bioreactor provided insight into platelet production mechanisms and generated functional platelets ex vivo for clinical use. Another bioreactor design using innovative microgravity technology (to minimise shear forces and turbulence) for a dynamic ex vivo tissue culture allowed study of the vascular compartment of the multiple myeloma niche, which plays a critical role in the progression and response to therapy of this cancer [44]. This bioreactor also proved suitable for long-term culture of healthy bone marrow explants. A microfluidics approach has also been used to study multiple myeloma biology, in a 3D ossified tissue system [166]. This culture system consisted of components of the bone/bone marrow microenvironment such as the endosteum-resembling surface (i.e., 3D osteoblasts) and their secreted ECM, as well as growth factors and cytokines from the multiple myeloma patient's plasma, and other putative patient-derived bone marrow stromal cells, all within a microperfused environment. This approach was able to optimise the maintenance of otherwise difficult-to-grow cancer cells, and is a step towards studying the mechanisms responsible for drug resistance and relapse for the development of personalised therapeutics. More recently, another microfluidic 3D HA ceramic-based scaffold system showed the advantage of this approach for freshly-isolated BMSC cultivation [109]. Thanks to ongoing perfusion, the BMSCs preserved better their early progenitor properties in terms of a higher clonogenicity and a superior multilineage differentiation capacity, and displayed reduced inter-donor variability and consistent upregulation of multipotency-related pathways, as assessed by transcriptomic analysis.

4.7 Future Approaches

Stem cell niches are complex and dynamic systems, therefore fully recapitulating the microenvironment with all its interactions and tightly regulated signalling events is not an easy task. Recreating all of the key factors crucial for formation of the bone marrow niche would also involve extensive understanding of different cell types and their interactions, both with each other and with their extracellular matrix. The spatially confined soluble and immobilised signalling molecules, ionic calcium, 3D mechanical forces and systemically regulated metabolite and oxygen gradients should also be considered. The provisional ECM constructs should have defined degradability, stiffness and adhesive properties, all closely resembling the bone marrow ECM. Therefore, reconstituting the in vivo stem cell microenvironment requires understanding of mesenchymal and haematopoietic stem cell biology, which is still not fully complete, as well as the ability to implement the desired 3D architectures, with appropriate physicochemical and biological cues.

Fortunately, advances in stem cell research as well as in the material science field are increasing, bringing a more complete in vitro reconstruction of stem cell niches within closer reach. Solutions are emerging at the interface between biology and engineering, such as advanced multicomponent biomaterials combined with cellular probes and physicochemical manipulation to study the nature of the bone marrow niche in a dynamic setting. These dynamic microenvironments are incorporating so-called four-dimensional (4D) biology: cell-laden matrices engineered to recapitulate tissue and organ function in a 3D space over time [146]. Thanks to novel bio-click reactions developed to add or remove biomolecules in a spatial and temporal manner, these biomimetic matrices can be locally and dynamically modified.

In order to achieve their full biological potential, the scaffold-microbioreactor systems should serve as an in vitro mimic of the milieu of development, regeneration, or disease as required. With the capability to generate spatial gradients of regulatory signals, to subject cells to dynamic changes in their environment, and to offer insight into cellular responses in real time, these new technologies are setting the stage for an entirely new approach to stem cell research.

5 General Conclusions and Challenges Ahead

Cutting-edge stem cell bioengineering platforms can provide strict control of the culture environment along with regulation of signalling molecules, oxygen tension and shear stress thereby enabling the study of tissue development, regeneration and disease, under conditions that can predict the human in vivo context. Unfortunately, to date human stem and progenitor cells are mostly studied in 2D plastic cultures, which lack the structural and signalling blueprints of native tissues. It is difficult to study the molecular regulatory factors in a spatiotemporal manner, and impossible to study the mechanical forces and systemic factors provided by blood circulation. Animal models to study human cell interactions are not entirely representative of the interactions taking place in human tissues. What is more, there is limited control and insight into the experiment itself and analysis of the outcomes is more complicated and easily confounded. Therefore, these studies often fail to predict the outcomes of human clinical studies, increasing time and cost, and decreasing the effectiveness of any translational therapeutic strategies. A better understanding of how different tissues develop and how different cell types relate to each other, for example by using legitimate and verified genetic lineage tracing models in mice, may facilitate more precise definition of the 3D architectural aspects of organ development and repair. To reconstruct tissues ex vivo, detailed information on cell position and reaction to particular conditions is needed. Hence, stem cell research in the coming decades will focus on interdisciplinary stem cell bioengineering as a new regenerative medicine approach, enabling precise study of the complex biology of tissues and progressing it towards translational applications.

References

1. Adams GB, Chabner KT, Alley IR, Olson DP, Szczepiorkowski ZM, Poznansky MC, et al. Stem cell engraftment at the endosteal niche is specified by the calcium-sensing receptor. Nature. 2006;439(7076):599–603.
2. Antebi B, Zhang Z, Wang Y, Lu Z, Chen XD, Ling J. Stromal-cell-derived extracellular matrix promotes the proliferation and retains the osteogenic differentiation capacity of mesenchymal stem cells on three-dimensional scaffolds. Tissue Eng Part C Methods. 2015;21(2):171–81.
3. Ara T, Tokoyoda K, Sugiyama T, Egawa T, Kawabata K, Nagasawa T. Long-term hematopoietic stem cells require stromal cell-derived factor-1 for colonizing bone marrow during ontogeny. Immunity. 2003;19(2):257–67.
4. Arai F, Hirao A, Ohmura M, Sato H, Matsuoka S, Takubo K, et al. Tie2/angiopoietin-1 signaling regulates hematopoietic stem cell quiescence in the bone marrow niche. Cell. 2004;118(2):149–61.
5. Arnsdorf EJ, Tummala P, Kwon RY, Jacobs CR. Mechanically induced osteogenic differentiation–the role of RhoA, ROCKII and cytoskeletal dynamics. J Cell Sci. 2009;122(Pt 4):546–53.
6. Banfi A, Muraglia A, Dozin B, Mastrogiacomo M, Cancedda R, Quarto R. Proliferation kinetics and differentiation potential of ex vivo expanded human bone marrow stromal cells: implications for their use in cell therapy. Exp Hematol. 2000;28(6):707–15.
7. Barcellos-de-Souza P, Gori V, Bambi F, Chiarugi P. Tumor microenvironment: bone marrow-mesenchymal stem cells as key players. Biochim Biophys Acta. 2013;1836(2):321–35.
8. Barker JE. Sl/Sld hematopoietic progenitors are deficient in situ. Exp Hematol. 1994;22 (2):174–7.
9. Barker N, van Es JH, Kuipers J, Kujala P, van den Born M, Cozijnsen M, et al. Identification of stem cells in small intestine and colon by marker gene Lgr5. Nature. 2007;449 (7165):1003–7.
10. Baron R, Kneissel M. WNT signaling in bone homeostasis and disease: from human mutations to treatments. Nat Med. 2013;19(2):179–92.
11. Bellantuono I, Aldahmash A, Kassem M. Aging of marrow stromal (skeletal) stem cells and their contribution to age-related bone loss. Biochim Biophys Acta. 2009;1792(4):364–70.
12. Bianco P, Cao X, Frenette PS, Mao JJ, Robey PG, Simmons PJ, et al. The meaning, the sense and the significance: translating the science of mesenchymal stem cells into medicine. Nat Med. 2013;19(1):35–42.
13. Braccini A, Wendt D, Jaquiery C, Jakob M, Heberer M, Kenins L, et al. Three-dimensional perfusion culture of human bone marrow cells and generation of osteoinductive grafts. Stem Cells. 2005;23(8):1066–72.
14. Brigger I, Dubernet C, Couvreur P. Nanoparticles in cancer therapy and diagnosis. Adv Drug Deliv Rev. 2002;54(5):631–51.
15. Calvi LM, Adams GB, Weibrecht KW, Weber JM, Olson DP, Knight MC, et al. Osteoblastic cells regulate the haematopoietic stem cell niche. Nature. 2003;425(6960):841–6.
16. Carrion B, Huang CP, Ghajar CM, Kachgal S, Kniazeva E, Jeon NL, et al. Recreating the perivascular niche ex vivo using a microfluidic approach. Biotechnol Bioeng. 2010;107 (6):1020–8.
17. Case N, Sen B, Thomas JA, Styner M, Xie Z, Jacobs CR, et al. Steady and oscillatory fluid flows produce a similar osteogenic phenotype. Calcif Tissue Int. 2011;88(3):189–97.
18. Castillo AB, Jacobs CR. Mesenchymal stem cell mechanobiology. Curr Osteoporos Rep. 2010;8(2):98–104.
19. Chan CK, Lindau P, Jiang W, Chen JY, Zhang LF, Chen CC, et al. Clonal precursor of bone, cartilage, and hematopoietic niche stromal cells. Proc Natl Acad Sci U S A. 2013;110 (31):12643–8.

20. Chen XD, Shi S, Xu T, Robey PG, Young MF. Age-related osteoporosis in biglycan-deficient mice is related to defects in bone marrow stromal cells. J Bone Miner Res. 2002;17(2):331–40.
21. Chen XD, Dusevich V, Feng JQ, Manolagas SC, Jilka RL. Extracellular matrix made by bone marrow cells facilitates expansion of marrow-derived mesenchymal progenitor cells and prevents their differentiation into osteoblasts. J Bone Miner Res. 2007;22(12):1943–56.
22. Choi NW, Cabodi M, Held B, Gleghorn JP, Bonassar LJ, Stroock AD. Microfluidic scaffolds for tissue engineering. Nat Mater. 2007;6(11):908–15.
23. Chow A, Lucas D, Hidalgo A, Mendez-Ferrer S, Hashimoto D, Scheiermann C, et al. Bone marrow CD169+ macrophages promote the retention of hematopoietic stem and progenitor cells in the mesenchymal stem cell niche. J Exp Med. 2011;208(2):261–71.
24. Claycombe K, King LE, Fraker PJ. A role for leptin in sustaining lymphopoiesis and myelopoiesis. Proc Natl Acad Sci U S A. 2008;105(6):2017–21.
25. Cohen A, Dempster DW, Stein EM, Nickolas TL, Zhou H, McMahon DJ, et al. Increased marrow adiposity in premenopausal women with idiopathic osteoporosis. J Clin Endocrinol Metab. 2012;97(8):2782–91.
26. Crisan M, Deasy B, Gavina M, Zheng B, Huard J, Lazzari L, et al. Purification and long-term culture of multipotent progenitor cells affiliated with the walls of human blood vessels: myoendothelial cells and pericytes. Methods Cell Biol. 2008;86:295–309.
27. Cuddihy MJ, Wang Y, Machi C, Bahng JH, Kotov NA. Replication of bone marrow differentiation niche: comparative evaluation of different three-dimensional matrices. Small. 2013;9(7):1008–15.
28. da Silva Meirelles L, Caplan AI, Nardi NB. In search of the in vivo identity of mesenchymal stem cells. Stem Cells. 2008;26(9):2287–99.
29. Dalby MJ, Gadegaard N, Tare R, Andar A, Riehle MO, Herzyk P, et al. The control of human mesenchymal cell differentiation using nanoscale symmetry and disorder. Nat Mater. 2007;6(12):997–1003.
30. Dalby MJ, Gadegaard N, Oreffo RO. Harnessing nanotopography and integrin-matrix interactions to influence stem cell fate. Nat Mater. 2014;13(6):558–69.
31. Dawson JI, Wahl DA, Lanham SA, Kanczler JM, Czernuszka JT, Oreffo RO. Development of specific collagen scaffolds to support the osteogenic and chondrogenic differentiation of human bone marrow stromal cells. Biomaterials. 2008;29(21):3105–16.
32. de Barros AP, Takiya CM, Garzoni LR, Leal-Ferreira ML, Dutra HS, Chiarini LB, et al. Osteoblasts and bone marrow mesenchymal stromal cells control hematopoietic stem cell migration and proliferation in 3D in vitro model. PLoS ONE. 2010;5(2):e9093.
33. Denhardt DT, Noda M. Osteopontin expression and function: role in bone remodeling. J Cell Biochem Suppl. 1998;30–31:92–102.
34. Dexter TM, Wright EG, Krizsa F, Lajtha LG. Regulation of haemopoietic stem cell proliferation in long term bone marrow cultures. Biomedicine. 1977;27(9–10):344–9.
35. Di Maggio N, Piccinini E, Jaworski M, Trumpp A, Wendt DJ, Martin I. Toward modeling the bone marrow niche using scaffold-based 3D culture systems. Biomaterials. 2011;32(2):321–9.
36. Ding L, Saunders TL, Enikolopov G, Morrison SJ. Endothelial and perivascular cells maintain haematopoietic stem cells. Nature. 2012;481(7382):457–62.
37. Ding L, Morrison SJ. Haematopoietic stem cells and early lymphoid progenitors occupy distinct bone marrow niches. Nature. 2013;495(7440):231–5.
38. Doetsch F, Caille I, Lim DA, Garcia-Verdugo JM, Alvarez-Buylla A. Subventricular zone astrocytes are neural stem cells in the adult mammalian brain. Cell. 1999;97(6):703–16.
39. Dominici M, Le Blanc K, Mueller I, Slaper-Cortenbach I, Marini F, Krause D, et al. Minimal criteria for defining multipotent mesenchymal stromal cells. The international society for cellular therapy position statement. Cytotherapy. 2006;8(4):315–7.
40. Ehninger A, Trumpp A. The bone marrow stem cell niche grows up: mesenchymal stem cells and macrophages move in. J Exp Med. 2011;208(3):421–8.

41. Eng G, Lee BW, Parsa H, Chin CD, Schneider J, Linkov G, et al. Assembly of complex cell microenvironments using geometrically docked hydrogel shapes. Proc Natl Acad Sci U S A. 2013;110(12):4551–6.
42. Engler AJ, Sen S, Sweeney HL, Discher DE. Matrix elasticity directs stem cell lineage specification. Cell. 2006;126(4):677–89.
43. Etheridge SL, Spencer GJ, Heath DJ, Genever PG. Expression profiling and functional analysis of wnt signaling mechanisms in mesenchymal stem cells. Stem Cells. 2004;22 (5):849–60.
44. Ferrarini M, Steimberg N, Ponzoni M, Belloni D, Berenzi A, Girlanda S, et al. Ex-vivo dynamic 3-D culture of human tissues in the RCCS bioreactor allows the study of multiple myeloma biology and response to therapy. PLoS ONE. 2013;8(8):e71613.
45. Ferreira MS, Jahnen-Dechent W, Labude N, Bovi M, Hieronymus T, Zenke M, et al. Cord blood-hematopoietic stem cell expansion in 3D fibrin scaffolds with stromal support. Biomaterials. 2012;33(29):6987–97.
46. Fleming HE, Janzen V, Lo Celso C, Guo J, Leahy KM, Kronenberg HM, et al. Wnt signaling in the niche enforces hematopoietic stem cell quiescence and is necessary to preserve self-renewal in vivo. Cell Stem Cell. 2008;2(3):274–83.
47. Florian MC, Nattamai KJ, Dorr K, Marka G, Uberle B, Vas V, et al. (2013) A canonical to non-canonical Wnt signalling switch in haematopoietic stem-cell ageing. Nature.
48. Friedenstein AJ, Chailakhjan RK, Lalykina KS. The development of fibroblast colonies in monolayer cultures of guinea-pig bone marrow and spleen cells. Cell Tissue Kinet. 1970;3 (4):393–403.
49. Fuchs E, Tumbar T, Guasch G. Socializing with the neighbors: stem cells and their niche. Cell. 2004;116(6):769–78.
50. Fuchs E, Chen T. A matter of life and death: self-renewal in stem cells. EMBO Rep. 2013;14 (1):39–48.
51. Gattazzo F, Urciuolo A, Bonaldo P. Extracellular matrix: a dynamic microenvironment for stem cell niche. Biochim Biophys Acta. 2014;1840(8):2506–19.
52. Ghia P, Granziero L, Chilosi M, Caligaris-Cappio F. Chronic B cell malignancies and bone marrow microenvironment. Semin Cancer Biol. 2002;12(2):149–55.
53. Gordon MY, Lewis JL, Marley SB. Of mice and men…and elephants. Blood. 2002;100 (13):4679–80.
54. Greenbaum A, Hsu YM, Day RB, Schuettpelz LG, Christopher MJ, Borgerding JN, et al. CXCL12 in early mesenchymal progenitors is required for haematopoietic stem-cell maintenance. Nature. 2013;495(7440):227–30.
55. Greenbaum AM, Revollo LD, Woloszynek JR, Civitelli R, Link DC. N-cadherin in osteolineage cells is not required for maintenance of hematopoietic stem cells. Blood. 2012;120(2):295–302.
56. Guerriero A, Worford L, Holland HK, Guo GR, Sheehan K, Waller EK. Thrombopoietin is synthesized by bone marrow stromal cells. Blood. 1997;90(9):3444–55.
57. Guo G, Luc S, Marco E, Lin TW, Peng C, Kerenyi MA, et al. Mapping cellular hierarchy by single-cell analysis of the cell surface repertoire. Cell Stem Cell. 2013;13(4):492–505.
58. Gurkan UA, Akkus O. The mechanical environment of bone marrow: a review. Ann Biomed Eng. 2008;36(12):1978–91.
59. Isern J, Martin-Antonio B, Ghazanfari R, Martin AM, Lopez JA, del Toro R, et al. Self-renewing human bone marrow mesenspheres promote hematopoietic stem cell expansion. Cell Rep. 2013;3(5):1714–24.
60. Jones DL, Wagers AJ. No place like home: anatomy and function of the stem cell niche. Nat Rev Mol Cell Biol. 2008;9(1):11–21.
61. Joseph C, Quach JM, Walkley CR, Lane SW. Lo Celso C, Purton LE. Deciphering hematopoietic stem cells in their niches: a critical appraisal of genetic models, lineage tracing, and imaging strategies. Cell Stem Cell. 2013;13(5):520–33.

62. Katayama Y, Battista M, Kao WM, Hidalgo A, Peired AJ, Thomas SA, et al. Signals from the sympathetic nervous system regulate hematopoietic stem cell egress from bone marrow. Cell. 2006;124(2):407–21.
63. Kiel MJ, Yilmaz OH, Iwashita T, Yilmaz OH, Terhorst C, Morrison SJ. SLAM family receptors distinguish hematopoietic stem and progenitor cells and reveal endothelial niches for stem cells. Cell. 2005;121(7):1109–21.
64. Kiel MJ, Morrison SJ. Maintaining hematopoietic stem cells in the vascular niche. Immunity. 2006;25(6):862–4.
65. Kiel MJ, Morrison SJ. Uncertainty in the niches that maintain haematopoietic stem cells. Nat Rev Immunol. 2008;8(4):290–301.
66. Kiel MJ, Acar M, Radice GL, Morrison SJ. Hematopoietic stem cells do not depend on N-cadherin to regulate their maintenance. Cell Stem Cell. 2009;4(2):170–9.
67. Kimble JE, White JG. On the control of germ cell development in Caenorhabditis elegans. Dev Biol. 1981;81(2):208–19.
68. Kimura S, Roberts AW, Metcalf D, Alexander WS. Hematopoietic stem cell deficiencies in mice lacking c-Mpl, the receptor for thrombopoietin. Proc Natl Acad Sci U S A. 1998;95 (3):1195–200.
69. Kingham E, White K, Gadegaard N, Dalby MJ, Oreffo RO. Nanotopographical cues augment mesenchymal differentiation of human embryonic stem cells. Small. 2013;9(12):2140–51.
70. Klein G. The extracellular matrix of the hematopoietic microenvironment. Experientia. 1995;51(9–10):914–26.
71. Kollet O, Dar A, Shivtiel S, Kalinkovich A, Lapid K, Sztainberg Y, et al. Osteoclasts degrade endosteal components and promote mobilization of hematopoietic progenitor cells. Nat Med. 2006;12(6):657–64.
72. Konopleva MY, Jordan CT. Leukemia stem cells and microenvironment: biology and therapeutic targeting. J Clin Oncol. 2011;29(5):591–9.
73. Kunisaki Y, Bruns I, Scheiermann C, Ahmed J, Pinho S, Zhang D, et al. Arteriolar niches maintain haematopoietic stem cell quiescence. Nature. 2013;502(7473):637–43.
74. Lai Y, Sun Y, Skinner CM, Son EL, Lu Z, Tuan RS, et al. Reconstitution of marrow-derived extracellular matrix ex vivo: a robust culture system for expanding large-scale highly functional human mesenchymal stem cells. Stem Cells Devel. 2010;19(7):1095–107.
75. Lander AD, Kimble J, Clevers H, Fuchs E, Montarras D, Buckingham M, et al. What does the concept of the stem cell niche really mean today? BMC Biol. 2012;10:19.
76. Leisten I, Kramann R, Ventura Ferreira MS, Bovi M, Neuss S, Ziegler P, et al. 3D co-culture of hematopoietic stem and progenitor cells and mesenchymal stem cells in collagen scaffolds as a model of the hematopoietic niche. Biomaterials. 2012;33(6):1736–47.
77. Li L, Clevers H. Coexistence of quiescent and active adult stem cells in mammals. Science. 2010;327(5965):542–5.
78. Lin CC, Anseth KS. PEG hydrogels for the controlled release of biomolecules in regenerative medicine. Pharm Res. 2009;26(3):631–43.
79. Ling L, Nurcombe V, Cool SM. Wnt signaling controls the fate of mesenchymal stem cells. Gene. 2009;433(1–2):1–7.
80. Lo Celso C, Fleming HE, Wu JW, Zhao CX, Miake-Lye S, Fujisaki J, et al. Live-animal tracking of individual haematopoietic stem/progenitor cells in their niche. Nature. 2009;457 (7225):92–96.
81. Lord BI, Hendry JH. The distribution of haemopoietic colony-forming units in the mouse femur, and its modification by x rays. Br J Radiol. 1972;45(530):110–5.
82. Lu Y, Gong Y, Lian J, Wang L, Kretlow JD, Zhou G, et al. Expansion of endothelial progenitor cells in high density dot culture of rat bone marrow cells. PLoS ONE. 2014;9(9): e107127.
83. Luis TC, Naber BA, Roozen PP, Brugman MH, de Haas EF, Ghazvini M, et al. Canonical wnt signaling regulates hematopoiesis in a dosage-dependent fashion. Cell Stem Cell. 2011;9 (4):345–56.

84. Luu YK, Pessin JE, Judex S, Rubin J, Rubin CT. Mechanical signals as a non-invasive means to influence mesenchymal stem cell fate, promoting bone and suppressing the fat phenotype. Bonekey Osteovision. 2009;6(4):132–49.
85. Mahadik BP, Wheeler TD, Skertich LJ, Kenis PJ, Harley BA. Microfluidic generation of gradient hydrogels to modulate hematopoietic stem cell culture environment. Adv Healthc Mater. 2014;3(3):449–58.
86. Mann AP, Tanaka T, Somasunderam A, Liu X, Gorenstein DG, Ferrari M. E-selectin-targeted porous silicon particle for nanoparticle delivery to the bone marrow. Adv Mater. 2011;23(36):H278–82.
87. Mata A, Kim EJ, Boehm CA, Fleischman AJ, Muschler GF, Roy S. A three-dimensional scaffold with precise micro-architecture and surface micro-textures. Biomaterials. 2009;30(27):4610–7.
88. McMurray RJ, Gadegaard N, Tsimbouri PM, Burgess KV, McNamara LE, Tare R, et al. Nanoscale surfaces for the long-term maintenance of mesenchymal stem cell phenotype and multipotency. Nat Mater. 2011;10(8):637–44.
89. Mendelson A, Frenette PS. Hematopoietic stem cell niche maintenance during homeostasis and regeneration. Nat Med. 2014;20(8):833–46.
90. Mendez-Ferrer S, Lucas D, Battista M, Frenette PS. Haematopoietic stem cell release is regulated by circadian oscillations. Nature. 2008;452(7186):442–7.
91. Mendez-Ferrer S, Michurina TV, Ferraro F, Mazloom AR, Macarthur BD, Lira SA, et al. Mesenchymal and haematopoietic stem cells form a unique bone marrow niche. Nature. 2010;466(7308):829–34.
92. Mendez-Ferrer S, Scadden DT, Sanchez-Aguilera A. Bone marrow stem cells: current and emerging concepts. Ann N Y Acad Sci. 2015;1335(1):32–44.
93. Metzger S, Lienemann PS, Ghayor C, Weber W, Martin I, Weber FE, et al. Modular poly(ethylene glycol) matrices for the controlled 3D-localized osteogenic differentiation of mesenchymal stem cells. Adv Healthc Mater. 2014;4(4):550–8.
94. Mizoguchi T, Pinho S, Ahmed J, Kunisaki Y, Hanoun M, Mendelson A, et al. Osterix marks distinct waves of primitive and definitive stromal progenitors during bone marrow development. Dev Cell. 2014;29(3):340–9.
95. Moghimi SM. Exploiting bone marrow microvascular structure for drug delivery and future therapies. Adv Drug Deliv Rev. 1995;17:61–73.
96. Mohyeldin A, Garzon-Muvdi T, Quinones-Hinojosa A. Oxygen in stem cell biology: a critical component of the stem cell niche. Cell Stem Cell. 2010;7(2):150–61.
97. Moore KA, Lemischka IR. Stem cells and their niches. Science. 2006;311(5769):1880–5.
98. Morrison SJ, Scadden DT. The bone marrow niche for haematopoietic stem cells. Nature. 2014;505(7483):327–34.
99. Nagasawa T, Hirota S, Tachibana K, Takakura N, Nishikawa S, Kitamura Y, et al. Defects of B-cell lymphopoiesis and bone-marrow myelopoiesis in mice lacking the CXC chemokine PBSF/SDF-1. Nature. 1996;382(6592):635–8.
100. Nakamura-Ishizu A, Okuno Y, Omatsu Y, Okabe K, Morimoto J, Uede T, et al. Extracellular matrix protein tenascin-C is required in the bone marrow microenvironment primed for hematopoietic regeneration. Blood. 2012;119(23):5429–37.
101. Naveiras O, Nardi V, Wenzel PL, Hauschka PV, Fahey F, Daley GQ. Bone-marrow adipocytes as negative regulators of the haematopoietic microenvironment. Nature. 2009;460(7252):259–63.
102. Ng CP, Sharif AR, Heath DE, Chow JW, Zhang CB, Chan-Park MB, et al. Enhanced ex vivo expansion of adult mesenchymal stem cells by fetal mesenchymal stem cell ECM. Biomaterials. 2014;35(13):4046–57.
103. Nichols JE, Cortiella J, Lee J, Niles JA, Cuddihy M, Wang S, et al. In vitro analog of human bone marrow from 3D scaffolds with biomimetic inverted colloidal crystal geometry. Biomaterials. 2009;30(6):1071–9.

104. Nilsson SK, Johnston HM, Whitty GA, Williams B, Webb RJ, Denhardt DT, et al. Osteopontin, a key component of the hematopoietic stem cell niche and regulator of primitive hematopoietic progenitor cells. Blood. 2005;106(4):1232–9.
105. Nombela-Arrieta C, Pivarnik G, Winkel B, Canty KJ, Harley B, Mahoney JE, et al. Quantitative imaging of haematopoietic stem and progenitor cell localization and hypoxic status in the bone marrow microenvironment. Nat Cell Biol. 2013;15(5):533–43.
106. Olson TS, Caselli A, Otsuru S, Hofmann TJ, Williams R, Paolucci P, et al. Megakaryocytes promote murine osteoblastic HSC niche expansion and stem cell engraftment after radioablative conditioning. Blood. 2013;121(26):5238–49.
107. Owen M, Friedenstein AJ. Stromal stem cells: marrow-derived osteogenic precursors. Ciba Found Symp. 1988;136:42–60.
108. Pallotta I, Lovett M, Kaplan DL, Balduini A. Three-dimensional system for the in vitro study of megakaryocytes and functional platelet production using silk-based vascular tubes. Tissue Eng Part C Methods. 2011;17(12):1223–32.
109. Papadimitropoulos A, Piccinini E, Brachat S, Braccini A, Wendt D, Barbero A, et al. Expansion of human mesenchymal stromal cells from fresh bone marrow in a 3D scaffold-based system under direct perfusion. PLoS ONE. 2014;9(7):e102359.
110. Papayannopoulou T, Scadden DT. Stem-cell ecology and stem cells in motion. Blood. 2008;111(8):3923–30.
111. Park D, Spencer JA, Koh BI, Kobayashi T, Fujisaki J, Clemens TL, et al. Endogenous bone marrow MSCs are dynamic, fate-restricted participants in bone maintenance and regeneration. Cell Stem Cell. 2012;10(3):259–72.
112. Parmar K, Mauch P, Vergilio JA, Sackstein R, Down JD. Distribution of hematopoietic stem cells in the bone marrow according to regional hypoxia. Proc Natl Acad Sci U S A. 2007;104 (13):5431–6.
113. Pittenger MF, Mackay AM, Beck SC, Jaiswal RK, Douglas R, Mosca JD, et al. Multilineage potential of adult human mesenchymal stem cells. Science. 1999;284(5411):143–7.
114. Plikus MV, Baker RE, Chen CC, Fare C, de la Cruz D, Andl T, et al. Self-organizing and stochastic behaviors during the regeneration of hair stem cells. Science. 2011;332 (6029):586–9.
115. Potier E, Noailly J, Ito K. Directing bone marrow-derived stromal cell function with mechanics. J Biomech. 2010;43(5):807–17.
116. Puri MC, Bernstein A. Requirement for the TIE family of receptor tyrosine kinases in adult but not fetal hematopoiesis. Proc Natl Acad Sci U S A. 2003;100(22):12753–8.
117. Qian H, Buza-Vidas N, Hyland CD, Jensen CT, Antonchuk J, Mansson R, et al. Critical role of thrombopoietin in maintaining adult quiescent hematopoietic stem cells. Cell Stem Cell. 2007;1(6):671–84.
118. Quarto R, Thomas D, Liang CT. Bone progenitor cell deficits and the age-associated decline in bone repair capacity. Calcif Tissue Int. 1995;56(2):123–9.
119. Raic A, Rodling L, Kalbacher H, Lee-Thedieck C. Biomimetic macroporous PEG hydrogels as 3D scaffolds for the multiplication of human hematopoietic stem and progenitor cells. Biomaterials. 2014;35(3):929–40.
120. Rattis FM, Voermans C, Reya T. Wnt signaling in the stem cell niche. Curr Opin Hematol. 2004;11(2):88–94.
121. Reagan MR, Mishima Y, Glavey SV, Zhang Y, Manier S, Lu ZN, et al. Investigating osteogenic differentiation in multiple myeloma using a novel 3D bone marrow niche model. Blood. 2014;124(22):3250–9.
122. Reya T, Clevers H. Wnt signalling in stem cells and cancer. Nature. 2005;434(7035):843–50.
123. Rinker TE, Hammoudi TM, Kemp ML, Lu H, Temenoff JS. Interactions between mesenchymal stem cells, adipocytes, and osteoblasts in a 3D tri-culture model of hyperglycemic conditions in the bone marrow microenvironment. Integr Biol (Camb). 2014;6(3):324–37.

124. Rosland GV, Svendsen A, Torsvik A, Sobala E, McCormack E, Immervoll H, et al. Long-term cultures of bone marrow-derived human mesenchymal stem cells frequently undergo spontaneous malignant transformation. Cancer Res. 2009;69(13):5331–9.
125. Sacchetti B, Funari A, Michienzi S, Di Cesare S, Piersanti S, Saggio I, et al. Self-renewing osteoprogenitors in bone marrow sinusoids can organize a hematopoietic microenvironment. Cell. 2007;131(2):324–36.
126. Saleh FA, Whyte M, Genever PG. Effects of endothelial cells on human mesenchymal stem cell activity in a three-dimensional in vitro model. Eur Cell Mater. 2011;22:242–57; Discussion 57.
127. Scadden DT. Nice neighborhood: emerging concepts of the stem cell niche. Cell. 2014;157 (1):41–50.
128. Schettini DA, Ribeiro RR, Demicheli C, Rocha OG, Melo MN, Michalick MS, et al. Improved targeting of antimony to the bone marrow of dogs using liposomes of reduced size. Int J Pharm. 2006;315(1–2):140–7.
129. Schofield R. The relationship between the spleen colony-forming cell and the haemopoietic stem cell. Blood Cells. 1978;4(1–2):7–25.
130. Sharma MB, Limaye LS, Kale VP. Mimicking the functional hematopoietic stem cell niche in vitro: recapitulation of marrow physiology by hydrogel-based three-dimensional cultures of mesenchymal stromal cells. Haematologica. 2012;97(5):651–60.
131. Shi S, Gronthos S. Perivascular niche of postnatal mesenchymal stem cells in human bone marrow and dental pulp. J Bone Miner Res. 2003;18(4):696–704.
132. Sou K, Goins B, Takeoka S, Tsuchida E, Phillips WT. Selective uptake of surface-modified phospholipid vesicles by bone marrow macrophages in vivo. Biomaterials. 2007;28 (16):2655–66.
133. Sou K, Goins B, Oyajobi BO, Travi BL, Phillips WT. Bone marrow-targeted liposomal carriers. Expert Opin Drug Deliv. 2011;8(3):317–28.
134. Spencer JA, Ferraro F, Roussakis E, Klein A, Wu J, Runnels JM, et al. Direct measurement of local oxygen concentration in the bone marrow of live animals. Nature. 2014;508 (7495):269–73.
135. Steiner D, Lampert F, Stark GB, Finkenzeller G. Effects of endothelial cells on proliferation and survival of human mesenchymal stem cells and primary osteoblasts. J Orthop Res. 2012;30(10):1682–9.
136. Stenderup K, Justesen J, Clausen C, Kassem M. Aging is associated with decreased maximal life span and accelerated senescence of bone marrow stromal cells. Bone. 2003;33(6):919–26.
137. Stier S, Ko Y, Forkert R, Lutz C, Neuhaus T, Grunewald E, et al. Osteopontin is a hematopoietic stem cell niche component that negatively regulates stem cell pool size. J Exp Med. 2005;201(11):1781–91.
138. Stolzing A, Jones E, McGonagle D, Scutt A. Age-related changes in human bone marrow-derived mesenchymal stem cells: consequences for cell therapies. Mech Ageing Dev. 2008;129(3):163–73.
139. Sugimura R, He XC, Venkatraman A, Arai F, Box A, Semerad C, et al. Noncanonical Wnt signaling maintains hematopoietic stem cells in the niche. Cell. 2012;150(2):351–65.
140. Sugiyama T, Kohara H, Noda M, Nagasawa T. Maintenance of the hematopoietic stem cell pool by CXCL12-CXCR4 chemokine signaling in bone marrow stromal cell niches. Immunity. 2006;25(6):977–88.
141. Sungaran R, Markovic B, Chong BH. Localization and regulation of thrombopoietin mRNa expression in human kidney, liver, bone marrow, and spleen using in situ hybridization. Blood. 1997;89(1):101–7.
142. Swami A, Reagan MR, Basto P, Mishima Y, Kamaly N, Glavey S, et al. Engineered nanomedicine for myeloma and bone microenvironment targeting. Proc Natl Acad Sci U S A. 2014;111(28):10287–92.
143. Takubo K, Goda N, Yamada W, Iriuchishima H, Ikeda E, Kubota Y, et al. Regulation of the HIF-1alpha level is essential for hematopoietic stem cells. Cell Stem Cell. 2010;7(3):391–402.

144. Tan J, Liu T, Hou L, Meng W, Wang Y, Zhi W, et al. Maintenance and expansion of hematopoietic stem/progenitor cells in biomimetic osteoblast niche. Cytotechnology. 2010;62(5):439–48.
145. Tan SH, Senarath-Yapa K, Chung MT, Longaker MT, Wu JY, Nusse R. Wnts produced by Osterix-expressing osteolineage cells regulate their proliferation and differentiation. Proc Natl Acad Sci U S A. 2014;111(49):E5262–71.
146. Tibbitt MW, Anseth KS. Dynamic microenvironments: the fourth dimension. Sci Transl Med. 2012;4(160):160ps24.
147. Tokalov SV, Gruner S, Schindler S, Wolf G, Baumann M, Abolmaali N. Age-related changes in the frequency of mesenchymal stem cells in the bone marrow of rats. Stem Cells and Development. 2007;16(3):439–46.
148. Travlos GS. Normal structure, function, and histology of the bone marrow. Toxicol Pathol. 2006;34(5):548–65.
149. Tumbar T, Guasch G, Greco V, Blanpain C, Lowry WE, Rendl M, et al. Defining the epithelial stem cell niche in skin. Science. 2004;303(5656):359–63.
150. Unwin RD, Smith DL, Blinco D, Wilson CL, Miller CJ, Evans CA, et al. Quantitative proteomics reveals posttranslational control as a regulatory factor in primary hematopoietic stem cells. Blood. 2006;107(12):4687–94.
151. Villa-Diaz LG, Ross AM, Lahann J, Krebsbach PH. Concise review: The evolution of human pluripotent stem cell culture: from feeder cells to synthetic coatings. Stem Cells. 2013;31(1):1–7.
152. Von Drygalski A, Alespeiti G, Ren L, Adamson JW. Murine bone marrow cells cultured ex vivo in the presence of multiple cytokine combinations lose radioprotective and long-term engraftment potential. Stem Cells Devel. 2004;13(1):101–11.
153. Vunjak-Novakovic G, Scadden DT. Biomimetic platforms for human stem cell research. Cell Stem Cell. 2011;8(3):252–61.
154. Wagner W, Horn P, Bork S, Ho AD. Aging of hematopoietic stem cells is regulated by the stem cell niche. Exp Gerontol. 2008;43(11):974–80.
155. Wagner W, Horn P, Castoldi M, Diehlmann A, Bork S, Saffrich R, et al. Replicative senescence of mesenchymal stem cells: a continuous and organized process. PLoS ONE. 2008;3(5):e2213.
156. Walenda T, Bork S, Horn P, Wein F, Saffrich R, Diehlmann A, et al. Co-culture with mesenchymal stromal cells increases proliferation and maintenance of haematopoietic progenitor cells. J Cell Mol Med. 2010;14(1–2):337–50.
157. Wang LD, Wagers AJ. Dynamic niches in the origination and differentiation of haematopoietic stem cells. Nat Rev Mol Cell Biol. 2011;12(10):643–55.
158. Watt FM, Huck WT. Role of the extracellular matrix in regulating stem cell fate. Nat Rev Mol Cell Biol. 2013;14(8):467–73.
159. Whitesides GM. The origins and the future of microfluidics. Nature. 2006;442(7101):368–73.
160. Willert K, Brown JD, Danenberg E, Duncan AW, Weissman IL, Reya T, et al. Wnt proteins are lipid-modified and can act as stem cell growth factors. Nature. 2003;423(6938):448–52.
161. Wilson A, Oser GM, Jaworski M, Blanco-Bose WE, Laurenti E, Adolphe C, et al. Dormant and self-renewing hematopoietic stem cells and their niches. Ann N Y Acad Sci. 2007;1106:64–75.
162. Wilson A, Laurenti E, Oser G, van der Wath RC, Blanco-Bose W, Jaworski M, et al. Hematopoietic stem cells reversibly switch from dormancy to self-renewal during homeostasis and repair. Cell. 2008;135(6):1118–29.
163. Wright DE, Wagers AJ, Gulati AP, Johnson FL, Weissman IL. Physiological migration of hematopoietic stem and progenitor cells. Science. 2001;294(5548):1933–6.
164. Yoshihara H, Arai F, Hosokawa K, Hagiwara T, Takubo K, Nakamura Y, et al. Thrombopoietin/MPL signaling regulates hematopoietic stem cell quiescence and interaction with the osteoblastic niche. Cell Stem Cell. 2007;1(6):685–97.
165. Zhang J, Niu C, Ye L, Huang H, He X, Tong WG, et al. Identification of the haematopoietic stem cell niche and control of the niche size. Nature. 2003;425(6960):836–41.

166. Zhang W, Lee WY, Siegel DS, Tolias P, Zilberberg J. Patient-specific 3D microfluidic tissue model for multiple myeloma. Tissue Eng Part C Methods. 2014;20(8):663–70.
167. Zhong Z, Akkus O. Effects of age and shear rate on the rheological properties of human yellow bone marrow. Biorheology. 2011;48(2):89–97.
168. Zou YR, Kottmann AH, Kuroda M, Taniuchi I, Littman DR. Function of the chemokine receptor CXCR4 in haematopoiesis and in cerebellar development. Nature. 1998;393(6685):595–9.

The Male Stem Cell Niche: Insights from *Drosophila* and Mammalian Model Systems

Fani Papagiannouli and Ingrid Lohmann

Abbreviations

Abd-B	Abdominal-B
AJ	Adherens junction
CySCs	Somatic cyst stem cells
ECM	Extracellular matrix
GSCs	Germline stem cells
SCCs	Somatic cyst cells
SJ	Septate junction
SGPs	Somatic gonadal precursors
SSC	Spermatogonial stem cell
TJ	Tight junction
wt	Wild type
L3	3rd instar *Drosophila* larvae

1 Stem Cells and Their Niches

Stem cells are undifferentiated cells, present in all multicellular organisms, with a remarkable potential to develop into many different cell types. Depending on the source, they have the potential to form one, many or all cell types of an organism. During early mammalian embryogenesis, pluripotent stem cells propagate and give rise to every cell type that built up our adult body architecture (embryonic stem cells). During adulthood, tissue-specific stem cells serve as a sort of internal repair

F. Papagiannouli (✉) · I. Lohmann (✉)
Centre for Organismal Studies (COS) Heidelberg, Cell Networks – Cluster of Excellence, University of Heidelberg, Heidelberg 69120, Germany
e-mail: fani.papagiannouli@cos.uni-heidelberg.de

I. Lohmann
e-mail: ilohmann@flydev.org

K. Turksen (ed.), *Tissue-Specific Stem Cell Niche*,
Stem Cell Biology and Regenerative Medicine,
DOI 10.1007/978-3-319-21705-5_12

system, dividing to replenish other cells as long as the person or animal is still alive (adult stem cells). Such examples are the germline stem cells in the mammalian testis producing one type of differentiated cell, the spermatozoon and the hematopoietic stem cells producing erythrocytes and all white blood cell types [1]. Therefore, adult stem cells are present in many tissues and are important in homeostasis and tissue repair.

Along with the production of specialized cells types, stem cells have the capacity to self-renew. Elucidating the basic mechanisms that stem cells use to accomplish self-renewal vs. differentiation can provide fundamental insights into the origin and design of multicellular organisms [2]. On a practical level, the stem cell field has dramatically changed over the last decades by the success in culturing human embryonic stem cells, the manipulation of their differentiation in vitro and the evidence that adult stem cells have a much higher plasticity than initially thought. Therefore, these cells not only open a new window for understanding the embryonic development of our species but also represent an incredible source for the repair of diseased and damaged tissues in our bodies [1].

Can the knowledge we have on stem cells be used to develop cell-based medicine and repair malformed, damaged or aging tissues? Scientists now face the task of bringing the stem cell therapy to the clinic. Yet, in doing so scientists need to better understand stem cell function at the molecular level and how these cells behave in their biological context. A critical point in order to discern how stem cells perform their tasks is to discover their regulation and specification. In other words, the factors that maintain stem cells in a multipotent, proliferative state or drive them to create differentiated daughter cells in vivo and in vitro need to be identified. Nowadays we know that several of the specialized functions required to ensure proper stem cell function are vested in the neighboring differentiated cells. By signals and other intercellular interactions, these cells control the behavior of the stem cells that are initially unspecialized. This local tissue microenvironment, called the stem cell niche, homes the stem cells and regulates their balance between self-renewal and differentiation [3].

In 1978, Ray Schofield proposed the "niche" hypothesis to describe the physiologically limited microenvironment that supports the hematopoietic stem cells (HSCs) [4]. While defining the stem cell niche in mammals has been difficult due to their complex anatomic structures, the stem cell niches in other genetic model systems, including *Drosophila* and *C. elegans*, were among the first to be characterized. In 2000, the germarial tip adjacent to germline stem cells (GSCs) was defined as the niche in the *Drosophila* ovary [5] while the hub, located at the apical end of the *Drosophila* testis, fulfills this function in the testis [6, 7]. In *C. elegans*, it is the distal tip cell (DTC), located at the distal end of the gonad, that functions as the niche [8]. In mammalian systems, the location of the niche is defined largely based on its proximity to the stem cells. The niche or stem cell regulatory microenvironment is defined by the cellular components and extracellular matrix (ECM) in proximity to the stem cells, and the signals emanating from the support cells [9, 10]. Significant progress regarding stem cells and their surrounding microenvironment has been made in different mammalian tissue types like the nervous system, the endothelial cells and hematopoietic system, the skin and hair follicles as well as the intestine [11–18]. Therefore, it is of crucial importance to determine how different cell types adjacent to stem cells in mammalian systems contribute to niche function and stem cell regulation.

A strong support of the niche-based stem cell regulation proposed by Schonfield [4] evolved by studying spermatogenesis [2]. In adult mammalian testes, GSCs, which lie in contact with the basement membrane of the seminiferous tubule, have the remarkable ability to both self-renew and differentiate, ensuring that a continuous population of mature spermatozoa is produced throughout the animal life-time. The existence of niches was demonstrated when GSCs were implanted into the seminiferous tubules of host males whose GSCs had been depleted [19]. However, further analysis was limited for several years by the inability to mark the stem cells and the niche in vivo with precision. In *Drosophila*, on the other hand, it has been possible to tag individual stem cells, genetically modify them and analyze their functional requirements over time. For this reason, studies in the *Drosophila* testis opened the way for understanding basic mechanisms and principles governing the male stem cell niches and the GSCs. Nowadays the *Drosophila* male stem cell niche is considered to be one of the best-characterized ones. In this book chapter, we review the current state-of-the art in the field of male stem cell niches, compare the differences and similarities of *Drosophila* to the mammalian male stem cell niches and finally we discuss important future topics to be addressed in male stem cell niches.

2 *Drosophila* Testis and the Male Stem Cell Niche

In all adult tissues harboring stem cells, the stem cell niche has a critical function as an organizer, which recruits the stem cells and provides the microenvironment required for stem cell maintenance. Much of the knowledge we have on testis stem cells and their niche comes from studies in *Drosophila*, a well-characterized system to study the biology of the stem cell niche, the GSCs and spermatogenesis [3]. Organogenesis of the *Drosophila* testis, a structure first made by the coalescence of germ cells and somatic gonadal cells at stage 14 of embryogenesis, continues throughout embryonic and larval stages, and goes through a second wave of organ shaping in the pupae, to reach maturation in adult stages. The *Drosophila* male stem cell niche, called the hub, is a cluster of non-dividing cells specified in the anterior most somatic gonadal cells already before gonad coalescensce [20–25].

The first signs of testis organogenesis are already detected in late embryogenesis (stages 14–17), once the specified hub cells recruit the anterior-most germ cells to become the germline stem cells (GSCs) [26]. A testis with a mature stem cell niche and all pre-meiotic stages is detected at 3rd instar larvae (L3) (Fig. 1). The *Drosophila* testis contains two types of stem cells: the GSCs and the somatic cyst stem cells (CySCs). Each GSC is flanked by two somatic cyst stem cells (CySCs) and both types of stem cells are maintained through their association to the hub

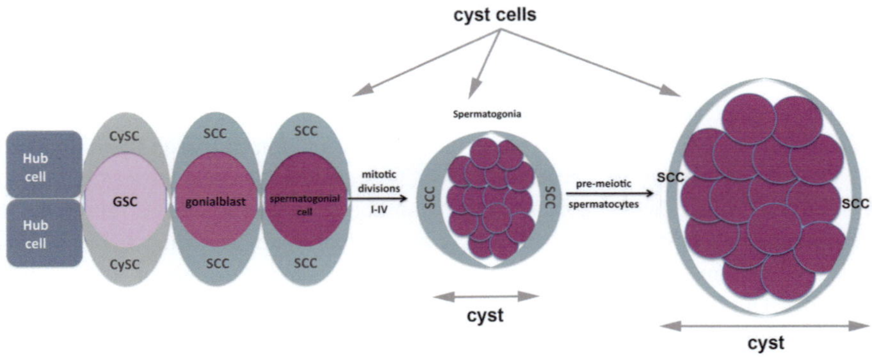

Fig. 1 Diagram depicting early spermatogenesis in *Drosophila*. *GSC* germline stem cell, *CySC* somatic cyst stem cell, *SCC* somatic cyst cell. For simplicity reasons CySCs and SCCs are collectively called cyst cells. Testicular cysts comprise of a pair of cyst cells flanking the germline (GSCs, spermatogonia or spermatocytes)

cells, a cluster of non-dividing cells forming the niche organizer. Upon asymmetric cell division, each GSC produces a new GSC attached to the hub and a distally located gonialblast. The CySCs also divide asymmetrically to generate a CySC remaining associated with the hub and a distally located post-mitotic daughter somatic cyst cell (SCC) [27]. Two SCCs enclose each gonialblast forming a testicular cyst (Fig. 1) "sealed" from the outside by ECM (Fig. 2) [28]. For simplicity reasons, CySCs and SCCs are collectively referred to in this chapter as "cyst cells". The gonialblast divides mitotically four more times to give rise to 16 interconnected spermatogonial cells, which then undergo pre-meiotic DNA replication, become spermatocytes, turn on the transcription program for terminal differentiation and undergo meiosis. During pupal stages testis morphogenesis is completed with the addition of the acto-myosin sheath originating from the genital disc [29]. The SCCs co-differentiate with the germ cells they enclose, grow enormously in size, elongate and accompany them throughout their differentiation steps up to individualization and sperm production in the adult testis [30].

2.1 Specification of the Male Stem Cell Niche

Specification of the hub cells is a prerequisite for establishment of the testis stem cell niche per se. Hub cells are somatic cells specified before gonad formation from a subpopulation of the lateral mesoderm, the somatic gonadal precursor cells (SGPs), in bilateral clusters of the abdominal parasegments 10–13 [20, 22, 31–34]. The different SGP populations joining the embryonic male gonad orchestrate testis morphogenesis at this initial stage, since the germ cells represent a uniform population at this time. *zinc-finger homeodomain 1* (*zfh-1*), a key player in SGPs specification, is initially expressed in cell clusters of the lateral mesoderm (PS2-14)

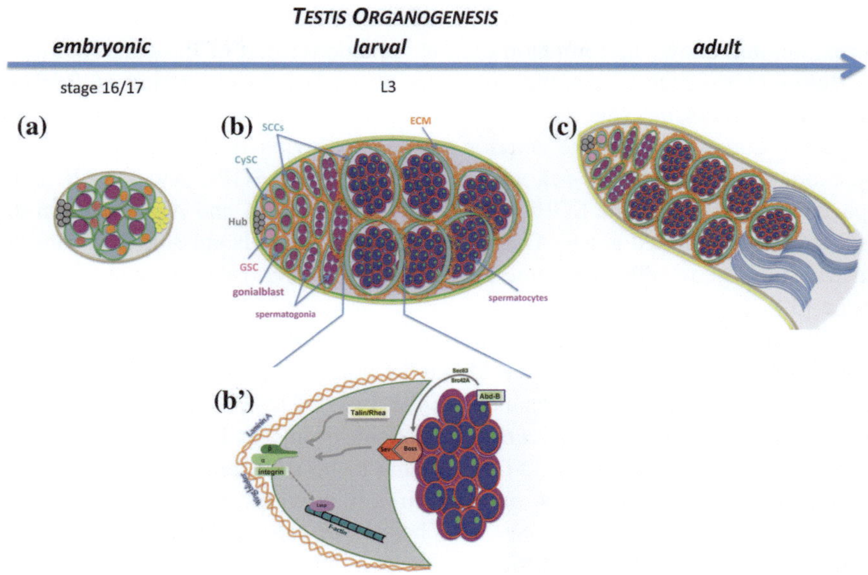

Fig. 2 Diagram showing the *Drosophila* male stem cell niche during testis organogenesis. *Upper panel* represents a schematic diagram of the embryonic male gonad (**a**), larval (**b**) and adult (**c**) testis during *Drosophila* testis organogenesis. *GSC* germline stem cell, *CySC* somatic cyst stem cell, *SCC* somatic cyst cell, *ECM* extra-cellular matrix. Germ cells are shown in *purple*. *Gray circles* represent hub cells in the anterior of male gonads and testes. The somatic lineage (somatic gonadal cells or CySCs and SCCs) is shown in *green*. **a** Within the male gonad, *abd-A* is expressed in *red-colored* nuclei, *abd-A* and *Abd-B* are co-expressed in orange-colored nuclei. Male-specific somatic gonadal cells expressing *Abd-B* are shown in *yellow*. **b**, **c** Integrin localization is indicated in *green*. ECM molecules (*orange*) surround the cyst cells and testicular cysts. Within the spermatocytes, the *red line* indicates the nuclear membrane, the green dots resemble *Abd-B* distribution in the nucleolus and blue represents the nucleus. **b** Diagram showing key players involved in larval niche positioning. Schematic diagram of a spermatocyte cyst depicting local germline-soma signaling and key players involved in niche positioning. For simplicity only one SCC is shown

whereas at a later stage *zfh-1* expression in parasegments 10–13 correlates with the specification of these cells as SGPs [34–36].

However, not only the hub cells but also the cyst cells are specified from the SGPs. The common origin between the hub and CySCs has been shown by lineage tracing experiments [21]. Hub cell fate vs. cyst cell fate is specified prior to gonad coalescence in a subset of SGPs upon Notch signaling activation [21]. Specification of CySCs vs. hub cell fate is further shaped by the antagonistic function of the cytoplasmic protein Lines (Lin) and the transcription factor Brother of odd with entrails limited (Bowl) [25, 37]. Bowl promotes hub cell fate and Lin CySCs fate, evidenced by fewer hub cells in *bowl* mutant gonads and increased number of hub cells in *lin* mutant gonads. Also, *lin* depleted CySCs acquired some hub-like properties and markers [21]. This is further supported by the fact that both cell types can be traced with the same cell markers such as Zfh-1 and Traffic Jam (TJ) [37].

In the posterior SGPs, the Epidermal growth factor receptor (EGFR) represses hub formation and allows its formation only at the anterior part of the gonad [38].

Before gonad coalescence, the *Hox* genes *abdominal-A* (*abd-A*) and *Abdominal-B* (*Abd-B*) pattern the anterior-posterior (A/P) axis of the male embryonic gonad (Fig. 2a): *abd-A* specifies the anterior most SGPs giving rise to the hub, a combination of *abd-A* and *Abd-B* specifies the posterior SGPs, and *Abd-B* alone specifies the male-specific SGPs [20, 22–24]. Thus, *abd-A* and *Abd-B* pattern the A/P axis of the formed gonad. Once specified, the hub cells are able to recruit the anterior-most germ cells to become the GSCs [26], giving rise to the male stem cell niche [39].

Although the separated fate of hub cells and CySCs is already set up during embryogenesis, a recent study showed that the quiescent post-mitotic hub cells can still give rise to CySCs upon challenge [40]. Loss of CySCs by genetic ablation triggers hub cells to exit quiescence, delaminate from the hub and convert into functional CySCs, and Cyclin D-Cdk4 seems to be sufficient to trigger this fate switch. This is interesting not only because it provides a working model on understanding oncogenic processes via changes in cell fate but because it also proves that the *Drosophila* male niche can respond to environmental changes and is in reality way more flexible than initially thought.

2.2 Positioning the Male Stem Cell Niche

Stem cell niche and subsequent testis morphogenesis is a stepwise process based on the physical contact and diffusible signals exchanged between the germline and somatic cell populations [41]. In order to ensure normal niche function, the hub cells of the *Drosophila* testis not only need to be properly specified but also need to be correctly placed and the architectural integrity of the system has to be maintained. Proper niche function in terms of hub positioning and integrity is tightly coupled to adhesion and cell communication, with βPS-Integrin [encoded by the *myospheroid* (*mys*) gene] and the Bride of Sevenless (Boss)/Sevenless (Sev) signaling pathway playing key roles in embryonic [42, 43] and adult stages [28, 42–44]. Integrin-mediated adhesion is important for maintaining the correct position of the embryonic hub cells during gonad morphogenesis. In the absence of integrin-mediated adhesion, the hub cells still form a cluster, but instead of remaining at the anterior part of the gonad they migrate to the middle part of the developing gonad [42]. Disruption of integrin-mediated adhesion in adult testes, for example by knocking down *talin/rhea,* a gene coding for an integrin-binding and essential focal adhesion protein of the integrin-cytoskeleton [45, 46], results in gradual hub disappearance, a phenotype which becomes more severe as adult males age [42]. As the hub is progressively lost in *talin*-depleted adult testis, the signals that normally emanate from the hub to instruct stem cell renewal are absent, driving the balance between stem cell maintenance and differentiation towards more differentiation. As a result the GSCs are progressively lost [42]. A similar hub displacement phenotype is observed by reducing Lasp levels, an actin-binding protein, in the adult testis [44]. It is known from vertebrate systems that Lasp interacts genetically with Integrin [47] and in blood platelets Lasp requires integrin for its proper localization to the cytoskeleton [48]. In a few cases, loss of Lasp leads to hub integrity defects in which the hub cell arrangement is disturbed or double hubs are observed. Genetic interaction studies showed that βPS-integrin and Lasp proteins are active in different pathways, which cooperate to position the hub in adult testes [44].

The Boss/Sev pathway plays an important role in hub positioning and integrity in the *Drosophila* male gonads by preventing ectopic niche differentiation in the posterior gonadal somatic cells. Sev is activated by the Boss ligand emanating from the primordial germ cells to represses ectopic hub differentiation [43]. Upstream of this cascade, *Abd-B* activates *sevenless* (*sev*) in the posterior male-specific SGPs [43]. Consistent with this observation, weak *Abd-B* mutant alleles result in hub expansion and hub integrity defects in embryonic gonads [20]. Boss and Sev are required for hub positioning and integrity in the adult testis, but the mechanism of action remains so far unknown. Taken together, hub positioning and integrity relies on the employment of the same players during testis organogenesis from embryonic up to adult stages of *Drosophila* development.

Recent work revealed a new role for the posterior *Hox* gene *Abd-B* in niche positioning and integrity during larval stages (Fig. 2b). In addition to its described role in the male embryonic gonads, *Abd-B* present in the pre-meiotic germline spermatocytes of the larval testis acts upstream of the Boss/Sev pathway to regulate hub positioning and integrity, which finally leads to loss of Integrin and Actin localization in the neighboring cyst cells [28]. Cell-type specific knockdown of *Abd-B* in larval *Drosophila* germline spermatocytes, using an *Abd-B*^{RNAi} transgene, leads to hub mispositioning, hub integrity defects and less frequently to the formation of two independent niches. The incorrect placement of the niche in *Abd-B* depleted testes, results in cell non-autonomous centrosome mispositioning and reduced GSC divisions, leading to a dramatic reduction of the pre-meiotic stages of the adult testis, a hallmark of aging in testis [49, 50].

In the *Drosophila* larval testis, the atypical G-protein coupled receptor Boss is found in the germline spermatocytes, primarily in vesicles, whereas Sev localizes in the cyst cells enclosing them. *Abd-B* performs its function by affecting Boss internalization in the germline, as Boss is lost from internalized vesicles in *Abd-B* depleted testes [28]. Expression of activated Sev in cyst cells of *Abd-B* depleted testes can fully rescue the phenotype, meaning that Boss exerts its function via Sev activation. In order to elucidate how the Hox transcription factor *Abd-B* affects Boss localization, genes directly regulated by *Abd-B* in the *Drosophila* testis were identified by mapping *Abd-B* binding sites in vivo using the DNA adenine methyltransferase identification (DamID) technology [51–54]. Two genes, one encoding the non-receptor tyrosine kinase Src oncogene at 42A (Src42A) and another one encoding the putative signal recognition binding protein Sec63, were identified as potential mediators of Boss function in the larval testis (Fig. 2b′) [28]. *src42A* and *sec63* mRNA levels were significantly reduced in spermatocytes of *Abd-B* depleted testes [28]. Functional analysis confirmed that *src42A* and *sec63* depleted testes mimic the loss of *Abd-B* function as Boss protein was not detected in vesicles, the hub was mispositioned and βPS-Integrin was not properly localized in SCCs. In the adult testis, *Abd-B* is expressed additionally in the nuclei of the acto-myosin sheath [28], which surrounds the adult testis and fuses it to the seminal vesicle. Recent work provided evidence that in contrast to the larval stages, *Abd-B* from the adult testis spermatocytes no longer affects integrin localization in the neighboring SCCs [28], meaning that *Abd-B* has most likely other germline stage-specific functions in the adult testis spermatocytes.

It becomes clear that male stem cell niche function in *Drosophila melanogaster*, from initial specification to its continuous maintenance during testis organogenesis and adult life, is a dynamic process that relies on a combination of cell-type and stage- specific regulators. All these studies evidence that the same players, AbdB, Boss, Sev and Integrin are used, often in an alternative way, to preserve the male stem cell niche positioning and integrity during subsequent steps of testis organogenesis. *Abd-B* acts as an upstream regulator of the Boss/Sev pathway, by controlling *sev* expression in the embryonic male gonad [43] and Boss function in the larval testis via *sec63* and *src42A* expression [28]. This means that the switch of *Abd-B* expression from the embryonic male-specific somatic cells to the larval spermatocyte germ cells correlates with a change in the *Abd-B* dependent mechanism of hub positioning between embryonic and larval stages [55]. Therefore, it seems that the occurrence of new cell types and cell interactions in the course of testis organogenesis made it necessary to adapt the whole stem cell system to the new cellular conditions by reusing the same main players of niche positioning in an alternative manner. Notably, correct niche positioning and architecture is a prerequisite for stem cell niche function in order to prevent the accumulation of aging-related defects in testes at adult stages when reproduction starts. This underlines the vital importance to protect the niche in order to preserve adult germline stem cell function, protect spermatogenesis and produce healthy gametes and progeny critical for organismal function [55].

2.3 Signaling Regulation of Stemness Versus Differentiation

Gamete development requires a coordinated soma-germ line interaction that keeps the balance between germline stem cell renewal and differentiation. The balance between stem cell identity and differentiation at the *Drosophila* testicular niche results from signals exchanged among the hub, GSCs and CySCs. The Janus-kinase transducer and activator of transcription (JAK-STAT) pathway was the first signaling pathway found to regulate GSC and CySC maintenance in the *Drosophila* testis [6, 7]. The hub cells secrete the ligand Unpaired (Upd), which activates the JAK-STAT pathway in adjacent GSCs and CySCs [6, 7, 56]. In the absence of JAK-STAT signaling the GSCs differentiate and are unable to self-renew, whereas ectopic expression of *upd* in the germline greatly expands the population of GSCs and CySCs in adult as well as in larval testes [6, 7]. In GSCs, STAT is required so that E-cadherin (E-cad) maintains the connection of the GSC to the hub and ectopic E-cad partially rescues the stem cell identity of the STAT-depleted GSCs by maintaining the adhesion of STAT-depleted GSCs to the hub [57]. Another STAT target in GSCs is *chickadde (chic)*, the homologue of the *Drosophila* profilin. Chic is required cell autonomously to maintain GSCs by facilitating GSC-hub contact possibly via E-cad whereas Chic in the SCCs is affecting germ cell enclosure and restricting trans-amplifying (TA) spermatogonial divisions [58]. When GSCs divide, their daughter cells displaced from the hub are thought to receive lower levels of hub-derived signals and therefore differentiate. In CySCs, STAT is critical for maintaining their stem cell character and the activation of targets essential for their identity such as *zfh-1* and *chinmo* [57, 59]. *zfh-1* is expressed predominantly in CySCs and their immediate SCCs, and ectopic expression in late SCCs outside the niche leads in accumulation of GSC- and CySCs-like cells which fill in the whole testis. Similarly, *chinmo* is expressed in comparable levels in CySCs and early SCCs, is required for CySCs and not GSC renewal, and ectopic expression causes accumulation of GSCs- and CySCs-like cells. Furthermore, *zfh-1* and *chinmo* are not expressed in GSCs meaning that STAT can activate distinct downstream cascades in the in GSC vs. CySCs. *ken* and *barbie* (*ken*) is another gene necessary and sufficient to promote CySC identity, yet in a STAT independent manner and with similar ectopic phenotypes like *zfh-1* and *chinmo* [60]. Critical for CySC and GSC survival and maintenance is the primary steroid 20-hydroxyecdyson (20E) [61]. Ecdyson receptor (EcR) and its partner *ultraspiracle* (*usp*) are expressed in CySCs and EcR is required cell-autonomously in the CySCs and cell non-autonomously in GSCs to promote their survival and maintenance [61].

Interestingly, very recent findings revealed that the Hedgehog (Hh) ligand secreted from the hub cells activates the Hh signaling in CySCs (and not in the GSCs) with critical function in CySC maintenance [62–65]. Hh overexpression leads in increased number of CySCs, identified as Zfh-1 positive cyst cells outside the niche, which can still proliferate in contrast to the normal post-mitotic SCCs. Furthermore, rescue of STAT depleted testis by Hh signaling activation in the CySCs can rescue the CySCs but GSC and germline maintenance is still impaired, as these Zfh-1 positive CySCs are not able to induce the GSC overproliferation phenotype observed in SCCs ectopic Zfh-1 activation [63]. This suggests that (1) *zfh-1* expression relies on inputs from both Hh and JAK-STAT signaling pathways and that (2) apart from Zfh-1 other STAT regulated factors are necessary for allowing the CySC-to-GSC communication, which promotes GSC maintenance.

Notably, BMP seems to be the primary pathway leading to GSC self-renewal in the *Drosophila* testis [66–69]. BMP ligands and the BMP modulator *magu*, are expressed in the hub and CySCs that serve as the GSC niche and their loss results in reduced GSC numbers and *bam* de-repression, whereas the hub and CySCs remain unaffected [67–69]. This could also suggest that expansion of the GSC population by the JAK-STAT signaling could be due to its activation in the CySCs that consequently leads to enhanced expression of BMP ligands from CySCs [57] that finally drive GSC expansion. The BMP pathway is also negatively regulated in the course of testis morphogenesis along embryonic-larval-adult stages via Smurf (SMAD ubiquitination regulatory factor) [70]. High BMP levels are required at the initial steps of niche establishment when the hub cells attract the nearby germ cells to become GSCs in late embryogenesis up to early 3rd instar larval stages. Apparently, BMP signaling is spatially and temporally downregulated in stem cells and early germline cells in late 3rd instar larval and pupal testes through Smurf proteolytic activity. The described BMP downregulation seems to be critical for the normal decrease in stem cell number during pupal development, for restricting TA spermatogonia proliferation and control of the testis size. This dynamic regulation indicates the requirement for fine trimming the BMP signaling intensity during subsequent developmental stages and might even suggest a difference in establishment vs. maintenance of certain cell populations across different stages. Yet, another recent story revealed that GSC characteristics can be maintained over time even after ablating the CySC and SCCs [71]. Without CySCs and SCCs, early germ cells away from the hub failed to initiate differentiation and maintained their GSC-like characteristics. Therefore, it becomes evident that the interactions between different stem cell populations and how one stem cell population influences the other can be indeed very complex. Finally, antagonistic functions between the *Drosophila* β-catenin Armadillo (Arm) and the microRNAs- (miR-) 310–313 suggest that modulation of the Wingless signaling activity is important to buffer germ cell and somatic differentiation in the *Drosophila* testis [72].

The fine-tuning of signaling pathways can regulate competition between GSCs and CySCs around the hub niche cells. Suppressor of cytokine signaling 36E (Socs36E) suppresses Jak-Stat signaling in the CySCs preventing them from outcompeting the GSCs and thereby maintains the proper balance of GSCs and CySCs, in a manner that depends on the adhesion protein integrin [56, 73]. Downstream of the JAK-STAT pathway in the *Drosophila* testis is the recently identified Slit-Roundabout 2 (Robo2) signaling pathway, which provides new information on how CySCs compete for occupying the niche [74]. The ligand Slit from the hub cells, signals to its receptor Robo2 present at the CySCs, which together with the Abelson kinase (Abl), also at the CySCs, balance the adhesion levels and thereby prevent the over-adhesion of CySCs to the niche. Robo2 and Abl modulate adherens junctions components such as E-cadherin (E-cad) and Arm, since Robo2 null CySCs are rescued by E-cad overexpression and Arm is required for Abl-mediated stem cell competition [74]. More insights on CySC-GSC competition round the niche emerged by studying neutral competition in that individual stem cell cells can be lost and replaced by their neighbors stochastically. CySCs follow this principle while the Hh and the Hippo pathways active in these cells affect neutral competition, independent of one another, by affecting the CySC proliferation rates [75].

Critical for germ cell differentiation is the expression of *bag of marbles* (*bam*) and *benign gonial cell neoplasm* (*bgcn*) in dividing spermatogonial cells in order to regulate their proliferation [76]. *bam* transcription is negatively regulated by the cooperation of the Glass bottom boat (Gbb) and Decapentaplegic (Dpp) signaling pathways emanating from the hub and CySCs to maintain the GSC identity [67].

Bam is required cell autonomously in TA spermatogonia to stop proliferation and enter the spermatocyte differentiation program [77]. The switch from TA proliferation to differentiation is mediated by translational control: Mei-P26 facilitates the accumulation of Bam in TA cells whereas Bam and Bcgn bind *mei-P26* 3' untranslated region and repress translation of mei-P26 in late TA cells. Thus, germ cells progress through subsequent regulatory states that is: from a "Mei-P26 on/Bam off" to a "Bam on/Mei-P26 off" state.

Another signaling pathway critical for spermatogenesis and testis homeostasis is the Epidermal Growth Factor Receptor (EGFR) pathway, whose inactivation in SCCs leads to an expansion of male GSCs [78]. In *Drosophila* testis, the major ligand of the EGFR pathway, Spitz (Spi) is secreted from the germline cells to stimulate the EGFR on cyst cells (CySCs and SCCs) [37]. Removal of either *spi* or *stet* from the germline cells, or removal of the EGFR from the cyst cells results in increased division frequencies of GSCs but does not affect the division frequencies of CySCs, suggesting that EGF signaling downregulates GSC divisions. Detailed follow-up studies showed that EGFR pathway has stage-specific and dose-specific effects in the cyst cells. More precisely, the EGFR pathway controls the GSCs division frequency, with GSCs dividing faster in adult but not in larval testes [79]. Furthermore, EGFR pathway has a critical role in guiding the differentiating germline into spermatogenesis. This is achieved in a stepwise manner by inducing different responses depending on its dose [80]. In fact, the cyst cells develop a temporal "signature" of EGF signaling created by the coordinated increase of the production of active ligands by the germline and the amount of available receptor molecules on the cyst cells [80]. In early spermatogonial cysts, low dose of EGF signaling reduces GSC characteristics in the germline spermatogonia that enter to synchronous TA divisions. In late cysts, high dose of EGF signaling induces the spermatogonia to end the TA divisions and become spermatocytes. This is supported by the observation that SCCs expressing a constitutively active EGFR exit the TA division prematurely and become spermatocytes prior to the 16-cell stage. A similar phenotype was also seen in testes with germline cells depleted of *nucleoporin 98-96* (*nup98-96*), where the germline differentiates to spermatocytes prematurely at the 2-, 4-, and 8- germ cell cysts stage [81]. All these data point at a mechanism which prevents the entry of spermatogonia into the spermatocyte stage before the completion of exactly four rounds of TA-divisions Likewise, Raf, an EGFR downstream component, is required in SCCs to limit GSC expansion [82–84]. In testes mutated for the *rhomboid* homologue *stet*, the germ cells fail to associate with SCCs. Furthermore, germ cells recruit CySCs via the ligand Spitz, which binds to EGFR, and acts through the nucleotide exchange factor Vav to regulate the activity of Rac1, a downstream component of the EGFR pathway. Taken together, EGF signaling from the germline cells produces differential Rac- and Rho- activities across the cyst cells that leads to a directional growth of the cyst cells around the germline cells [37]. Finally, Zero population growth (Zpg), the *Drosophila* gap junction Innexin 4, is localized to the spermatogonia surface, primarily on the sides adjacent to SCCs [85] and is required for the survival and differentiation of early germ cells in both sexes [86, 87]. All these studies advance our understanding on stem cell dynamics in the *Drosophila* testis, uncover the plasticity of the system and reveal the great signaling complexity underlying testis homeostasis, in order to secure what is supposed to be the most precious thing for the organism: the production of healthy gametes as a prerequisite for successful sexual reproduction.

2.4 Cyst Stem Cells and Somatic Cyst Cells: The Supportive Cells of the Germline

Critical for testis differentiation and morphogenesis is the cyst microenvironment created by (i) the CySCs enclosing the GSCs and (ii) the SCCs enclosing the differentiating germ cells and accompanying them throughout their differentiation steps up to sperm individualization. Consequently, CySCs and SCCs maintain the integrity and architecture of the testicular cysts, and in a broader sense they act as "niche" cells for the germ cells they encapsulate [88, 89]. During terminal differentiation, the two cyst cells of the same cyst acquire different identities followed by morphological changes [37]: the forward SCC becomes the "head cyst cell" (HCC) onto which all 64 spermatid heads are anchored shortly after meiosis, and the posterior one becomes the much larger "tail cyst" (TCC) that surrounds the spermatid tails of 1.8 mm length [90]. This results in polarized cysts across the testis anterior-posterior (A-P) axis and towards the direction (A $\rightarrow$ P) of differentiation. The HCC finally is engulfed by cells of the terminal epithelium to allow coiling of the spermatid bundles towards the testis base [91].

Although it is well established that soma-germline physical contact is critical for the cell communication and for promoting their mutual development and differentiation [41], it remains so far elusive how these tightly packed cysts coordinate adhesion and cell shape changes with signaling and how they can grow enourmously on a mechanistic level. The thin and squamous cyst cells lack the columnar epithelial structure of e.g. the ovarian follicular epithelium, which caught the attention of scientists analyzing apico-basal polarity many years ago. So far the main evidence for cyst cell (CySCs and SCCs) function came from the analysis of individual signal transduction pathways that establish a cross talk between the soma and the germline, as outlined in the previous part. In this part, recent findings affecting germline-soma coordination will be highlighted, with emphasis on the role of cytoskeletal, junctional and scaffolding components.

Critical cytoskeletal and polarity components localize at cyst cells, such as Rho1, Rac1 [84], Profilin (encoded by the *chickadee/chic* gene) [58, 92], βPS-Integrin (encoded by the *myospheroid* gene) [28], Talin (encoded by the *rhea* gene) [42], as well as the septate junction components Neurexin-IV (Nrx-IV), Coracle (Cora) [92], Discs large (Dlg), Scribble (Scrib) and Lethal (2) giant larvae (Lgl) [88, 93]. Septate junctions (SJ) (the equivalent of vertebrate Tight Junctions) are primary candidates for cyst integrity and coordination, as apart from acting as sealing junctions in epithelia and neurons by mediating cell-cell adhesion, they act as scaffolding networks together with multiple pathways to promote organ morphogenesis [94]. The last years a number of studies addressed the role of septate junction components in the testicular cyst integrity, spermatogenesis and testis homeostasis [58, 88, 92, 93]. These include studies analyzing the critical role of the *Drosophila* Profilin and septate junctional core components Nrx-IV and Cora [58, 92] in germline encapsulation by the cyst cells. Soma-germline interaction proceeds through two subsequent steps during early spermatogenesis: an "encapsulation" stage in which the cyst cells wrap the germ cells, and an "occlusion" stage in which a permeability barrier is established around the germline [92]. Permeability assays in the *Drosophila* testis cysts revealed that this permeability barrier is established by the function of septate junction components in the cyst cells and is critical for restricting soma-germline communication within the cysts, keeping the differentiating germline isolated of from the niche signals and allowing spermatogenesis to proceed.

The critical requirement of germline encapsulation by cyst cells has been shown by studies on Dlg, Scrib and Lgl. These genes have been identified as tumor suppressor genes in *Drosophila* leading to neoplastic transformation [95–98] which is characterized by overproliferating epithelial cells that lose their apico-basal polarity and their ability to terminally differentiate [99]. Dlg and Scrib localize at the cytoplasmic side of SJs underlying the membrane cortical side of adjacent epithelial cells, neublasts and neuromuscular junctions (NMJs). Yet, Dlg and Scrib are not part of the highly stable core protein complex that builds the SJs in *Drosophila* [100]. *scrib* is expressed in the somatic gonadal cells of the newly formed embryonic *Drosophila* gonads [101, 102] and analysis of agametic gonads and pseudo-gonads made of aggregated germ cells confirmed this observation [102]. Analysis of *scrib* and *dlg* mutant gonads revealed that the gonadal mesodermal cells are not able to extend projections between the germ cells suggesting a role in establishing the intimate contacts of the gonadal mesoderm to the germ cells [93, 102]. At larval stages, *dlg*, *scrib* and *lgl* expression in the somatic lineage is indispensable for testis homeostasis and spermatogenesis, as depletion of these genes results in extremely small testes with reduced number of GSCs, increased number of early cyst cells (positive for early cyst cell marker Traffic-Jam), impaired differentiation and infertility. Late cyst cells and their corresponding spermatocyte cysts are lost via apoptosis in *dlg* testes [88]. Similar to *dlg*, *lgl* testes also lose late cyst cells, whereas in *scrib* testes late cyst cells are still present albeit with a significantly reduced nucleus size [93]. Rescue experiments expressing a *dlg* transgene in cyst cells could restore normal testis morphology and function. Expression of a *dlg* transgene in late cyst cells could restore only partially the development of the cysts since the cyst cells could not perfectly elongate and ensheath the germline. *dlg* overexpression in cyst cells leads to the formation of wavy and ruffled plasma membrane, suggesting that Dlg may affect the process through which cyst cells grow enormously, elongate and ensheath the germline. Such a function would be in agreement with previous findings showing that Dlg regulates membrane proliferation in a subset of NMJs and is an important player in the process of polarized membrane insertion during cellularization [99, 103–105].

Taken together, *dlg* is required (i) in early cyst cells to properly ensheath the GSCs and spermatogonia, and establish normal testicular cyst architecture and (ii) in late cyst cells flanking the germline spermatocytes, for cyst cell survival, growth, expansion and maintaining the integrity of the cysts [88].

All these studies on the role of septate junction components in the *Drosophila* testis, underline the importance of the CySCs and SCCs to encapsulate the germline and establish the cyst integrity, which is necessary for germline differentiation and normal testis function to finally produce healthy sperm and ensure fertility [10, 89, 93]. Interestingly, the *Drosophila* testis cyst cells show striking similarities with the Sertoli cells, the supportive cells of the mammalian germline, in terms of cytoskeletal and scaffolding components [10]. Since several of these genes show high degree of conservation to their vertebrate homologues [28, 89], what we learn about cyst cell function, soma-germline coordination and the underlying regulatory logic in the *Drosophila* testis can be directly tested in other organisms and stem cell systems in other tissues.

3 Mammalian Testes and the Spermatogonial Stem Cell Niche

Morphogenesis of the mammalian testis begins shortly after birth and continues until puberty when the first round of spermatogenesis is completed. Spermatogenesis is a highly regulated and complex process, which proceeds in three phases [106]. First is the proliferative phase, in which spermatogonia undergo a series of amplifying divisions and differentiate into primary spermatocytes. Second is the meiotic phase, in which the germline goes through meiosis and genetic recombination, resulting in the formation of haploid spermatids. Finally during spermiogenesis, the round germ cells are transformed through rearrangement of the cytoskeletal structure, to specialized spermatozoa [106, 107].

At the anatomical level, the mammalian testis does not possesses an overall "polarity" the way it is described for other model systems such as *Drosophila*. The entire developmental process from spermatogonia to spermatozoa occurs in the seminiferous tubules (Fig. 3). Seminiferous tubules show a simple structure composed of Sertoli cells and peritubular myoid cells, with the basal membrane made by contributions of both Sertoli and myoid cells. Normal spermatogenesis in mammals, like in *Drosophila*, depends on functional interactions between the Sertoli cells and the reproductive germ cells. The Sertoli cells are the supporting somatic cell type of the mammalian testis, like nursery cells, which maintain the spermatogonial stem cells niche. Each Sertoli cell supports $\sim$ 50 germ cells in the epithelium, produces a number of factors important for stem cell self-renewal and plays a role in germ cell maturation and balance between self-renewal, proliferation and differentiation [108–110]. Sertoli cells form the blood-testis barrier, which divides the seminiferous epithelium into a basal compartment and an adluminal compartment (Fig. 3b).

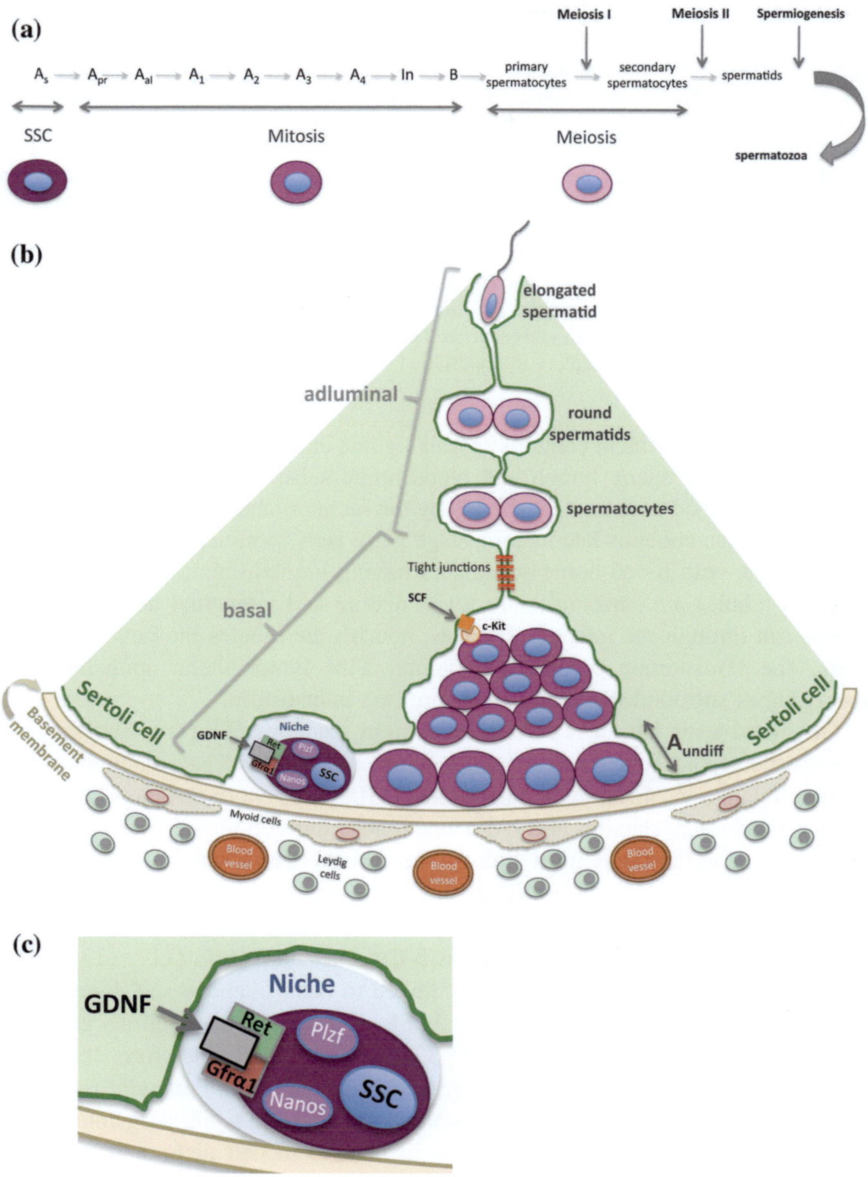
(a)
Meiosis I
Meiosis II
Spermiogenesis
A_s → A_{pr} → A_{al} → A_1 → A_2 → A_3 → A_4 → In → B → primary spermatocytes → secondary spermatocytes → spermatids
SSC
Mitosis
Meiosis
spermatozoa
(b)
elongated spermatid
adluminal
round spermatids
spermatocytes
Tight junctions
SCF
c-Kit
basal
Sertoli cell
Basement membrane
Niche
GDNF
Ret
Gfrα1
Plzf
Nanos
SSC
A_{undiff}
Sertoli cell
Myoid cells
Blood vessel
Leydig cells
Blood vessel
Blood vessel
(c)
Niche
GDNF
Ret
Gfrα1
Plzf
Nanos
SSC

◀ **Fig. 3** Diagram depicting the spermatogonial stem cell niche and spermatogenesis in the mammalian testis. **a** Schematic diagram showing the stages of spermatogenesis in mice, which is the sum of all germ cell divisions and differentiation steps, beginning with the spermatogonial stem cell (SSC) differentiation and ending with the generation of spermatozoa. **b** Diagram of a seminiferous epithelium, showing the germ cells, the Sertoli cells touching the basal membrane and the underlying myoid, Leyding and blood cells. The basal compartment of the Sertoli cells, including spermatogonia and preleptotene spermatocytes, are exposed to growth factors and signals that emanate from the lymph and blood cells or signals captured at the basement membrane. The SSC niche is formed on the basis of both architectural support and secreted signals produced by the niche cells. The Sertoli cells provide the niche for the SSCs and secrete GDNF, while at later stages Sertoli cells secrete SCF and promote spermatogonia differentiation. **c** Diagram depicting a close up of the SSC niche from (**b**). SSC maintenance is promoted by signals emanating from the Sertoli cells such as the GDNF, but also by SSC intrinsic factors like Nanos, Plzf and the GDNF receptors Ret and Gfrα1

The basal compartment contains spermatogonia and preleptotene spermatocytes, and is exposed to many lymph and blood-borne substances. The adluminal is a structurally defined environment created by the secretory and endocytic activities of the Sertoli cells, contains late meiotic stage germ cells, postmeiotic spermatids and spermatozoa, with blood-borne substances having limited direct access. The seminiferous tubules are surrounded by a vasculature and interstitial network. Blood vessels run through the interstitial spaces, nourish the tubules but never penetrate them. The testosterone producing Leyding cells, lymphathetic epithelium and macrophages surround the vessels and form the interstisium.

The blood-testis barrier is composed of tight junctions (TJs) (the equivalent of invertebrate septate junctions), adherens junctions (AJs) and gap junctions with a different relative localization than in epithelia [108]. TJs consist of trans-membrane proteins (occludins, claudins and JAM/Junctional Adhesion Molecules), which bind scaffold proteins to build up a link to the cytoskeleton, vesicle trafficking and signaling pathways [111, 112]. The scaffold proteins consist mainly of Zonula Occludens (ZO) proteins, a family of MAGUK proteins, closely related to mammalian Dlg and adaptor proteins such as β-Catenin and Afadin [111, 113]. ZO-1, the backbone of the scaffold, is directly linked to actin filaments and binds to the transmembrane occludins, claudins and JAMs. ZO-1 binds also various gap junction proteins, among which connexin 43 (Con43) [108, 114]. Therefore, despite the morphological differences between mammalian Sertoli and *Drosophila* cyst cells, the high degree of conservation in terms of cytoskeletal and scaffolding components provides another level of commonalities in male stem cell niche and spermatogenesis to investigate between mammals and model system organisms.

3.1 *Spermatogonial Stem Cells and Germ Cell Differentiation*

The GSCs of the mammalian testis are the spermatogonial stem cells (SSCs), which, similar to all other stem cells, have the capacity to both self-renew and differentiate. In contrast to the *Drosophila* testis, where GSCs perform this dual function via asymmetric cell division, it is yet unknown whether the SSCs perform this function by using a symmetric, asymmetric or both division pathways in the mammalian testis [106, 115]. There are up to nine different spermatogonia populations in the mammalian testes, classified into three major groups: the type A, the intermediate (In) and the type B spermatogonia (Fig. 3a). The type A spermatogonia consist of A_{single} (A_s), A_{paired} (A_p), $A_{aligned}$ (A_{al}; chains of 4, 8, 16 and occasionally 32 germ cells), A_1, A_2, A_3 and A_4 spermatogonia. SSCs are considered to be the A_s, which are the most primitive spermatogonia containing no intercellular bridges. The A_s-A_{al} are primitive spermatogonia with minimal heterochromatic condensation, consist of less than 1 % of all testicular cells and are collectively called $A_{undifferentiated}$ (A_{undiff}). Due to incomplete cytokinesis, A_s progeny possess intercellular bridges, forming syncytial chains of interconnected germ cells. In that sense, A_{undiff} spermatogonia behave as a stem cell compartment, but since the individual germ cells are not all equivalent to stem cells, they are still a heterogeneous population. A_{undiff} give rise to A_1 spermatogonia, which go through six mitotic divisions (each forming A_2, A_3, A_4, In, B spermatogonia and preleptotene primary spermatocytes). A_4 spermatogonia mature into In and type B spermatogonia that enter two meiotic divisions, become primary and secondary spermatocytes, give rise to haploid spermatids and finally to spermatozoa [106, 107, 116, 117].

3.2 *The Mammalian Testis Niche: A Flexible Unit for Stem Cell Maintenance*

Stem cell self-renewal is tightly regulated by signals emanating from the stem cell microenvironment called the spermatogonial stem cell (SSC) niche. In contrast to *Drosophila*, the anatomical site of the SSC niche in mammals is less clearly defined. The mammalian testis contains seminiferous tubules lined by the seminiferous epithelium, which consists of germ cells that proliferate and differentiate into sperm in a direction towards the lumen. The niche microenvironment consists of the Sertoli cells, the major contributor of the SSC niche, but also of the basement membrane and other testicular somatic cell components of the interstitial space in between the seminiferous tubules (peritubular myoid cells, Leydig cells and vasculature) (Fig. 3b) [106, 118]. The niche extrinsic signals include growth factors produced by Sertoli cells, adhesion molecules linking the SSCs to basement membrane components such as the laminins, and stimuli from the vascular network and the interstitial cells.

Our understanding of the mammalian SSC niche has advanced since the discovery of the glial cell line-derived neurotrophic factor (GDNF), a TGFβ family member, secreted from the Sertoli cells (Fig. 3c). GDNF is so far the best described paracrine factor responsible for the maintenance and self-renewal of SSCs [118–122]. Indeed, GDNF overexpression in the mouse testis can sufficiently suppress SSC differentiation and lead to accumulation of stem cell-like undifferentiated germ cells. GDNF signals through a multicomponent receptor complex comprising of a glycosylphosphatidylinositol (GPI)-anchored cell surface molecule [the GDNF family receptor (Gfr)α-1] and Rearranged during Transfection (Ret) tyrosine kinase transmembrane protein [118, 123, 124]. The binding of GDNF triggers the activation of multiple signaling pathways in responsive cells. Interestingly, GDNF acts directly on the SSCs for controlling self-renewal and proliferation by activating Src and phosphatidylinositol 3-kinase (PI3 K)/Akt signaling activities [118, 125, 126]. In Sertoli cells, the production of GDNF is under the influence of the follicle-stimulating hormone (FSH), growth factors and cytokines such as the basic fibroblast growth factor (FGF2), the tumor necrosis factor (TNFα) and the interleucin-1β (Il-1β) [118, 127, 128]. This shows that SSC maintenance and self-renewal is controlled locally but also systemically.

The potential regulation of the SSC niche with systemic endocrine factors suggests that the physical location of SSCs to the niche requires close proximity to the vascular system. This hypothesis was tested with the use of a three-dimensional culture system and an image capturing system, and showed that A_{undiff} localize preferentially close to the vascular network and interstitial space surrounding the seminiferous tubules [107, 118, 129, 130]. These observations, suggest the presence of a rather "flexible" niche, which may be reversibly specified along with the vasculature pattern and its organization (Fig. 3b). This offers "robustness" to the SSC niche and may be of particular importance for mammals having a large body and a need for a long-lived functional stem cell system. Finally, it is important to stress that SSC niche function varies in a dynamic way during mammalian development (embryonic, post-natal and adult stages) [106, 131]. This is supported by studying niche stimulation via the gonadotropin-realising hormone (GnRH) [132], TGFβ signaling components [133, 134] and SSC transplantation assays showing that immature mouse pup testes can support the colonization and proliferation of transplanted SSCs much better than in adult testes [135]. These studies provide evidence that the male stem cell niche and spermatogenesis in mammals undergoes dramatic stage-specific changes during testis development and maturation.

3.3 Germ Cell Intrinsic Factors: Balancing Self-renewal Versus Differentiation

Apart from the extrinsic factors emanating from the SSC niche, germ cell intrinsic factors are also critical in regulating SSC maintenance or promoting germline differentiation [106, 131]. GDNF-regulated transcription within the germline is important for SSC self-renewal. A microarray-based gene expression profiling experiment, identified genes within the germline regulated by GDNF stimulation in cultures [136] such as the transcription-factor encoding genes *bcl6b* (B cell CLL/lymphoma 6, member B), *etv5* (Ets variant gene 5) and *lhx1* (Lim homeobox protein 1), which are also required for maintaining SSC self-renewing cultures. Analysis of disrupted spermatogenesis in null mutant mice uncovered the essential function of two other transcription factors in mouse SSC renewal: of Plzf (promyelotic leukemia zinc finger proteins) and of Taf4b [TATA box binding protein (TBF)-associated factor 4b]. Plzf is a transcriptional repressor that inhibits stem cell differentiation and helps maintain their presence in the niche. Plzf acts critically within the SSCs for self-renewal, since naturally occurring mutants or knockout mice for *plzf* lose spermatogonia as they progressively age and the balance is shifted towards differentiation at the expense of self-renewal (Fig. 3c) [118, 137, 138]. Plzf interacts with signaling pathways, in response to Sertoli-derived signals such as GDNF and Stem Cell Factor (SCF) encoded by the *Steel* (*Sl*) locus [2, 115, 139]. Taf4b is a germ cell specific component of the RNA polymerase II basal transcription apparatus. More precisely, Taf4b is expressed in gonocytes of post-natal testes, and in spermatogonia and spermatids of adult testes. Targeted disruption of the gene leads to a variety of phenotypes among which is SSC disappearance after 3 days of birth, progressive loss of germ cells and testis atrophy within 12 weeks time [118, 140]. Interesting is that GDNF does not influence the expression of either Plzf or Taf4b in SSC cultures.

Octomer-4 (Oct-4; also known as POU domain class 5 transcription factor 1/Pou5f1), localizes in proliferating gonocytes and after birth in A_{undiff} spermatogonia [115, 141]. Oct4 is used as a SSC marker and knockdown experiments revealed its importance for SSC maintenance. However, Plzf and Oct4 work in different pathways to SSC survival and self-renewal. A more recent study suggests that POU3F1 (but not Oct4) is an intrinsic regulator of GDNF-induced survival and self-renewal of mouse SSCs [115, 142]. The Zinc finger and Broad Complex/Tramtrack/bric-a-brac (ZBTB) 16 is a transcriptional repressor, expressed only in A_{diff}, with a critical role in SSC self-renewal [115, 138]. Other germ cell intrinsic factors include *nanos2* and *nanos3*, members of the zinc-finger motif containing conserved family of RNA-binding proteins [131, 143]. *nanos2* is exclusively expressed in A_s-A_{pr} and mutations lead to SSC loss while *nanos3* is expressed in A_s-A_{al} and differentiating A_1 spermatogonia, and mutations lead to germ cell loss in both males and females [131, 143, 144]. Up-regulation of *nanos3* results in accumulation of germ cells in G1 phase whereas treatment with retinoic acid results in dramatic reduction of Nanos3 [115, 145].

Several other germ cell intrinsic factors act at a later stage to promote the differentiation of spermatogonia. The transition from SSCs to differentiating spermatogonia is marked by the expression of *c-Kit* and the loss of *neurogenin3* (*ngn3*) expression [115, 146, 147]. Signals mediated by the Sertoli-produced SCF are received by the c-Kit tyrosine kinase receptor and allow germ cells to develop beyond type A spermatogonial stages (Fig. 3b) [2, 148, 149]. Plzf directly represses transcription of the receptor *c-kit* [118, 150] and accordingly homozygous *plzf* mutant mice show increased expression of *c-kit*. Ngn3, a basic helix loop helix transcription factor, is expressed in A_s-A_{al} spermatogonia that are *c-Kit*$^-$ to initiate spermatogonial differentiation [115, 129]. The SRY-box containing gene 3 (Sox3), member of a high mobility group family of transcription factors, is expressed in A_s, A_{pr} and A_{al} spermatogonia [115, 151] and regulates spermatogonia differentiation together with Ngn3 [115]. Finally, the RNA-binding protein Dazl, expressed in the primary spermatocytes and weakly in spermatogonia, regulates the transition of A_{al} to A_1 spermatogonia [115, 152].

3.4 *Spermatogonial Stem Cell Homing*

In stem cell biology, the term "homing" refers to the migration and retention of stem cells to their corresponding niche [153]. Niche architecture relies largely on the physical interactions of the stem cell membrane with tethering molecules on neighboring non-stem cells or surfaces that keep niche integrity and allow the exchange of signals shaping the niche [154]. The binding of adhesion molecules between cells promotes cell-to-cell communication and affects differentiation, growth and survival. Accordingly, proper SSC homing through adhesion is critical for spermatogenesis and successful sperm production. The basement membrane of the mammalian testis consists of ECM containing fibronectin, collagens and laminins [153]. Within the seminiferous epithelium, the Sertoli cells are anchored to the basement membrane and consequently spermatogonia reside in close proximity to it as well. One example is the interplay of integrins with focal adhesion proteins [155], laminins and other ECM components [156–159], pointing out once more that basic molecular features of niche integrity are found to be common in different stem cell and model systems. Within mammalian tissues, high levels of integrin expression is used as a marker for tissue stem cells, suggesting that the attachment to a basal lamina is important for holding stem cells within the niche [160]. In the mammalian testes, α6 integrin has been used as a surface marker for enrichment of SSCs [161]. During spermatogenesis, the moving of developing germ cells across the seminiferous epithelium is associated with extensive reconstructing of actin based AJs between Sertoli and germ cells. The regulation of AJ assembly and dynamics is regulated by the interplay of β1-integrin with focal adhesion complex associated proteins, such as vinculin and phosphorylated FAK [155]. On the other hand, the "homing efficiency" is critical for male fertility restoration and spermatogenesis regeneration [162]. Due to its requirement in both SSCs and Sertoli cells, β1-integrin seems to play an important role for SSC homing at the basal membrane, as this was shown in vivo and in vitro by transplanting SSCs into the lumen of the seminiferous tubules [163, 164]. Another example is the guanine nucleotide exchange factor Vav [165, 166], which is critical for retaining of SSCs in the mammalian testis [163]. Presumably, despite the differences in niche architecture across different stem cell systems, homologues or equivalent factors are frequently utilized for the execution of analogous tasks. Understanding integrin involvement in niche and stem cell regulation will not only provide the basic mechanisms underlying these processes but will also allow the development of a whole range of new stem cell based therapeutic approaches.

3.5 Bone Morphogenetic Factors: Germ Cell Maintenance, Differentiation and Spermatogenesis

Another group of TGFβ family members, with an important role in germline maintenance and differentiation, are the bone morphogenetic factor proteins (BMPs). Multiple BMPs are expressed in the testis germline, such as BMP7, BMP8a and BMP8b, whereas BMP4 is expressed in both the germline and the Sertoli cells. BMP4, BMP8a and BMP8b regulate primordial germ cell (PGC) specification, whereas BMP2 and BMP4 enhance juvenile spermatogonial proliferation in in vitro assays [167–169]. Although no evidence is yet available about BMP2 localization in young and adult testes, it has been shown that BMP2 increases spermatogonia proliferation without affecting the Sertoli mitotic activity in adult testes [169]. BMP7, 8a and 8b are expressed in spermatogonia and early spermatocytes in the juvenile testis, with expression shifting to round spermatids of the adult testis. Targeted activation of BMP8b revealed its importance for the onset of spermatogenesis in the juvenile testis, while BMP8a is necessary for maintaining spermatogenesis in the adult testis [133, 170]. BMP7 regulates Sertoli cell proliferation [169] and plays a role in spermatogenesis maintenance by exacerbating the function of BMP8a [171]. This further suggests, that BMP8a and BMP7 signal through similar, if not the same, receptors in the testis. Furthermore, targeted disruption of BMPs revealed that they all play redundant roles in maintaining the viability of germ cells including the SSCs [131, 169, 171]. BMP4 is required for sustained spermatogenesis in vivo [167, 172], since *BMP4* knockout mice die at birth with testis containing no germ cells [133, 173] and addition of BMP4 in cultured SSCs promotes differentiation and blocks SSC self-renewal by activating *c-kit* [115, 162, 168]. BMP signals are regulated by the inhibitory SMAD proteins, mediators of the TGFβ signaling, which show differential expression between early (fetal and postnatal] and adult spermatogenesis [167]. Interestingly, in co-cultured immature mouse spermatogonia and Sertoli cells, BMP2 and BMP4 stimulate the expression of Smad1, Smad5 and Smad8 in spermatogonia nuclei [115, 167]. As described before, the BMPs control the balanced growth of spermatogonial cells and their niches in a stage specific way during mammalian embryonic, post-natal and adult testis organogenesis [133, 134].

4 Male Stem Cell Niches: Common Themes and Future Challenges

The male stem cell niches, as highlighted in this book chapter, have a key-organizing role in proper spermatogenesis and healthy sperm production, which is the aim of all sexually reproductive organisms. Despite the structural and functional differences between e.g. *Drosophila* and mammalian testes, it becomes clear that the basic organizing principles of the male stem cell niches are similar and directly comparable. The male stem cell niche recruits and "homes" the GSCs and SSCs in *Drosophila* and mammalian testes respectively, and orchestrates the balance between stem cell maintenance and germ cell differentiation. Niche extrinsic signals control GSC and SSC self-renewal directly by activating stem cell maintaining factors and by suppressing cellular components promoting differentiation. TGFβ signaling components seem to be critical in both systems (GDNF versus Dpp and BMPs) whereas others, like the JAK-STAT pathway are so far identified only in *Drosophila*. Conversely, topics already known from mammals, like the progressive, step-wise development of germline stem cells and the male niche along different pathways during development [2, 131, 133, 134], now starts to emerge also in *Drosophila*: niche architecture and positioning is differentially regulated in a cell type and stage specific way [28, 55] and the EFGR signaling stage-specifically controls GSC division rate and germline differentiaton [79, 80].

Soma-germline coordination through regulated adhesion, junctional complexes and cytoskeletal components is also critical in male niches. This is impressively illustrated in the case of the SJs forming (i) the blood-testis barrier in the Sertoli cells that separates the differentiating spermatogonia at the adluminal part, and (ii) the permeability barrier in late SCCs that isolates the spermatocyte containing cysts from the outside environment. In both cases, the SJ barriers protect the germ cells from the signals of the niche once they irreversibly enter the differentiation program towards sperm production. However, one has to point out that the Sertoli cells perform the function of both the hub cells and the cyst cells (CySCs and SCCs) in the *Drosophila* testis. Sertoli cells provide the GDNF signal equivalent to the JAK-STAT and TGFβ signals emanating from the hub cells, while at the same time they act as supportive "escort" cells for the differentiating type A, In and type B spermatogonia (throughout the mitotic and meiotic divisions, spermatid and spermatozoa formation). Therefore, Sertoli cells support the germline in a dynamic way and it would exciting to elucidate in future how the Sertoli cells are programmed for this functional role. On the other hand, comparing the Sertoli cells not only to the *Drosophila* testis hub cells but also to the cyst cells fits very nicely with a newly emerging field in *Drosophila*, that cyst cells act as a "local niche" for the germ cells they encapsulate.

Male stem cell niche and testes of different organisms are regulated in a dynamic and very precise way, in which intrinsic and extrinsic factors in combination with adhesion and coordinated cell-cell cross-talk control the balance between stem cell self-renewal and germline differentiation. At the same time, the system is flexible enough to respond to environmental conditions e.g. nutrient supply, injury and other systemic influences. This is convincingly illustrated in the mammalian testes, where the niche seems to be a flexible unit, comprised of different cellular components. In *Drosophila* it was initially thought that the testis organization is rather stable, lacking the plasticity observed in mammals, with a well defined niche of post-mitotic hub cells confined in the anterior of the *Drosophila* testis. However, newly emerged studies on soma-germline competition, germline dedifferentiation and conversion of quiescent hub cells to cyst cells upon challenge, reveal that indeed the system is built upon a dynamic rather than static equilibrium.

Drosophila is a simple model system with genetic tools that allow to test in a straightforward way the role of signaling pathways and single genes homologues, where mammals usually possess several homologues with redundant functions. On the other side, mammals have far larger bodies with organs harboring many more cells, live much longer and have the need for a more robust system that protects sperm production over time and under changing conditions. The advantages of each model system, combined with the development of new tools and methodologies, open up the possibility to test concepts and transfer knowledge from one model system to the other. Finally, understanding the basic mechanisms regulating germline stem cells and their niches can ultimately be used in regenerative medicine to repair aging and damaged stem cells, treat male infertility and improve applications in agricultural species [2, 10, 130, 163, 174].

Acknowledgments We would like to thank the members of the Ingrid Lohmann lab for critically reading the manuscript. Finally, we thank Elsevier for licensed content use (Copyright Clearance Center's RightsLink service; License Number: 3570791468065). We apologize to all whose work was not cited due to space limitations. This work was supported by the DFG/SFB 873.

References

1. Donovan PJ, Gearhart J. The end of the beginning for pluripotent stem cells. Nature. 2001;414(6859):92–7.
2. Spradling A, Drummond-Barbosa D, Kai T. Stem cells find their niche. Nature. 2001;414 (6859):98–104.
3. Morrison SJ, Spradling AC. Stem cells and niches: mechanisms that promote stem cell maintenance throughout life. Cell. 2008;132(4):598–611.
4. Schofield R. The relationship between the spleen colony-forming cell and the haemopoietic stem cell. Blood Cells. 1978;4(1–2):7–25.
5. Xie T, Spradling AC. A niche maintaining germ line stem cells in the Drosophila ovary. Science. 2000;290(5490):328–30.
6. Kiger AA, Jones DL, Schulz C, Rogers MB, Fuller MT. Stem cell self-renewal specified by JAK-STAT activation in response to a support cell cue. Science. 2001;294(5551):2542–5.
7. Tulina N, Matunis E. Control of stem cell self-renewal in Drosophila spermatogenesis by JAK-STAT signaling. Science. 2001;294(5551):2546–9.
8. Crittenden SL, Bernstein DS, Bachorik JL, Thompson BE, Gallegos M, Petcherski AG, et al. A conserved RNA-binding protein controls germline stem cells in Caenorhabditis elegans. Nature. 2002;417(6889):660–3.

9. Li L, Xie T. Stem cell niche: structure and function. Annu Rev Cell Dev Biol. 2005;21:605–31.
10. Papagiannouli F, Lohmann I. Shaping the niche: Lessons from the Drosophila testis and other model systems. Biotechnol J. 2012.
11. Xie T. Germline stem cell niches. StemBook. Cambridge (MA)2008.
12. Kiel MJ, Yilmaz OH, Iwashita T, Yilmaz OH, Terhorst C, Morrison SJ. SLAM family receptors distinguish hematopoietic stem and progenitor cells and reveal endothelial niches for stem cells. Cell. 2005;121(7):1109–21.
13. Sugiyama T, Kohara H, Noda M, Nagasawa T. Maintenance of the hematopoietic stem cell pool by CXCL12-CXCR4 chemokine signaling in bone marrow stromal cell niches. Immunity. 2006;25(6):977–88.
14. Doetsch F. The glial identity of neural stem cells. Nat Neurosci. 2003;6(11):1127–34.
15. Shen Q, Goderie SK, Jin L, Karanth N, Sun Y, Abramova N, et al. Endothelial cells stimulate self-renewal and expand neurogenesis of neural stem cells. Science. 2004;304(5675):1338–40.
16. Cotsarelis G, Sun TT, Lavker RM. Label-retaining cells reside in the bulge area of pilosebaceous unit: implications for follicular stem cells, hair cycle, and skin carcinogenesis. Cell. 1990;61(7):1329–37.
17. Fuchs E, Tumbar T, Guasch G. Socializing with the neighbors: stem cells and their niche. Cell. 2004;116(6):769–78.
18. Barker N, van Es JH, Kuipers J, Kujala P, van den Born M, Cozijnsen M, et al. Identification of stem cells in small intestine and colon by marker gene Lgr5. Nature. 2007;449 (7165):1003–7.
19. Parreira GG, Ogawa T, Avarbock MR, Franca LR, Brinster RL, Russell LD. Development of germ cell transplants in mice. Biol Reprod. 1998;59(6):1360–70.
20. Le Bras S, Van Doren M. Development of the male germline stem cell niche in Drosophila. Dev Biol. 2006;294(1):92–103.
21. Dinardo S, Okegbe T, Wingert L, Freilich S, Terry N. lines and bowl affect the specification of cyst stem cells and niche cells in the Drosophila testis. Development. 2011;138(9):1687–96.
22. Boyle M, DiNardo S. Specification, migration and assembly of the somatic cells of the Drosophila gonad. Development. 1995;121(6):1815–25.
23. DeFalco T, Le Bras S, Van Doren M. Abdominal-B is essential for proper sexually dimorphic development of the Drosophila gonad. Mech Dev. 2004;121(11):1323–33.
24. DeFalco T, Camara N, Le Bras S, Van Doren M. Nonautonomous sex determination controls sexually dimorphic development of the Drosophila gonad. Dev Cell. 2008;14(2):275–86.
25. Hatini V, Green RB, Lengyel JA, Bray SJ, Dinardo S. The Drumstick/Lines/Bowl regulatory pathway links antagonistic Hedgehog and Wingless signaling inputs to epidermal cell differentiation. Genes Dev. 2005;19(6):709–18.
26. Sheng XR, Posenau T, Gumulak-Smith JJ, Matunis E, Van Doren M, Wawersik M. Jak-STAT regulation of male germline stem cell establishment during Drosophila embryogenesis. Dev Biol. 2009.
27. Fuller MT, Spradling AC. Male and female Drosophila germline stem cells: two versions of immortality. Science. 2007;316(5823):402–4.
28. Papagiannouli F, Schardt L, Grajcarek J, Ha N, Lohmann I. The hox gene abd-B controls stem cell niche function in the Drosophila testis. Dev Cell. 2014;28(2):189–202.
29. Kozopas KM, Samos CH, Nusse R. DWnt-2, a Drosophila Wnt gene required for the development of the male reproductive tract, specifies a sexually dimorphic cell fate. Genes Dev. 1998;12(8):1155–65.
30. Fuller MT. Spermatogenesis. Arias MBAM, editor. Cold Spring Harbor, New York: Cold Spring Harbor Laboratory Press; 1993. 71–147.
31. Okegbe TC, DiNardo S. The endoderm specifies the mesodermal niche for the germline in Drosophila via Delta-Notch signaling. Development. 2011;138(7):1259–67.
32. DeFalco TJ, Verney G, Jenkins AB, McCaffery JM, Russell S, Van Doren M. Sex-specific apoptosis regulates sexual dimorphism in the Drosophila embryonic gonad. Dev Cell. 2003;5 (2):205–16.

33. Riechmann V, Rehorn KP, Reuter R, Leptin M. The genetic control of the distinction between fat body and gonadal mesoderm in Drosophila. Development. 1998;125(4):713–23.
34. Moore LA, Broihier HT, Van Doren M, Lehmann R. Gonadal mesoderm and fat body initially follow a common developmental path in Drosophila. Development. 1998;125 (5):837–44.
35. Broihier HT, Moore LA, Van Doren M, Newman S, Lehmann R. zfh-1 is required for germ cell migration and gonadal mesoderm development in Drosophila. Development. 1998;125 (4):655–66.
36. Starz-Gaiano M, Lehmann R. Moving towards the next generation. Mech Dev. 2001;105(1–2):5–18.
37. Zoller R, Schulz C. The Drosophila cyst stem cell lineage: partners behind the scenes? Spermatogenesis. 2012;2(3):145–57.
38. Kitadate Y, Kobayashi S. Notch and Egfr signaling act antagonistically to regulate germ-line stem cell niche formation in Drosophila male embryonic gonads. Proc Natl Acad Sci USA. 2010;107(32):14241–6.
39. Casper A, Van Doren M. The control of sexual identity in the Drosophila germline. Development. 2006;133(15):2783–91.
40. Hetie P, de Cuevas M, Matunis E. Conversion of quiescent niche cells to somatic stem cells causes ectopic niche formation in the Drosophila testis. Cell Rep. 2014;7(3):715–21.
41. Walker MR, Patel KK, Stappenbeck TS. The stem cell niche. J Pathol. 2009;217(2):169–80.
42. Tanentzapf G, Devenport D, Godt D, Brown NH. Integrin-dependent anchoring of a stem-cell niche. Nat Cell Biol. 2007;9(12):1413–8.
43. Kitadate Y, Shigenobu S, Arita K, Kobayashi S. Boss/Sev signaling from germline to soma restricts germline-stem-cell-niche formation in the anterior region of Drosophila male gonads. Dev Cell. 2007;13(1):151–9.
44. Lee S, Zhou L, Kim J, Kalbfleisch S, Schock F. Lasp anchors the Drosophila male stem cell niche and mediates spermatid individualization. Mech Dev. 2008;125(9–10):768–76.
45. Cram EJ, Schwarzbauer JE. The talin wags the dog: new insights into integrin activation. Trends Cell Biol. 2004;14(2):55–7.
46. Tanentzapf G, Martin-Bermudo MD, Hicks MS, Brown NH. Multiple factors contribute to integrin-talin interactions in vivo. J Cell Sci. 2006;119(Pt 8):1632–44.
47. Suyama R, Jenny A, Curado S. Pellis-van Berkel W, Ephrussi A. The actin-binding protein Lasp promotes Oskar accumulation at the posterior pole of the Drosophila embryo. Development. 2009;136(1):95–105.
48. Traenka J, Hauck CR, Lewandrowski U, Sickmann A, Gambaryan S, Thalheimer P, et al. Integrin-dependent translocation of LASP-1 to the cytoskeleton of activated platelets correlates with LASP-1 phosphorylation at tyrosine 171 by Src-kinase. Thromb Haemost. 2009;102(3):520–8.
49. Boyle M, Wong C, Rocha M, Jones DL. Decline in self-renewal factors contributes to aging of the stem cell niche in the Drosophila testis. Cell Stem Cell. 2007;1(4):470–8.
50. Cheng J, Turkel N, Hemati N, Fuller MT, Hunt AJ, Yamashita YM. Centrosome misorientation reduces stem cell division during ageing. Nature. 2008;456(7222):599–604.
51. van Steensel B, Henikoff S. Identification of in vivo DNA targets of chromatin proteins using tethered dam methyltransferase. Nat Biotechnol. 2000;18(4):424–8.
52. van Steensel B, Delrow J, Henikoff S. Chromatin profiling using targeted DNA adenine methyltransferase. Nat Genet. 2001;27(3):304–8.
53. Vogel MJ, Peric-Hupkes D, van Steensel B. Detection of in vivo protein-DNA interactions using DamID in mammalian cells. Nat Protoc. 2007;2(6):1467–78.
54. Choksi SP, Southall TD, Bossing T, Edoff K, de Wit E, Fischer BE, et al. Prospero acts as a binary switch between self-renewal and differentiation in Drosophila neural stem cells. Dev Cell. 2006;11(6):775–89.
55. Papagiannouli F, Lohmann I. Stage-specific control of stem cell niche architecture in the Drosophila testis by the posterior Hox gene Abd-B. Comput Struct Biotechnol J. 2015.

56. Zeng X, Singh SR, Hou D, Hou SX. Tumor suppressors Sav/Scrib and oncogene Ras regulate stem-cell transformation in adult Drosophila malpighian tubules. J Cell Physiol. 2010;224(3):766–74.
57. Leatherman JL, Dinardo S. Germline self-renewal requires cyst stem cells and stat regulates niche adhesion in Drosophila testes. Nat Cell Biol. 2010;12(8):806–11.
58. Shields AR, Spence AC, Yamashita YM, Davies EL, Fuller MT. The actin-binding protein profilin is required for germline stem cell maintenance and germ cell enclosure by somatic cyst cells. Development. 2014;141(1):73–82.
59. Flaherty MS, Salis P, Evans CJ, Ekas LA, Marouf A, Zavadil J, et al. chinmo is a functional effector of the JAK/STAT pathway that regulates eye development, tumor formation, and stem cell self-renewal in Drosophila. Dev Cell. 2010;18(4):556–68.
60. Issigonis M, Matunis E. The Drosophila BCL6 homolog Ken and Barbie promotes somatic stem cell self-renewal in the testis niche. Dev Biol. 2012;368(2):181–92.
61. Li Y, Ma Q, Cherry CM, Matunis EL. Steroid signaling promotes stem cell maintenance in the Drosophila testis. Dev Biol. 2014;394(1):129–41.
62. Michel M, Kupinski AP, Raabe I, Bokel C. Hh signalling is essential for somatic stem cell maintenance in the Drosophila testis niche. Development. 2012;139(15):2663–9.
63. Amoyel M, Sanny J, Burel M, Bach EA. Hedgehog is required for CySC self-renewal but does not contribute to the GSC niche in the Drosophila testis. Development. 2013;140(1):56–65.
64. Zhang Z, Lv X, Jiang J, Zhang L, Zhao Y. Dual roles of Hh signaling in the regulation of somatic stem cell self-renewal and germline stem cell maintenance in Drosophila testis. Cell Res. 2013;23(4):573–6.
65. Zhang Z, Pan C, Zhao Y. Hedgehog in the Drosophila testis niche: what does it do there? Protein & Cell. 2013.
66. Shivdasani AA, Ingham PW. Regulation of stem cell maintenance and transit amplifying cell proliferation by tgf-beta signaling in Drosophila spermatogenesis. Curr Biol. 2003;13 (23):2065–72.
67. Kawase E, Wong MD, Ding BC, Xie T. Gbb/Bmp signaling is essential for maintaining germline stem cells and for repressing bam transcription in the Drosophila testis. Development. 2004;131(6):1365–75.
68. Schulz C, Kiger AA, Tazuke SI, Yamashita YM, Pantalena-Filho LC, Jones DL, et al. A misexpression screen reveals effects of bag-of-marbles and TGF beta class signaling on the Drosophila male germ-line stem cell lineage. Genetics. 2004;167(2):707–23.
69. Zheng Q, Wang Y, Vargas E, DiNardo S. magu is required for germline stem cell self-renewal through BMP signaling in the Drosophila testis. Dev Biol. 2011;357(1):202–10.
70. Chang YJ, Pi H, Hsieh CC, Fuller MT. Smurf-mediated differential proteolysis generates dynamic BMP signaling in germline stem cells during Drosophila testis development. Dev Biol. 2013;383(1):106–20.
71. Lim JG, Fuller MT. Somatic cell lineage is required for differentiation and not maintenance of germline stem cells in Drosophila testes. Proc Natl Acad Sci USA. 2012;109(45):18477–81.
72. Pancratov R, Peng F, Smibert P, Yang S Jr, Olson ER, Guha-Gilford C, et al. The miR-310/13 cluster antagonizes beta-catenin function in the regulation of germ and somatic cell differentiation in the Drosophila testis. Development. 2013;140(14):2904–16.
73. Issigonis M, Tulina N, de Cuevas M, Brawley C, Sandler L, Matunis E. JAK-STAT signal inhibition regulates competition in the Drosophila testis stem cell niche. Science. 2009;326 (5949):153–6.
74. Stine RR, Greenspan LJ, Ramachandran KV, Matunis EL. Coordinate regulation of stem cell competition by Slit-Robo and JAK-STAT signaling in the Drosophila testis. PLoS Genet. 2014;10(11):e1004713.
75. Amoyel M, Simons BD, Bach EA. Neutral competition of stem cells is skewed by proliferative changes downstream of Hh and Hpo. EMBO J. 2014;33(20):2295–313.
76. Gonczy P, Matunis E, DiNardo S. bag-of-marbles and benign gonial cell neoplasm act in the germline to restrict proliferation during Drosophila spermatogenesis. Development. 1997;124 (21):4361–71.

77. Insco ML, Bailey AS, Kim J, Olivares GH, Wapinski OL, Tam CH, et al. A self-limiting switch based on translational control regulates the transition from proliferation to differentiation in an adult stem cell lineage. Cell Stem Cell. 2012;11(5):689–700.
78. Kiger AA, White-Cooper H, Fuller MT. Somatic support cells restrict germline stem cell self-renewal and promote differentiation. Nature. 2000;407(6805):750–4.
79. Parrott BB, Hudson A, Brady R, Schulz C. Control of germline stem cell division frequency—a novel, developmentally regulated role for epidermal growth factor signaling. PLoS ONE. 2012;7(5):e36460.
80. Hudson AG, Parrott BB, Qian Y, Schulz C. A temporal signature of epidermal growth factor signaling regulates the differentiation of germline cells in testes of Drosophila melanogaster. PLoS ONE. 2013;8(8):e70678.
81. Parrott BB, Chiang Y, Hudson A, Sarkar A, Guichet A, Schulz C. Nucleoporin98-96 function is required for transit amplification divisions in the germ line of Drosophila melanogaster. PLoS ONE. 2011;6(9):e25087.
82. Tran J, Brenner TJ, DiNardo S. Somatic control over the germline stem cell lineage during Drosophila spermatogenesis. Nature. 2000;407(6805):754–7.
83. Schulz C, Wood CG, Jones DL, Tazuke SI, Fuller MT. Signaling from germ cells mediated by the rhomboid homolog stet organizes encapsulation by somatic support cells. Development. 2002;129(19):4523–34.
84. Sarkar A, Parikh N, Hearn SA, Fuller MT, Tazuke SI, Schulz C. Antagonistic roles of Rac and Rho in organizing the germ cell microenvironment. Curr Biol. 2007;17(14):1253–8.
85. Tazuke SI, Schulz C, Gilboa L, Fogarty M, Mahowald AP, Guichet A, et al. A germline-specific gap junction protein required for survival of differentiating early germ cells. Development. 2002;129(10):2529–39.
86. Gilboa L, Forbes A, Tazuke SI, Fuller MT, Lehmann R. Germ line stem cell differentiation in Drosophila requires gap junctions and proceeds via an intermediate state. Development. 2003;130(26):6625–34.
87. Bohrmann J, Zimmermann J. Gap junctions in the ovary of Drosophila melanogaster: localization of innexins 1, 2, 3 and 4 and evidence for intercellular communication via innexin-2 containing channels. BMC Dev Biol. 2008;8:111.
88. Papagiannouli F, Mechler BM. discs large regulates somatic cyst cell survival and expansion in Drosophila testis. Cell Res. 2009;19(10):1139–49.
89. Papagiannouli F, Mechler BM. discs large in the Drosophila testis: An old player on a new task. Fly (Austin). 2010;4(4).
90. Lindsley DL, Tokuyasu KT. Spermatogenesis. In: Ashburner M, Wright TRF, editors. Genetics and Biology of Drosophila. 2D. 2nd ed. New York: Academic Press; 1980. p. 225–94.
91. Tokuyasu KT, Peacock WJ, Hardy RW. Dynamics of spermiogenesis in Drosophila melanogaster. II. Coiling process. Z Zellforsch Mikrosk Anat. 1972;127(4):492–525.
92. Fairchild MJ, Smendziuk CM, Tanentzapf G. A somatic permeability barrier around the germline is essential for Drosophila spermatogenesis. Development. 2015;142(2):268–81.
93. Papagiannouli F. The internal structure of embryonic gonads and testis development in Drosophila melanogaster requires scrib, lgl and dlg activity in the soma. Int J Dev Biol. 2013;57(1):25–34.
94. Yi P, Johnson AN, Han Z, Wu J, Olson EN. Heterotrimeric G proteins regulate a noncanonical function of septate junction proteins to maintain cardiac integrity in Drosophila. Dev Cell. 2008;15(5):704–13.
95. Mechler BM, McGinnis W, Gehring WJ. Molecular cloning of lethal(2)giant larvae, a recessive oncogene of Drosophila melanogaster. EMBO J. 1985;4(6):1551–7.
96. Woods DF, Hough C, Peel D, Callaini G, Bryant PJ. Dlg protein is required for junction structure, cell polarity, and proliferation control in Drosophila epithelia. J Cell Biol. 1996;134(6):1469–82.
97. Bilder D, Li M, Perrimon N. Cooperative regulation of cell polarity and growth by Drosophila tumor suppressors. Science. 2000;289(5476):113–6.

98. Wodarz A, Nathke I. Cell polarity in development and cancer. Nat Cell Biol. 2007;9 (9):1016–24.
99. Elsum I, Yates L, Humbert PO, Richardson HE. The Scribble-Dlg-Lgl polarity module in development and cancer: from flies to man. Essays Biochem. 2012;53:141–68.
100. Oshima K, Fehon RG. Analysis of protein dynamics within the septate junction reveals a highly stable core protein complex that does not include the basolateral polarity protein Discs large. J Cell Sci. 2011;124(Pt 16):2861–71.
101. Li M, Marhold J, Gatos A, Torok I, Mechler BM. Differential expression of two scribble isoforms during Drosophila embryogenesis. Mech Dev. 2001;108(1–2):185–90.
102. Marhold J, Papagiannouli F, Li M, Patel A, Mechler BM. Requirements for scribble expression in newly formed gonads of Drosophila embryos. Gene Expr Patterns. 2003;3 (2):143–6.
103. Lee OK, Frese KK, James JS, Chadda D, Chen ZH, Javier RT, et al. Discs-Large and Strabismus are functionally linked to plasma membrane formation. Nat Cell Biol. 2003;5 (11):987–93.
104. Chen K, Featherstone DE. Discs-large (DLG) is clustered by presynaptic innervation and regulates postsynaptic glutamate receptor subunit composition in Drosophila. BMC Biol. 2005;3:1.
105. Gorczyca D, Ashley J, Speese S, Gherbesi N, Thomas U, Gundelfinger E, et al. Postsynaptic membrane addition depends on the Discs-Large-interacting t-SNARE Gtaxin. J Neurosci. 2007;27(5):1033–44.
106. Oatley JM, Brinster RL. Regulation of spermatogonial stem cell self-renewal in mammals. Annu Rev Cell Dev Biol. 2008;24:263–86.
107. Okajima M, Takahashi M, Higuchi M, Ohsawa T, Yoshida S, Yoshida Y, et al. Human T-cell leukemia virus type 1 Tax induces an aberrant clustering of the tumor suppressor Scribble through the PDZ domain-binding motif dependent and independent interaction. Virus Genes. 2008;37(2):231–40.
108. Fink C, Weigel R, Hembes T, Lauke-Wettwer H, Kliesch S, Bergmann M, et al. Altered expression of ZO-1 and ZO-2 in Sertoli cells and loss of blood-testis barrier integrity in testicular carcinoma in situ. Neoplasia. 2006;8(12):1019–27.
109. Siu MK, Cheng CY. Dynamic cross-talk between cells and the extracellular matrix in the testis. Bioessays. 2004;26(9):978–92.
110. Hess RA, Cooke PS, Hofmann MC, Murphy KM. Mechanistic insights into the regulation of the spermatogonial stem cell niche. Cell Cycle. 2006;5(11):1164–70.
111. Lui WY, Mruk D, Lee WM, Cheng CY. Sertoli cell tight junction dynamics: their regulation during spermatogenesis. Biol Reprod. 2003;68(4):1087–97.
112. Zahraoui A, Louvard D, Galli T. Tight junction, a platform for trafficking and signaling protein complexes. J Cell Biol. 2000;151(5):F31–6.
113. Lee NP, Cheng CY. Adaptors, junction dynamics, and spermatogenesis. Biol Reprod. 2004;71(2):392–404.
114. Giepmans BN, Moolenaar WH. The gap junction protein connexin43 interacts with the second PDZ domain of the zona occludens-1 protein. Curr Biol. 1998;8(16):931–4.
115. Pane A, Jiang P, Zhao DY, Singh M, Schupbach T. The Cutoff protein regulates piRNA cluster expression and piRNA production in the Drosophila germline. EMBO J. 2011;30 (22):4601–15.
116. de Rooij DG, Russell LD. All you wanted to know about spermatogonia but were afraid to ask. J Androl. 2000;21(6):776–98.
117. de Rooij DG. Proliferation and differentiation of spermatogonial stem cells. Reproduction. 2001;121(3):347–54.
118. Hofmann MC. Gdnf signaling pathways within the mammalian spermatogonial stem cell niche. Mol Cell Endocrinol. 2008;288(1–2):95–103.

119. Moore MW, Klein RD, Farinas I, Sauer H, Armanini M, Phillips H, et al. Renal and neuronal abnormalities in mice lacking GDNF. Nature. 1996;382(6586):76–9.
120. Trupp M, Ryden M, Jornvall H, Funakoshi H, Timmusk T, Arenas E, et al. Peripheral expression and biological activities of GDNF, a new neurotrophic factor for avian and mammalian peripheral neurons. J Cell Biol. 1995;130(1):137–48.
121. Hofmann MC, Braydich-Stolle L, Dym M. Isolation of male germ-line stem cells; influence of GDNF. Dev Biol. 2005;279(1):114–24.
122. Meng X, Lindahl M, Hyvonen ME, Parvinen M, de Rooij DG, Hess MW, et al. Regulation of cell fate decision of undifferentiated spermatogonia by GDNF. Science. 2000;287 (5457):1489–93.
123. Treanor JJ, Goodman L, de Sauvage F, Stone DM, Poulsen KT, Beck CD, et al. Characterization of a multicomponent receptor for GDNF. Nature. 1996;382(6586):80–3.
124. Jing S, Wen D, Yu Y, Holst PL, Luo Y, Fang M, et al. GDNF-induced activation of the ret protein tyrosine kinase is mediated by GDNFR-alpha, a novel receptor for GDNF. Cell. 1996;85(7):1113–24.
125. Oatley JM, Avarbock MR, Brinster RL. Glial cell line-derived neurotrophic factor regulation of genes essential for self-renewal of mouse spermatogonial stem cells is dependent on Src family kinase signaling. J Biol Chem. 2007;282(35):25842–51.
126. Lee J, Kanatsu-Shinohara M, Inoue K, Ogonuki N, Miki H, Toyokuni S, et al. Akt mediates self-renewal division of mouse spermatogonial stem cells. Development. 2007;134 (10):1853–9.
127. Simon L, Ekman GC, Tyagi G, Hess RA, Murphy KM, Cooke PS. Common and distinct factors regulate expression of mRNA for ETV5 and GDNF, Sertoli cell proteins essential for spermatogonial stem cell maintenance. Exp Cell Res. 2007;313(14):3090–9.
128. Tadokoro Y, Yomogida K, Ohta H, Tohda A, Nishimune Y. Homeostatic regulation of germinal stem cell proliferation by the GDNF/FSH pathway. Mech Dev. 2002;113(1):29–39.
129. Yoshida S, Sukeno M, Nabeshima Y. A vasculature-associated niche for undifferentiated spermatogonia in the mouse testis. Science. 2007;317(5845):1722–6.
130. Caires K, Broady J, McLean D. Maintaining the male germline: regulation of spermatogonial stem cells. J Endocrinol. 2010;205(2):133–45.
131. Wong MD, Jin Z, Xie T. Molecular mechanisms of germline stem cell regulation. Annu Rev Genet. 2005;39:173–95.
132. Kanatsu-Shinohara M, Morimoto T, Toyokuni S, Shinohara T. Regulation of mouse spermatogonial stem cell self-renewing division by the pituitary gland. Biol Reprod. 2004;70 (6):1731–7.
133. Itman C, Mendis S, Barakat B, Loveland KL. All in the family: TGF-beta family action in testis development. Reproduction. 2006;132(2):233–46.
134. Barakat B, Itman C, Mendis SH, Loveland KL. Activins and inhibins in mammalian testis development: new models, new insights. Mol Cell Endocrinol. 2012;359(1–2):66–77.
135. Shinohara T, Orwig KE, Avarbock MR, Brinster RL. Remodeling of the postnatal mouse testis is accompanied by dramatic changes in stem cell number and niche accessibility. Proc Natl Acad Sci USA. 2001;98(11):6186–91.
136. Oatley JM, Avarbock MR, Telaranta AI, Fearon DT, Brinster RL. Identifying genes important for spermatogonial stem cell self-renewal and survival. Proc Natl Acad Sci USA. 2006;103(25):9524–9.
137. Costoya JA, Hobbs RM, Barna M, Cattoretti G, Manova K, Sukhwani M, et al. Essential role of Plzf in maintenance of spermatogonial stem cells. Nat Genet. 2004;36(6):653–9.
138. Buaas FW, Kirsh AL, Sharma M, McLean DJ, Morris JL, Griswold MD, et al. Plzf is required in adult male germ cells for stem cell self-renewal. Nat Genet. 2004;36(6):647–52.
139. Kotaja N, Sassone-Corsi P. Plzf pushes stem cells. Nat Genet. 2004;36(6):551–3.
140. Falender AE, Freiman RN, Geles KG, Lo KC, Hwang K, Lamb DJ, et al. Maintenance of spermatogenesis requires TAF4b, a gonad-specific subunit of TFIID. Genes Dev. 2005;19 (7):794–803.

141. Pesce M, Wang X, Wolgemuth DJ, Scholer H. Differential expression of the Oct-4 transcription factor during mouse germ cell differentiation. Mech Dev. 1998;71(1–2):89–98.
142. Wu X, Oatley JM, Oatley MJ, Kaucher AV, Avarbock MR, Brinster RL. The POU domain transcription factor POU3F1 is an important intrinsic regulator of GDNF-induced survival and self-renewal of mouse spermatogonial stem cells. Biol Reprod. 2010;82(6):1103–11.
143. Tsuda M, Sasaoka Y, Kiso M, Abe K, Haraguchi S, Kobayashi S, et al. Conserved role of nanos proteins in germ cell development. Science. 2003;301(5637):1239–41.
144. Sada A, Suzuki A, Suzuki H, Saga Y. The RNA-binding protein NANOS2 is required to maintain murine spermatogonial stem cells. Science. 2009;325(5946):1394–8.
145. Lolicato F, Marino R, Paronetto MP, Pellegrini M, Dolci S, Geremia R, et al. Potential role of Nanos3 in maintaining the undifferentiated spermatogonia population. Dev Biol. 2008;313 (2):725–38.
146. Schrans-Stassen BH, van de Kant HJ, de Rooij DG, van Pelt AM. Differential expression of c-kit in mouse undifferentiated and differentiating type A spermatogonia. Endocrinology. 1999;140(12):5894–900.
147. Yamaguchi M, Hirose F, Inoue YH, Ohno K, Yoshida H, Hayashi Y, et al. Genetic link between p53 and genes required for formation of the zonula adherens junction. Cancer Sci. 2004;95(5):436–41.
148. Ohta H, Yomogida K, Dohmae K, Nishimune Y. Regulation of proliferation and differentiation in spermatogonial stem cells: the role of c-kit and its ligand SCF. Development. 2000;127(10):2125–31.
149. Yoshinaga K, Nishikawa S, Ogawa M, Hayashi S, Kunisada T, Fujimoto T, et al. Role of c-kit in mouse spermatogenesis: identification of spermatogonia as a specific site of c-kit expression and function. Development. 1991;113(2):689–99.
150. Filipponi D, Hobbs RM, Ottolenghi S, Rossi P, Jannini EA, Pandolfi PP, et al. Repression of kit expression by Plzf in germ cells. Mol Cell Biol. 2007;27(19):6770–81.
151. Raverot G, Weiss J, Park SY, Hurley L, Jameson JL. Sox3 expression in undifferentiated spermatogonia is required for the progression of spermatogenesis. Dev Biol. 2005;283 (1):215–25.
152. Schrans-Stassen BH, Saunders PT, Cooke HJ, de Rooij DG. Nature of the spermatogenic arrest in Dazl -/- mice. Biol Reprod. 2001;65(3):771–6.
153. Oatley JM, Brinster RL. The germline stem cell niche unit in mammalian testes. Physiol Rev. 2012;92(2):577–95.
154. Scadden DT. The stem-cell niche as an entity of action. Nature. 2006;441(7097):1075–9.
155. Siu MK, Mruk DD, Lee WM, Cheng CY. Adhering junction dynamics in the testis are regulated by an interplay of beta 1-integrin and focal adhesion complex-associated proteins. Endocrinology. 2003;144(5):2141–63.
156. Seki M, Kameoka J, Takahashi S, Harigae H, Yanai N, Obinata M, et al. Identification of tenascin-C as a key molecule determining stromal cell-dependent erythropoiesis. Exp Hematol. 2006;34(4):519–27.
157. Garcion E, Faissner A. ffrench-Constant C. Knockout mice reveal a contribution of the extracellular matrix molecule tenascin-C to neural precursor proliferation and migration. Development. 2001;128(13):2485–96.
158. Garcion E, Halilagic A, Faissner A, ffrench-Constant C. Generation of an environmental niche for neural stem cell development by the extracellular matrix molecule tenascin C. Development. 2004;131(14):3423–32.
159. Garzon-Muvdi T, Quinones-Hinojosa A. Neural stem cell niches and homing: recruitment and integration into functional tissues. ILAR J. 2009;51(1):3–23.
160. Voog J, D'Alterio C, Jones DL. Multipotent somatic stem cells contribute to the stem cell niche in the Drosophila testis. Nature. 2008;454(7208):1132–6.
161. Shinohara T, Avarbock MR, Brinster RL. beta1- and alpha6-integrin are surface markers on mouse spermatogonial stem cells. Proc Natl Acad Sci USA. 1999;96(10):5504–9.
162. Nagano MC. Homing efficiency and proliferation kinetics of male germ line stem cells following transplantation in mice. Biol Reprod. 2003;69(2):701–7.

163. Kanatsu-Shinohara M, Takehashi M, Takashima S, Lee J, Morimoto H, Chuma S, et al. Homing of mouse spermatogonial stem cells to germline niche depends on beta1-integrin. Cell Stem Cell. 2008;3(5):533–42.
164. de Rooij DG, Repping S, van Pelt AM. Role for adhesion molecules in the spermatogonial stem cell niche. Cell Stem Cell. 2008;3(5):467–8.
165. Hornstein I, Alcover A, Katzav S. Vav proteins, masters of the world of cytoskeleton organization. Cell Signal. 2004;16(1):1–11.
166. Sanchez-Aguilera A, Lee YJ, Lo Celso C, Ferraro F, Brumme K, Mondal S, et al. Guanine nucleotide exchange factor Vav1 regulates perivascular homing and bone marrow retention of hematopoietic stem and progenitor cells. Proc Natl Acad Sci USA. 2011;108 (23):9607–12.
167. Itman C, Loveland KL. SMAD expression in the testis: an insight into BMP regulation of spermatogenesis. Dev Dyn. 2008;237(1):97–111.
168. Pellegrini M, Grimaldi P, Rossi P, Geremia R, Dolci S. Developmental expression of BMP4/ALK3/SMAD5 signaling pathway in the mouse testis: a potential role of BMP4 in spermatogonia differentiation. J Cell Sci. 2003;116(Pt 16):3363–72.
169. Puglisi R, Montanari M, Chiarella P, Stefanini M, Boitani C. Regulatory role of BMP2 and BMP7 in spermatogonia and Sertoli cell proliferation in the immature mouse. Eur J Endocrinol/Eur Fed Endocrine Soc. 2004;151(4):511–20.
170. Ying Y, Liu XM, Marble A, Lawson KA, Zhao GQ. Requirement of Bmp8b for the generation of primordial germ cells in the mouse. Mol Endocrinol. 2000;14(7):1053–63.
171. Zhao GQ, Chen YX, Liu XM, Xu Z, Qi X. Mutation in Bmp7 exacerbates the phenotype of Bmp8a mutants in spermatogenesis and epididymis. Dev Biol. 2001;240(1):212–22.
172. Hu J, Chen YX, Wang D, Qi X, Li TG, Hao J, et al. Developmental expression and function of Bmp4 in spermatogenesis and in maintaining epididymal integrity. Dev Biol. 2004;276 (1):158–71.
173. Lawson KA, Dunn NR, Roelen BA, Zeinstra LM, Davis AM, Wright CV, et al. Bmp4 is required for the generation of primordial germ cells in the mouse embryo. Genes Dev. 1999;13(4):424–36.
174. Oatley JM, Reeves JJ, McLean DJ. Biological activity of cryopreserved bovine spermatogonial stem cells during in vitro culture. Biol Reprod. 2004;71(3):942–7.

The Male Germinal Stem Cell Niche in Mammals

Unai Silván, Pablo Moreno and Juan Aréchaga

Abbreviations

BTB	Blood-testis barrier
CAM	Cell adhesion molecule
CSF-1	Colony stimulating factor-1
GDNF	Glial cell line-derived neurotrophic factor
GFR-1	GDNF-family receptor α-1
LIF	Leukemia inhibitory factor
NCAM	Neural cell adhesion molecule
PGC	Primordial germ cell
PMC	Peritubular myoid cell
SCF	Stem cell factor
SSC	Spermatogonial stem cell
TGCT	Testicular germ cell tumor
VEGF	Vascular endothelial growth factor

U. Silván · P. Moreno · J. Aréchaga
Faculty of Medicine and Dentistry, Laboratory of Stem Cells, Development and Cancer, Department of Cell Biology and Histology, University of the Basque Country, E-48942 Leioa, Spain

J. Aréchaga (✉)
Faculty of Medicine and Dentistry, Analytical and High Resolution Biomedical Microscopy Core Facility, University of the Basque Country, E-48942 Leioa, Spain
e-mail: juan.arechaga@ehu.es

Present Address:
U. Silván
Uniklinik Balgrist, Institute for Biomechanics, ETH Zürich, Forchstrasse 340, 8008 Zürich, Switzerland

K. Turksen (ed.), *Tissue-Specific Stem Cell Niche*,
Stem Cell Biology and Regenerative Medicine,
DOI 10.1007/978-3-319-21705-5_13

1 Establishment of the Spermatogonial Stem Cell (SSC) Niche

Mammalian primordial germ cells (PGCs) arise from the embryonic epiblast, proximal to the extra-embryonic ectoderm, but are first detected during gastrulation at 7.5 dpc mouse embryos, inside the extra-embryonic mesoderm, as a small cluster of non-specific alkaline phosphatase expressing cells at the level of the base of the allantoid pediculum. PGCs passively penetrate inside the embryo body during the invagination of the visceral endoderm to form the hindgut, proliferating and migrating later along the hindgut wall and the dorsal mesentery to reach finally the genital ridges (Fig. 1). The somatic cells present in these protuberances drive the differentiation of PGCs to follow a process in which female PGCs enter meiosis and become arrested in prophase I to form primordial follicles and male ones incorporate into the seminiferous cords, and proliferate giving rise to mitotic prospermatogonia (also known as gonocytes). Some time after birth, ranging from days in the case of rodents to months in primates, gonocytes migrate from the center of the seminiferous cords to the basal lamina and remodel their morphology, forming a population of undifferentiated spermatogonia, composed of SSCs and non-stem cell progenitors.

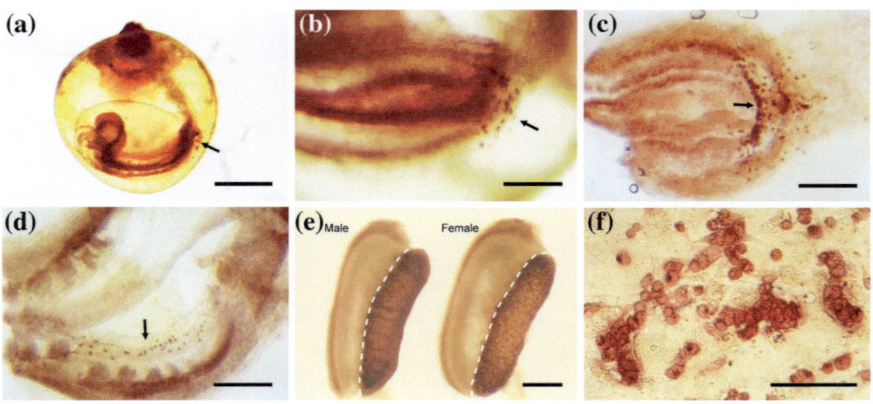

Fig. 1 Cultured, postimplanted, murine embryos showing phosphatase-alkaline positive primordial germ cell (PGCs) origin, migration and colonization of the genital ridges. **a** At 7.5 dpc PGCs (stained dark-red for endogenous alkaline phosphatase) become visible in the extra-embryonic mesoderm of the allantois base. **b** This initial founding population of PGCs is composed of a small cluster of only 20–40 cells, which can be seen at higher magnification. **c-d** During the next developmental stages, PGCs proliferate and migrate towards the genital ridges. **e** By 12.5 dpc all PGCs have already reached the embryonic gonads, ceasing to proliferate and adopting a different pattern of organization (staggered in males and disperse in females); whereas PGCs in females enter into meiosis, PGCs in males undergo mitotic arrest. **f** Isolated PGCs (alkaline phosphatase stained) cultivated *in vitro* on a feeder layer of embryonic fibroblasts. Scale bars represent 500 μm in **a**, 80 μm in **b** and **c**, 100 μm in **d** and **f**, and 200 μm in **e**

With the beginning of puberty a number of hormonal changes trigger the initiation of spermatogenesis. It is then when the testicular cords, which until that moment are solid, develop a lumen. In the first steps of the establishment of the germinal epithelium, self-renewal and proliferation of the stem cells is favored over differentiation to secure a stable pool of SSCs that will support homeostasis during adult life. In the testis, as in other organs containing a stem cell population, such as bone marrow and intestinal epithelium [2, 9], resident stem cells are most of the time quiescent, and in case required by tissue homeostasis, they give rise to a transient population of amplifying progenitors, which proliferate and differentiate [27]. In fact, during steady state conditions, both self-renewal events are reduced and only occur when new progenitor cells are required.

2 Adult Histological Organization of the SSC Niche Inside the Germinal Epithelium

The seminiferous tubules represent the vast majority of the volume of the testicles, forming a dense network with both their ends connected to a structure called the rete testis (Fig. 2a). In mice, the rete testis (from Latin, meaning testicular network) consists of a collection chamber with variable dimensions between subjects [38], while in humans it is a complex network of small ducts, which gave the structure its original name [32]. From this compartment, the newly formed spermatozoa leave the testes and are conducted to the epididymis, where they finish their maturation and acquire fertilization potential.

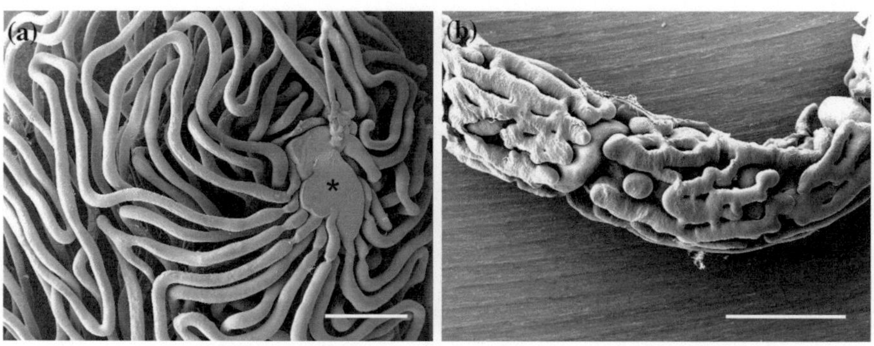

Fig. 2 Seminiferous tubule network and vascularization of the epididymis. **a** Corrosion cast of the murine seminiferous network reveals the complex organization of this organ, with all seminiferous tubules connected to the rete testis (asterisk). This chamber drives the formed gametes through the efferent ducts to the epididymis. **b** Corrosion cast of the vascular pampiniform plexus of the epididymis. Scale bar represents 500 μm in **a** and 400 μm in **b**

During early embryonic development male gonads are located inside the abdomen. In many mammalian species, including human and mouse, the testes descend to their definitive scrotal position through a two-stage process of trans-abdominal and inguino-scrotal migration at later stages of pregnancy [16]. This final position of the testicles defines one of the characteristics of the mammalian SSC niche and, in general, the testis, which is a typically low temperature (varying from 2 to 6°C less than the temperature of the body, depending on the species). Reduced pressure has been proposed as the main evolutionary goal of descendant testicles, with the temperature change being a spillover effect [7]. Nevertheless, reduced temperature has become crucial for proper functioning of the testis. In fact, the anomalous disruption of normal testicular descent is clinically known as cryptorchidism, which has been reported to be a prognostic factor for infertility and an increased risk factor for testicular cancer in human and other species [42], at least partially due to elevated testicular temperature [13]. Reduced temperature of the testis is achieved not only by its anatomical location outside the body, but also by the irrigation of the testis with blood that has been previously cooled down from abdominal to testicular temperature. This refrigeration takes place in the pampiniform plexus, an intricate network of small vessels that surround the testicular artery as it enters the testis (Fig. 2b). Vascular connections between the testicular artery and the pampiniform plexus confer on the system the possibility of bypassing testicular blood flow, and could possibly represent a mechanism to regulate testicular temperature [40].

Histological sections of the mammalian adult testis reveal a densely packed mass of seminiferous tubules, which are surrounded by the tunica albuginea, a thick fibromuscular connective tissue capsule (Fig. 3). The tubules contain the germinal (or seminiferous) epithelium, which is where SCCs differentiate to generate male gametes. The vast majority of the adult seminiferous epithelium is occupied by

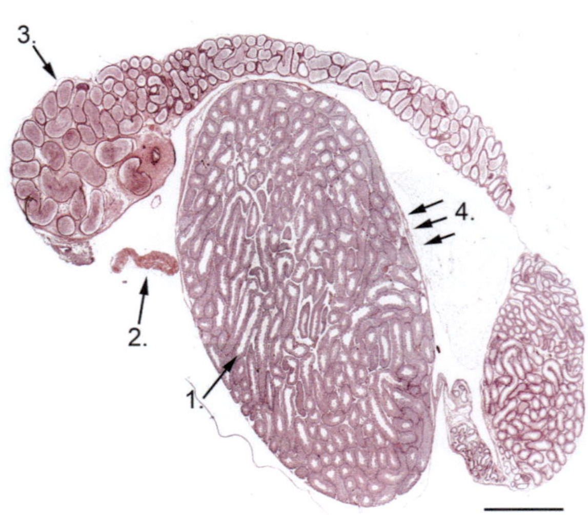

Fig. 3 Histology of the male mouse gonad. Longitudinal section stained with hematoxylin/eosin in which the most important structures can be observed, including seminiferous tubules *1*, testicular artery *2*, epididymis *3* and tunica albuginea *4*. Scale bar represents 500 μm

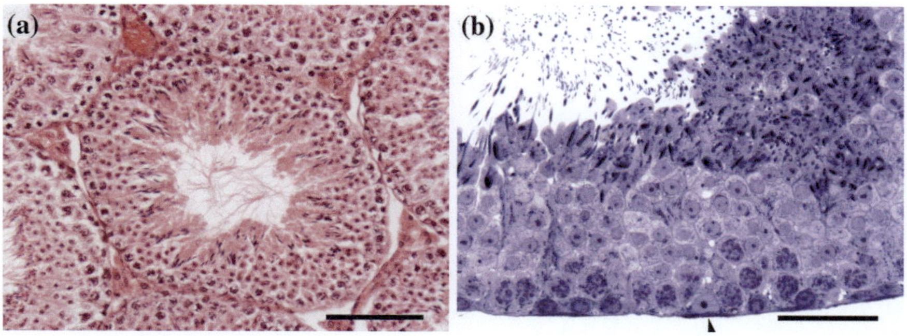

Fig. 4 Testis germinal epithelium. **a** Horizontal section of seminiferous tubules *in situ* stained with hematoxylin-eosin. **b** A detail of the germinal epithelium in a toluidine blue stained semithin section (arrowhead points the nucleus of a Sertoli Cell and a PMC underneath). Scale bars: 50 μm **a** and 20 μm **b**

germ cells at different stages of differentiation but, as in other organs, the size of the cell population with stem cell characteristics is small. SSCs are located in the basal region of the seminiferous epithelium, confined between the basal membrane of the seminiferous tubules and the Sertoli cells (Fig. 4a, b). It is nowadays accepted that type As (single) mouse SSCs can divide to generate either SSCs or a couple of Ap (A paired) spermatogonia. SSCs assure self-renewal, while Ap spermatogonia remain interconnected by intracellular bridges and further divide to form 'chains' of Aal (A aligned) spermatogonia composed of 4, 8 or 16 cells. In turn, these aligned spermatogonia continue the process and by serial divisions produce A1 to A4 spermatogonia, which further give rise to intermediate spermatogonia, B spermatogonia, and then preleptotene spermatocytes (reviewed in Meng et al. [25]). **Sertoli cells**, the only non-germinal cell type present in the seminiferous epithelium, extend from the basal membrane to the luminal space of the tubule. Their intimate association with spermatogonia points to them as essential regulators of spermatogenesis. These cells have been classically described as 'nurse cells of the testis', and as we will see, they are responsible as well for different aspects of the establishment of the SSC niche, including the formation and maintenance of the blood-testis barrier (BTB).

Peritubular myoid cells (PMCs) are myofibroblast-like cells that surround the seminiferous tubules and, together with Sertoli cells, enclose the SSC niche (Fig. 5). PMCs are characterized by their cytoplasmic system of α-smooth muscle actin (α-SMA) and myosin filament bundles that are distributed in two independent layers, orthogonal to each other [22]. Using this cytoskeletal machinery, they participate in the propulsion of the tubular fluid and spermatozoa towards the rete testis [45]. In addition, and together with the basement membrane, PMCs constitute a physical barrier that provides the structural support of the seminiferous tubules. **Leydig cells** represent a population of cells localized within the loose connective tissue in the intratubular space. Histologically these interstitial cells have a polyhedral shape and

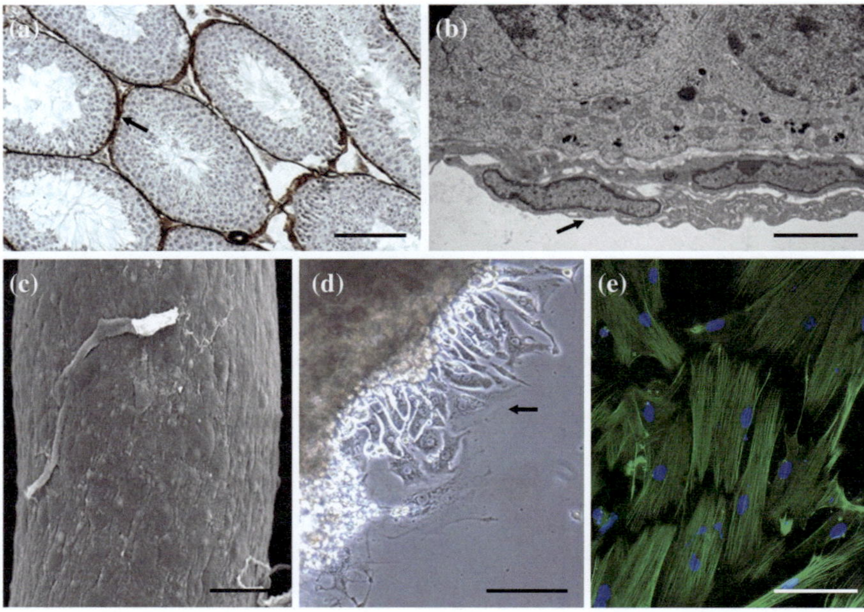

Fig. 5 Peritubular myoid cell (PMC) population of the testis. **a** Immunoperoxidase staining with anti-smooth muscle actin reveals the presence of a thin layer of PMCs surrounding mouse seminiferous tubules. **b** Transmission electron microscopy reveals the elongated and flat shape of PMCs. **c** Surface image of a single seminiferous tubule covered with a continuous layer of PMCs as seen with scanning electron microscopy. **d** PMCs growing out from an explant of a single seminiferous tubule (phase contrast microscopy). **e** A pure population of PMCs derived from these explants reveals an extensive actin cytoskeleton, as observed with immunofluorescence microscopy (*Phalloidin* staining in green and nuclear staining with *DAPI*). Scale bars represent 50 μm in **a**, 5 μm in **b**, 30 μm in **c**, 200 μm in **d**, and 150 μm in **e**

frequently contain 'crystals of Reinke', rod-shaped structures with variable sizes and unknown function. To facilitate their hormone-related functions, Leydig cells are in close proximity to blood capillaries.

3 Extracellular Matrix and Adhesion Molecules Related to the SSC Niche

Transplantation of cells into the seminiferous tubules is a very powerful approach to study the regeneration of the germinal epithelium in mammals and also to study earliest stages of germinal tumor invasion. Brinster and Zimmermann described this technique in [5] as microinjections of cell suspensions into the lumen of individual seminiferous tubules, but later we improved it [31, 38, 39] (Fig. 6). This experimental approach revealed that isolated SSCs located in the lumen of the seminiferous tubules retain their ability to return home to their niche and subsequently

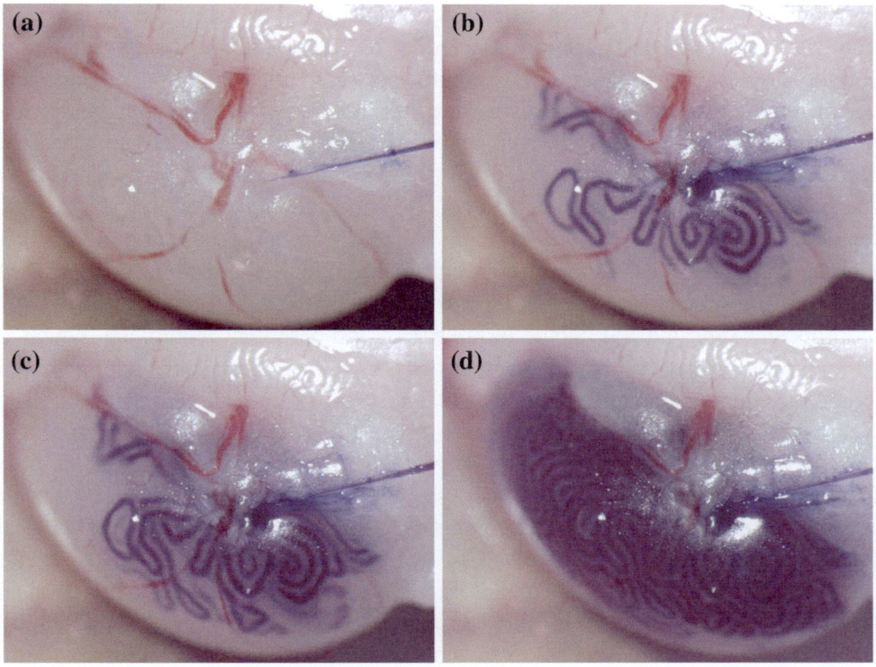

Fig. 6 Cell transplantation into the seminiferous tubules. **a–d** Cells were transplanted into the seminiferous tubules by microinjection through the testis efferent ducts. A small a quantity of bromophenol blue was added to the cell suspension to facilitate visualization. Figure adapted from Silván et al. [39] with permission from *Cognizant Communication Corporation*

recover spermatogenesis in previously infertile mice [4]. In addition to SSCs, other cell types including testicular somatic cells [36], cancer cells [21], and embryonic stem cells [39] have been transplanted into the seminiferous tubules. This transplantation procedure further represents a functional assay to test the impact of different characteristics of the transplanted SSCs on migration into the niche and physical confinement in that location. For instance, transplantation experiments have revealed that the homing and retention events are dictated by cell adhesion molecules (CAMs) expressed by both SSC and Sertoli cells. CAMs are a large group of membrane proteins that belong to different families, including cadherins, integrins, selectins, and members of the immunoglobulin superfamily. As occurs in other tissues, expression of CAM molecules by SSCs and adjacent cells is a key factor in the establishment of the niche.

One of the peculiarities of the SSC niche is its well-defined physical localization between the basal membrane of the seminiferous tubules and the BTB, one of the tightest blood-tissue barriers in the mammalian body, which physically divides the seminiferous epithelium into a basal compartment, where the most undifferentiated germinal cells are localized, and an apical (or adluminal) one, that contains the most differentiated cells, including fully differentiated spermatozoa. The BTB is formed by adjacent Sertoli cells, which are physically connected to each other by means of tight junctions, adherent junctions (including the basal ectoplasmic specialization and the basal tubulobulbar complexes, both specific to this testicular structure), and desmosome-like junctions. This structure has been suggested to assume an immunoprotective role [35]. However, its precise mechanism of functioning and the associated biological implications are still a matter of discussion [18]. During spermatogenesis, the BTB can disassemble to permit the transit of preleptotene and leptotene spermatocytes to the luminal side of the seminiferous tubules. As previously mentioned, this journey, although in the opposite direction, occurs when SSCs are transplanted into the adluminal side of the seminiferous tubules [4]. On the opposite side, the SSC niche is limited by the basement membrane, which is composed of proteins secreted by Sertoli cells, PMCs and germ cells, with its major structural components being laminin, collagens (mostly type IV), and fibronectin. Cells located on the basal membrane bind to it via integrins, a large family of cell-matrix and cell-cell transmembrane proteins that combine into a large number of alternative dimers. The subunit composition of these complexes dictates their binding affinity and functions in the integrity of the tissues. Besides their structural function, integrins are also involved in cell signaling and participate in several cellular events such as cell death, proliferation, differentiation, and migration in processes such as embryonic development, homeostasis and immune response [23].

Gene expression analysis revealed the expression of numerous β-integrin subunits in SSCs; the expression levels of β1 and β5 were predominant [17]. The β1-integrin is known to participate in hematopoietic stem cell homing, as well as to be required for PGC colonization of the genital ridges during embryonic development. Transplantation experiments in which integrin-β1-deficient SSCs were microinjected into wild type seminiferous tubules revealed that, although mutant SSCs retain competence to cross the blood-testis barrier, their inability to bind to the basal membrane hinders their confinement in their niche [17]. SSC homing also requires the expression of the β1-integrin in Sertoli cells, pointing to a mechanism in which this cell adhesion molecule is used to drive SSCs through the blood-testis barrier and later is needed to retain SSCs in their niche. Furthermore, it has been shown that this integrin plays an additional role in spermatogenesis, since the rare colonies of β1-deficient SSCs formed after transplantation do not achieve complete differentiation [17].

Besides cell-matrix binding, the formation and maintenance of intercellular junctions is central to the establishment and homeostasis of the stem cell niche; transmembrane cadherins are known to be the major architectural proteins at these sites [20]. The adhesion mediated by these molecules is generally homophilic (cadherin-cadherin) and homotypic (i.e. between the same cell types). Over 100 different types of cadherins have been reported so far, and at least three of them (E-, N-, and P-cadherin) are present in the testis [46]. N-cadherin has been found on the surface of endothelial cells, spermatogonia and primary spermatocytes, whereas it is not expressed in peritubular and Leydig cells [1]. In turn, in the adult testis, E-cadherin expression is restricted to undifferentiated spermatogonia, including As,

Apr, and Aal subtypes [28]. In fact, the chains of interconnected AaI spermatogonia it seems to be stabilized by E-cadherin [49]. Surprisingly, transplantation experiments in which the expression of this CAM was downregulated in SSCs found that its absence does not hinder SSC homing to their niche or spermatogenesis [17].

4 Growth Factors Involved in the Function and Maintenance of the SSC Niche

Testicular function, and subsequently the establishment and maintenance of the SSC niche, is largely governed by hormones. The hypothalamus, the pituitary gland and the gonads form the so-called hypothalamic-pituitary-gonadal axis, a system that governs the differentiation of SSCs into spermatozoa. This is achieved through the secretion of gonadotropin-releasing hormone (GnRH), a hormone produced by the hypothalamus that stimulates the anterior pituitary to secrete two gonadotropins, namely luteinizing hormone (LH) and follicle-stimulating hormone (FSH). Both are named after their effect on females, but they play different roles in males. In the latter, LH stimulates the production of testosterone in Leydig cells, while FSH acts on Sertoli cells, upregulating the expression and secretion of different growth factors [3]. Additionally, it has recently been reported that the mutation of the androgen receptor gene in PMCs causes the progressive loss of spermatogonia in mice carrying the modification [8], thereby manifesting the complexity of the interactions that regulate the SSC niche. Experiments, in which FSH stimulation was blocked, revealed a decrease in levels of glial cell line-derived neurotrophic factor (GDNF), along with a reduction in the proliferation rate of undifferentiated spermatogonia [44], probably through the downregulation of GDNF expression in PMCs and Sertoli cells [14, 41]. This neurotrophic cytokine operates through its binding to a receptor, named GDNF-family receptor α-1 (GFR1), that further recruits two receptor tyrosine kinase RET subunits, activating the intracellular signaling. Both molecules (GFR1 and RET) are present in SSCs and in different spermatogonia subtypes depending on the developmental stage of the testis [43]. The implication of Glial cell-Derived Neurotrophic Factor (GDNF) in SSC self-renewal was discovered using animals in which its expression had been altered. For instance, it was reported that seminiferous tubules of mice overexpressing GDNF contain an increased amount of undifferentiated spermatogonia, with cells in later differentiation stages being absent. On the other hand, mice in which one of the GDNF alleles had been disrupted did lose germ cells at older ages [25]. In fact, reduced expression of GDNF in aging testes has been proposed to be the cause of fertility loss in elderly males. Consequently, GDNF is a necessary component of the culture medium of these SSCs [26] and, although its mechanism of action has not been completely unraveled, it is known that SSCs cultured in the absence of this factor show a downregulation of E-cadherin expression, with the levels of β1 and β5 integrin remaining unaffected. It has also been reported that in mouse SSCs,

GDNF activates the phosphoinositide 3-kinase (PI3 K)-Akt pathway, which is necessary for SSC maintenance [29]. An autocrine mechanism of action on Sertoli cells is also possible, since these cells express NCAM (neural cell adhesion molecule), a receptor that has also been shown to bind members of the GDNF family, and is known to mediate the proliferative effect of GDNF on cultured immature Sertoli cells [47]. Besides GDNF, fibroblast growth factor-2 (FGF2; also known as basic fibroblast growth factor or bFGF), a cytokine, which is also expressed in Sertoli cells, is also known to play a role in SSC self-renewal. For instance, together with GDNF, FGF2 has been shown to synergistically enhance the expression of self-renewal genes in SSCs [15]. Although the precise mechanism by which this factor operates has not yet been completely characterized, in vitro experiments have revealed that its effects depend on MAP2K1 signaling.

As we have mentioned, Leydig cells, although being located relatively distant from the physical location of SSCs, represent a key cell population in the regulation of the SSC niche. Interestingly, experiments in which animals expressing fluorescent proteins under the control of a promoter differentially active in undifferentiated spermatogonia, revealed that these stem cells are preferentially located in those areas of the seminiferous tubules adjacent to the testicular stroma, and consequently close to the Leydig cells [48]. It has been suggested that Leydig cells, through the expression and secretion of factors such as colony-stimulating factor-1 (CSF-1), might guide SSCs to their final position [30]. Since the addition of CSF-1 to the SSC culture medium significantly enhances the ability of these cells to reestablish spermatogenesis upon their transplantation, without increasing the overall number of cells, it is likely that this factor also promotes SSC survival [30].

Other factors present in the testis, although not essential for SSC self-renewal, favor other processes important for the maintenance of the SSC pool, such as proliferation and survival of this cell type. For instance, SSCs express receptors for leukemia inhibitory factor (LIF) and stem cell factor (SCF) (Fig. 7). Both growth factors are expressed by Sertoli cells, and have been shown to promote the survival of gonocytes [19]. In turn, the expression levels of the vascular endothelial growth factor (VEGF) family in testicular tissue are extremely high, considering the stability of the testicular vasculature [37]. This, together with the expression pattern of VEGF receptors in undifferentiated spermatogonia, suggests a non-vascular function for these growth factors in the testis. In fact, studies in which the transplantation efficiency of neonatal isolated SSCs cultured with either VEGF-A 164 or VEGF-A 165b were compared and revealed that the former promotes self-renewal, whereas the latter stimulates differentiation [6]. Additionally, it has also been shown that VEGF-A knockout mice experience subfertility due to altered expression of genes that regulate SSC self-renewal [24]. The complexity of the regulatory mechanisms that define the SSC niche, together with the involvement of several cell types found in a specific physical arrangement, makes the study of this process intriguing. It is therefore likely that new mechanisms underlying the balance between SSC differentiation and self-renewal will be reported in the future, and that some of those reviewed here will be better understood.

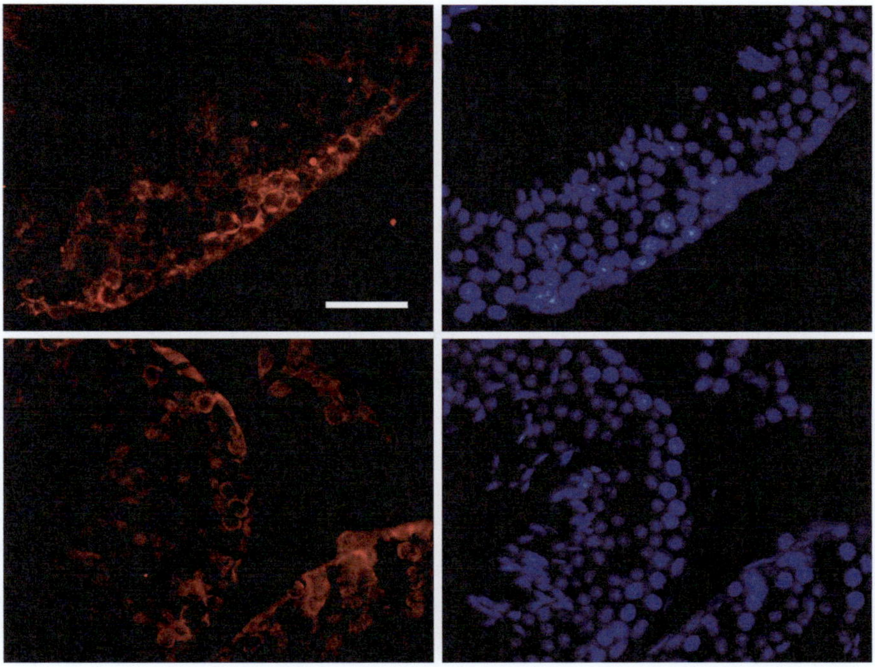

Fig. 7 cKITr (*upper row*) and LIFr (*lower row*) fluorescent immunohistochemistry revealed the presence of these receptors in the most undifferentiated cells of the seminiferous epithelium (in *red*), being practically absent in more differentiated cells (*blue signal* corresponds to nuclear DAPI staining). Scale bar: 25 μm

5 Concluding Remarks

Advances in stem cell biology bring the treatment of several degenerative medical conditions by means of stem cell therapy closer. However, the importance of the niche in which these stem cells are localized is frequently ignored. Serial transplantation of SSCs derived from old donors into the seminiferous tubules of young individuals has shown that fertility loss caused by aging is mainly caused by the impaired capacity of the SSC niche to sustain spermatogenesis, continuing the SSC self-renewal potential past the normal lifespan of the animal [33]. Similar mechanisms have been reported in other tissues; in which aged niches could be recovered after exposure to systemic factors from young individuals [10]. A better knowledge of the niche in which these stem cells differentiate may also contribute to improving current culture techniques for the production of functional gametes to be used in standard in vitro insemination techniques [12]. Furthermore, the characteristics of the SSC niche, including the presence of high concentrations of growth factors, such as VEGF and GDNF, are likely to play a central role in the development and progression of testicular germ cell tumors (TGCTs). For instance, the stromal population of PMCs has been hypothesized to give rise to cancer-associated myofibroblasts [11], a cell population known to participate in the development of tumors. It is therefore becoming increasingly important to understand the mechanisms by which the cell types present in the testis maintain and regulate SSC renewal and differentiation.

Acknowledgments The authors would like to thank Dr. David J. Fogarty for critical reading of the manuscript and Mr. Juan L. Vidaurrázaga for his editorial assistance. Research grants from the Spanish Ministry of Economy and Competitiveness (SAF2012-39773), the regional Basque Government Research Groups Program (T560-10) and the SAIOTEK Program (SOA11UN001) of the Basque Science, Technology & Innovation network (SPRI) to J. Aréchaga supported this work. U. Silván and P. Moreno enjoyed fellowships from the University of the Basque Country (UPV/EHU) and the Jesús De Gangoiti Barrera Foundation of Bilbao (Spain) respectively.

Bibliography

1. Andersson AM, Edvardsen K, Skakkebaek NE. Expression and localization of N- and E-cadherin in the human testis and epididymis. Int J Androl. 1994;17:174–80.
2. Barker N, Clevers H. Tracking down the stem cells of the intestine: strategies to identify adult stem cells. Gastroenterology. 2007;133:1755–60.
3. Boron WF, Boulpaep EL. Medical physiology. 2nd ed. Elsevier Health Sciences; 2012.
4. Brinster RL, Avarbock MR. Germline transmission of donor haplotype following spermatogonial transplantation. Proc Natl Acad Sci USA. 1994;91:11303–7.
5. Brinster RL, Zimmermann JW. Spermatogenesis following male germ-cell transplantation. Proc Natl Acad Sci USA. 1994;91:11298–302.
6. Caires KC, de Avila JM, Cupp AS, McLean DJ. VEGFA family isoforms regulate spermatogonial stem cell homeostasis in vivo. Endocrinology. 2012;153:887–900.
7. Chance MRA. Reason for externalization of the testis of mammals. J Zool. 1996;239:691–5.
8. Chen L-Y, Brown PR, Willis WB, Eddy EM. Peritubular myoid cells participate in male mouse spermatogonial stem cell maintenance. Endocrinology. 2014;155:en20141406.
9. Cheshier SH, Morrison SJ, Liao X, Weissman IL. In vivo proliferation and cell cycle kinetics of long-term self-renewing hematopoietic stem cells. Proc Natl Acad Sci USA. 1999;96:3120–5.
10. Conboy IM, Conboy MJ, Wagers AJ, Girma ER, Weissman IL, Rando TA. Rejuvenation of aged progenitor cells by exposure to a young systemic environment. Nature. 2005;433:760–4.
11. Díez-Torre A, Silván U, Moreno P, Gumucio J, Aréchaga J. Peritubular myoid cell-derived factors and its potential role in the progression of testicular germ cell tumours. Int J Androl. 2011;34:e252–64.
12. Eguizabal C, Montserrat N, Vassena R, Barragan M, Garreta E, Garcia-Quevedo L, Vidal F, Giorgetti A, Veiga A, Izpisúa-Belmonte JC. Complete meiosis from human induced pluripotent stem cells. Stem Cells. 2011;29:1186–95.
13. Garner MJ, Turner MC, Ghadirian P, Krewski D. Epidemiology of testicular cancer: an overview. Int J Cancer. 2005;116:331–9.
14. Hofmann M-C. Gdnf signaling pathways within the mammalian spermatogonial stem cell niche. Mol Cell Endocrinol. 2008;288:95–103.
15. Ishii K, Kanatsu-Shinohara M, Toyokuni S, Shinohara T. FGF2 mediates mouse spermatogonial stem cell self-renewal via upregulation of Etv5 and Bcl6b through MAP2K1 activation. Development. 2012;139:1734–43.
16. Kaleva M, Toppari J. Genetics and hormones in testicular descent. Hormones (Athens). 2003;2:211–216.
17. Kanatsu-Shinohara M, Takehashi M, Takashima S, Lee J, Morimoto H, Chuma S, et al. Homing of mouse spermatogonial stem cells to germline niche depends on beta1-integrin. Cell Stem Cell. 2008;3:533–42.

18. Kaur G, Mital P, Dufour JM. Testisimmune privilege—assumptions versus facts. Anim Reprod. 2013;10(1):3–15.
19. Kubota H, Avarbock MR, Brinster RL. Growth factors essential for self-renewal and expansion of mouse spermatogonial stem cells. Proc Natl Acad Sci USA. 2004;101:16489–94.
20. Leckband D, Sivasankar S. Cadherin recognition and adhesion. Curr Opin Cell Biol. 2012;24:620–7.
21. Li Y, Kido T, Luo J, Fukuda M, Dobrinski I. Lau Y-FC. Intratubular transplantation as a strategy for establishing animal models of testicular germ cell tumours. Int J Exp Pathol. 2008;89:342–9.
22. Losinno AD, Morales A, Fernández D, Lopez LA. Peritubular myoid cells from rat seminiferous tubules contain actin and myosin filaments distributed in two independent layers. Biol Reprod. 2012;86(150):1–8.
23. Lowell CA, Mayadas TN. Overview: studying integrins in vivo. Methods Mol Biol. 2012;757:369–97.
24. Lu N, Sargent KM, Clopton DT, Pohlmeier WE, Brauer VM, McFee RM, et al. Loss of vascular endothelial growth factor A (VEGFA) isoforms in the testes of male mice causes subfertility, reduces sperm numbers, and alters expression of genes that regulate undifferentiated spermatogonia. Endocrinology. 2013;154:4790–802.
25. Meng X, Lindahl M, Hyvönen ME, Parvinen M, De Rooij DG, Hess MW, et al. Regulation of cell fate decision of undifferentiated spermatogonia by GDNF. Science. 2000;287:1489–93.
26. Nagano M, Ryu B-Y, Brinster CJ, Avarbock MR, Brinster RL. Maintenance of mouse male germ line stem cells in vitro. Biol Reprod. 2003;68:2207–14.
27. Nakagawa T, Nabeshima Y-I, Yoshida S. Functional identification of the actual and potential stem cell compartments in mouse spermatogenesis. Dev Cell. 2007;12:195–206.
28. Nakagawa T, Sharma M, Nabeshima Y, Braun RE, Yoshida S. Functional hierarchy and reversibility within the murine spermatogenic stem cell compartment. Science. 2010;328:62–7.
29. Oatley JM, Avarbock MR, Brinster RL. Glial cell line-derived neurotrophic factor regulation of genes essential for self-renewal of mouse spermatogonial stem cells is dependent on Src family kinase signaling. J Biol Chem. 2007;282:25842–51.
30. Oatley JM, Oatley MJ, Avarbock MR, Tobias JW, Brinster RL. Colony stimulating factor 1 is an extrinsic stimulator of mouse spermatogonial stem cell self-renewal. Development. 2009;136:1191–9.
31. Ogawa T, Aréchaga JM, Avarbock MR, Brinster RL. Transplantation of testis germinal cells into mouse seminiferous tubules. Int J Dev Biol. 1997;41:111–22.
32. Roosen-Runge EC, Holstein AF. The human rete testis. Cell Tissue Res. 1978;189:409–33.
33. Ryu B-Y, Orwig KE, Oatley JM, Avarbock MR, Brinster RL. Effects of aging and niche microenvironment on spermatogonial stem cell self-renewal. Stem Cells. 2006;24:1505–11.
34. Schofield R. The relationship between the spleen colony-forming cell and the haemopoietic stem cell. Blood Cells. 1978;4:7–25.
35. Setchell BP, Uksila J, Maddocks S, Pöllänen P. Testis physiology relevant to immunoregulation. J Reprod Immunol. 1990;18:19–32.
36. Shinohara T, Orwig KE, Avarbock MR, Brinster RL. Restoration of spermatogenesis in infertile mice by Sertoli cell transplantation. Biol Reprod. 2003;68:1064–71.
37. Silván U. Díez-Torre a, Jiménez-Rojo L, Aréchaga J. Vascularization of testicular germ cell tumours: evidence from experimental teratocarcinomas. Int J Androl. 2010;33:765–74.
38. Silván U, Aréchaga J. Anatomical basis for cell transplantation into mouse seminiferous tubules. Reproduction. 2012;144:385–92.
39. Silván U, Díez-Torre A, Andrade R, Arluzea J, Silió M, Aréchaga J. Embryonic stem cell transplantation into seminiferous tubules: a model for the study of invasive germ cell tumors of the testis. Cell Transplant. 2011;20:637–42.
40. Skowroński A, Jedrzejewski K. The human testicular artery and the pampiniform plexus–where is the connection? Folia Morphol (Warsz). 2003;62:201–4.

41. Spinnler K, Köhn FM, Schwarzer U, Mayerhofer A. Glial cell line-derived neurotrophic factor is constitutively produced by human testicular peritubular cells and may contribute to the spermatogonial stem cell niche in man. Hum Reprod. 2010;25:2181–7.
42. Strader CH, Weiss NS, Daling JR. Vasectomy and the incidence of testicular cancer. Am J Epidemiol. 1988;128:56–63.
43. Suzuki H, Sada A, Yoshida S, Saga Y. The heterogeneity of spermatogonia is revealed by their topology and expression of marker proteins including the germ cell-specific proteins Nanos2 and Nanos3. Dev Biol. 2009;336:222–31.
44. Tadokoro Y, Yomogida K, Ohta H, Tohda A, Nishimune Y. Homeostatic regulation of germinal stem cell proliferation by the GDNF/FSH pathway. Mech Dev. 2002;113:29–39.
45. Tripiciano A, Palombi F, Ziparo E, Filippini A. Dual control of seminiferous tubule contractility mediated by ETA and ETB endothelin receptor subtypes. FASEB J. 1997;11:276–86.
46. Wu JC, Gregory CW, DePhilip RM. Expression of E-cadherin in immature rat and mouse testis and in rat Sertoli cell cultures. Biol Reprod. 1993;49:1353–61.
47. Yang Y, Han C. GDNF stimulates the proliferation of cultured mouse immature Sertoli cells via its receptor subunit NCAM and ERK1/2 signaling pathway. BMC Cell Biol. 2010;11:78.
48. Yoshida S, Sukeno M, Nabeshima Y-I. A vasculature-associated niche for undifferentiated spermatogonia in the mouse testis. Science. 2007;317:1722–6.
49. Zhang Y, Su H, Luo F, Wu S, Liu L, Liu T, et al. E-cadherin can be expressed by a small population of rat undifferentiated spermatogonia in vivo and in vitro. In Vitro Cell Dev Biol Anim. 2011;47:593–600.

Erratum to Chapter 10, 'Tendon Stem Cell Niche'

Raquel Costa-Almeida, A.I. Gonçalves, P. Gershovich, M.T. Rodrigues, R.L. Reis and M.E. Gomes

Erratum to Chapter 10, 'Tendon Stem Cell Niche' in *Tissue-Specific Stem Cell Niche* (K. Turksen, Ed.) in the series *Stem Cell Biology and Regenerative Medicine*.

In Chapter 10, the Table of Contents, and the Contributors List, the lead author's last name was misspelled 'Costa-Alameida'; it should have been spelled 'Costa-Almeida'.

The updated original online version for this chapter can be found at
DOI 10.1007/978-3-319-21705-5_10

R. Costa-Almeida · A.I. Gonçalves · P. Gershovich · M.T. Rodrigues · R.L. Reis · M.E. Gomes (✉)
3B's Research Group, Biomaterials, Biodegradables and Biomimetics, University of Minho, Headquarters of the European Institute of Excellence on Tissue Engineering and Regenerative Medicine, Avepark - Parque de Ciência e Tecnologia, Zona Industrial da Gandra, Barco GMR, 4805-017 Guimarães, Portugal
e-mail: megomes@dep.uminho.pt

R. Costa-Almeida
e-mail: raquel.almeida@dep.uminho.pt

A.I. Gonçalves
e-mail: ana.goncalves@dep.uminho.pt

K. Turksen (ed.), *Tissue-Specific Stem Cell Niche*,
Stem Cell Biology and Regenerative Medicine,
DOI 10.1007/978-3-319-21705-5_14

P. Gershovich
e-mail: pm.gershovich@gmail.com

M.T. Rodrigues
e-mail: mrodrigues@dep.uminho.pt

R.L. Reis
e-mail: rgreis@dep.uminho.pt

R. Costa-Almeida · A.I. Gonçalves · P. Gershovich · M.T. Rodrigues · R.L. Reis · M.E. Gomes
ICVS/3B's Government Associate Laboratory, University of Minho, Braga, Portugal

Index

K. Turksen (ed.), *Tissue-Specific Stem Cell Niche*,
Stem Cell Biology and Regenerative Medicine,
DOI 10.1007/978-3-319-21705-5

MIX
Papier aus verantwortungsvollen Quellen
Paper from responsible sources
FSC® C105338

If you have any concerns about our products,
you can contact us on
ProductSafety@springernature.com

In case Publisher is established outside the EU,
the EU authorized representative is:
Springer Nature Customer Service Center GmbH
Europaplatz 3, 69115 Heidelberg, Germany

Printed by Libri Plureos GmbH
in Hamburg, Germany